VOLUME ONE HUNDRED AND THIRTY

INTERNATIONAL REVIEW OF NEUROBIOLOGY

Nanotechnology and the Brain

INTERNATIONAL REVIEW OF NEUROBIOLOGY

VOLUME 130

SERIES EDITOR

PETER JENNER
Division of Pharmacology and Therapeutics
GKT School of Biomedical Sciences
King's College, London, UK

EDITORIAL BOARD

VOLUME ONE HUNDRED AND THIRTY

INTERNATIONAL REVIEW OF NEUROBIOLOGY

Nanotechnology and the Brain

Edited by

KHULOUD T. AL-JAMAL

Institute of Pharmaceutical Science,
Faculty of Life Sciences and Medicine,
King's College London, London, United Kingdom

ELSEVIER

AMSTERDAM • BOSTON • HEIDELBERG • LONDON
NEW YORK • OXFORD • PARIS • SAN DIEGO
SAN FRANCISCO • SINGAPORE • SYDNEY • TOKYO
Academic Press is an imprint of Elsevier

Academic Press is an imprint of Elsevier
50 Hampshire Street, 5th Floor, Cambridge, MA 02139, United States
525 B Street, Suite 1800, San Diego, CA 92101-4495, United States
The Boulevard, Langford Lane, Kidlington, Oxford OX5 1GB, United Kingdom
125 London Wall, London, EC2Y 5AS, United Kingdom

First edition 2016

Notices
Knowledge and best practice in this field are constantly changing. As new research and experience broaden our understanding, changes in research methods, professional practices, or medical treatment may become necessary.

Practitioners and researchers must always rely on their own experience and knowledge in evaluating and using any information, methods, compounds, or experiments described herein. In using such information or methods they should be mindful of their own safety and the safety of others, including parties for whom they have a professional responsibility.

To the fullest extent of the law, neither the Publisher nor the authors, contributors, or editors, assume any liability for any injury and/or damage to persons or property as a matter of products liability, negligence or otherwise, or from any use or operation of any methods, products, instructions, or ideas contained in the material herein.

ISBN: 978-0-12-804636-4
ISSN: 0074-7742

Publisher: Zoe Kruze
Acquisition Editor: Kirsten Shankland
Editorial Project Manager: Hannah Colford
Production Project Manager: Surya Narayanan Jayachandran
Cover Designer: Christian Bilbow

Typeset by SPi Global, India

CONTENTS

CONTRIBUTORS

K.T. Al-Jamal
Institute of Pharmaceutical Science, Faculty of Life Sciences and Medicine, King's College London, London, United Kingdom

K. Andrieux
Université Paris Descartes, Université Paris-Sorbonne, UTCBS, UMR CNRS 8258, UE1022 INSERM, Paris, France

E. Araya
Advanced Center for Chronic Diseases (ACCDiS); Facultad de Ciencias Exactas, Universidad Andrés Bello, Santiago, Chile

E. Barbu
School of Pharmacy and Biomedical Sciences, University of Portsmouth, Portsmouth, United Kingdom

G. Battaglia
University College London, London, United Kingdom

D. Brambilla
Institute of Pharmaceutical Sciences, ETH Zurich, Zurich, Switzerland

A.L. Cardoso
CNC – Center for Neuroscience and Cell Biology, University of Coimbra, Coimbra, Portugal

A.M. Cardoso
CNC – Center for Neuroscience and Cell Biology, University of Coimbra, Coimbra, Portugal

D. Carradori
Advanced Drug Delivery and Biomaterials, Louvain Drug Research Institute, Université catholique de Louvain, Bruxelles, Belgium

R. Costa
CNC – Center for Neuroscience and Cell Biology, University of Coimbra, Coimbra, Portugal

P. Cunha
CNC – Center for Neuroscience and Cell Biology, University of Coimbra, Coimbra, Portugal

G. Fullstone
University College London, London, United Kingdom

E. Gallardo-Toledo
Facultad de Ciencias Químicas y Farmacéuticas, Universidad de Chile; Advanced Center for Chronic Diseases (ACCDiS), Santiago, Chile

A. Gaudin
Yale University, New Haven, CT, United States

R. Gromnicova
The Open University, Milton Keynes, United Kingdom

J.R. Guedes
CNC – Center for Neuroscience and Cell Biology, University of Coimbra, Coimbra, Portugal

A. Jurado
CNC – Center for Neuroscience and Cell Biology; Department of Life Sciences, University of Coimbra, Coimbra, Portugal

H. Kafa
Institute of Pharmaceutical Science, Faculty of Life Sciences and Medicine, King's College London, London, United Kingdom

M.J. Kogan
Facultad de Ciencias Químicas y Farmacéuticas, Universidad de Chile; Advanced Center for Chronic Diseases (ACCDiS), Santiago, Chile

A. Lalatsa
School of Pharmacy and Biomedical Sciences, University of Portsmouth, Portsmouth, United Kingdom

D. Male
The Open University, Milton Keynes, United Kingdom

C. McQuaid
The Open University, Milton Keynes, United Kingdom

C. Morais
CNC – Center for Neuroscience and Cell Biology; Department of Life Sciences, University of Coimbra, Coimbra, Portugal

S. Nyberg
University College London, London, United Kingdom; Biological Sciences, Sunnybrook Research Institute, Toronto, ON, Canada

M.C. Pedroso de Lima
CNC – Center for Neuroscience and Cell Biology, University of Coimbra, Coimbra, Portugal

A.C. Sintov
Faculty of Engineering Sciences, Ben Gurion University of the Negev, Be'er Sheva, Israel

X. Tian
School of Life Sciences, Anhui University, Hefei, People's Republic of China

C. Velasco-Aguirre
Facultad de Ciencias Químicas y Farmacéuticas, Universidad de Chile; Advanced Center for Chronic Diseases (ACCDiS), Santiago, Chile

A.T. Viegas
CNC – Center for Neuroscience and Cell Biology; Inter-University Doctoral Programme in Ageing and Chronic Disease, Faculty of Medicine, University of Coimbra, Coimbra, Portugal

J.T.-W. Wang
Institute of Pharmaceutical Science, Faculty of Life Sciences and Medicine, King's College London, London, United Kingdom

PREFACE

Central nervous system (CNS) diseases, and more specifically the chronic age-related neurodegenerative disorders, constitute a set of challenging pathological conditions, from both diagnostic and therapeutic point of view. This is particularly true for most of these disorders, due to the lack of early diagnostic biomarkers to allow proper follow-up of disease progression and effective therapeutic strategies to allow a persistent cure. Once the disease is accurately diagnosed, treatment of CNS diseases constitutes another challenge mainly due to the physically and chemically protected brain and spinal cord, when compared to peripheral organs. The blood–brain barrier (BBB) maintains essential brain homeostasis but significantly restricts the delivery of most therapeutic agents to the brain parenchyma. This is further complexed by the lack of tissue regenerative properties of the brain cells. The recent advances in nanotechnology have allowed new fields of research to investigate cutting-edge brain-specific therapies and to tackle the complex brain-related disorders. Nanoparticles (NPs) provide a flexible platform for conjugating drugs and targeting ligands and have been extensively researched to facilitate BBB crossing and effective delivery to the brain.

This Special Issue sought to bring together researchers and the NPs they use to overcome the big challenge of crossing the BBB. Because the focus was on establishing knowledge to experts from other fields, selective NP examples and targeting strategies are being introduced. Some of the results are novel and not discussed elsewhere. We believe that bringing together experts in nanotechnology, each working on a different class of NPs, makes this work distinctive in nature. In addition, this volume presents some in vitro models to be used for assessing NPs transcytosis, which is distinctive from transport from small molecules due to the need for specialized transport mechanisms, that is lacking in many existing in vitro BBB testing models. Another topic that needs to be addressed and is introduced in this issue is the different transcytosis routes exploited by a range of NPs.

The format of the book is that of stand-alone chapters in the form of a compendium. This way, each chapter can be used as a reference on a particular type of NPs, containing the most updated literature on that topic at the time of publication, in the majority of the chapters. The first chapter *Recent Trends in Nanotechnology Toward CNS Diseases: Lipid-Based Nanoparticles and Exosomes for Targeted Therapeutic Delivery*, by Drs. Cardoso, Guedes, Cardoso, Morais, Cunha, Viegas, Costa, Jurado, and Pedroso de

Lima, starts with an introduction on structural and functional units of CNS composition and the CNS microvessels. It then discusses the latest advances employing lipid-based NPs and cell-produced exosomes, as drug and nucleic acid delivery systems. It summarizes some examples of their applications in the context of neurological diseases, with emphasize on success stories in preclinical disease models. Findings from this chapter provide new prospects for further developments using this class of biocompatible nanocarriers, thus enabling researchers to move from the research realm to the clinical arena.

Designing NPs that effectively enter the CNS rapidly and without alteration is one of the major challenges in the use of nanotechnology for the brain. Transcytosis, a receptor-mediated transport pathway, permits endogenous macromolecules to enter the CNS by crossing the BBB. Transcytosis across the BBB involves a number of distinct stages, including receptor binding, endocytosis into a transport vesicle, trafficking of the vesicle to the opposite side of the cell, and finally exocytosis and release of cargo. The second chapter *From the Blood to the Central Nervous System: A Nanoparticle's Journey Through the Blood–Brain Barrier by Transcytosis*, by Drs. Fullstone, Nyberg, Tian, and Battaglia, discusses the current knowledge on biological, physiological, and physical factors that influence NP transit through that stage of transcytosis, with implications for NP design. This chapter concludes with current progress made in designing NPs that exploit transcytosis for CNS delivery.

It is already recognized that the integrity of the BBB is compromised in some CNS conditions such as brain tumor and inflammatory disorders. The third chapter *Application of Nanomedicine to the CNS Diseases*, by Drs. Carradori, Gaudin, Brambilla, and Andrieux, discusses in a systematic manner, and in a particular order, the pathophysiology of glioblastoma, Alzheimer's disease and stroke, conventional treatment options, and the limitations encountered in a disease by disease fashion. This is followed by insights on the preclinical and clinical NP-based approaches for treatment of these three conditions, detailing NP advantages and disadvantages. The examples given in this chapter contribute to demonstrate that delivering drugs into the brain is one of the promising applications of nanotechnology in clinical neuroscience.

Carbohydrate-based NPs have received interest as BBB delivery carriers due to their hydrophilicity, biodegradability, biocompatibility, and ability for large-scale manufacture. Chitosan-based nanocarriers, for example, have shown ability to deliver drugs, peptides/proteins, and genetic drugs after intravenous, nasal, and oral administration in preclinical proof-of-concept studies resulting in enhanced brain levels and pharmacodynamics responses. The fourth chapter *Carbohydrate Nanoparticles for Brain Delivery*, by Drs. Lalatsa and Barbu, discussed methods of preparation of carbohydrate-based NPs and the associated preclinical studies with main findings tabulated. The chapter touches upon some of the novel emerging NPs types which belong to this chemical class. The authors proposes that pullulan, dextran, and cellulose nanocrystals are emerging technologies that warrant further research to elicit novel delivery systems for overcoming the BBB.

Some types of metallic NPs hold great potential for delivery of therapies into the CNS, with gold NPs being the most studied type. The fifth chapter *Gold Nanoparticles for Imaging and Drug Transport to the CNS*, by Drs. Male, Gromnicova, and McQuaid, discusses how gold NPs with a core size of 2 nm, depending on their surface coat, can cross the brain endothelium in vitro by the plasma membrane/cytosolic route (passive transport) or by vesicular transcytosis (active transport). Gold NPs are relatively easy to synthesize and can be modified with various types of surface ligands. Their relatively small sizes allow them to pass through brain endothelium in vitro and in vivo and move rapidly through the brain parenchyma, in neurons and glia, within minutes of infusion. It is considered crucial to improve on the biodistribution of the NPs for CNS drug delivery; smaller gold NPs are removed rapidly via the kidney, while larger NPs are taken up by mononuclear phagocytes in various tissues. The sixth chapter *Metal Nanoparticles as Targeted Carriers Circumventing the Blood–Brain Barrier*, by Drs. Sintov Velasco-Aguirre, Gallardo-Toledo, Araya, and Kogan, proposes the intranasal administration route as a way to circumvent the BBB. Despite exploiting this route for nose-to-brain delivery, there is little known about fate of metallic NPs following this route of administration which is one of the aspects this chapter addresses.

NPs explored over the past decades are spherical in nature. Recent advances in nanotechnology allow fabrication NPs of high aspect ratios (>3) such as rod-shaped gold NPs and carbon nanotubes (CNTs). CNTs exhibit several attractive characteristics allowing their use in the brain environment. The properties include the ability to efficient crossing of biological membranes, multiple cell internalization pathways, and their intrinsic physical properties, eg, their infrared absorption and self-heating properties. The seventh chapter *Current Perspective of Carbon Nanotubes Application in Neurology*, by Drs. Kafa, Wang, and Al-Jamal, discusses major advances in using CNTs for treating brain tumor and degenerative diseases with special focus on their abilities to cross the BBB following systemic administration, which is the major obstacle for most other NPs.

Collectively, this Special Issue provides some examples, and is not exclusive, of the types of the NPs that have been explored preclinically in brain delivery. It is clear that from the material presented here that there is real progress in the understanding of the physiology of the BBB and the types of nanotechnology-based carriers that need to be used to overcome this barrier. The advances in chemical modification approaches of these nanocarriers helped to improve their in vivo performance. The choice of the in vitro BBB models is critical; there is a need for a tight polarized model. The coculture with cells that could serve in induction of polarization and act as a target cell is reasonable idea. We need to find powerful imaging tools, ie, real time to understand mechanisms for transcytosis. Correlation of in vitro findings with in vivo distribution and function needs to be considered as one of the far-reaching objectives during the study design. There is clearly a future perspective for a wider range of drug delivery options.

Finally, despite the promise nanotechnology holds in brain delivery and imaging, it is important not to forget addressing other issues that may arise as a consequence. Some are the fate of these NPs and the challenges encountered before their clinical translation can be achieved. It is our sincere hope that the readers will be motivated to continue searching for new technology tools or even optimize on the existing ones to overcome the barriers to brain delivery. Some technologies such as sonoporation, which relies on temporarily opening the tight junctions of the BBB, although not covered in this volume, hold great promise. The number of CNS disorders and brain cancer is on a steep increase and success in the field of brain disease diagnosis and therapy will no doubt create a huge societal and economic impact.

I would like to thank all of the contributing authors for their excellent contributions to this volume. I also thank my research team members, namely, Drs. Pedro Costa and Julie Wang, and Mrs. Kuo-Ching Mei and Izzat Suffian, for their assistance and helping me editing this book. I also thank Prof. Peter Jenner, an editor of the *International Review of Neurobiology* book series, for his kind invitation. We also thank Ms. Poppy Garraway and Ms. Hannah Colford and all of the other fine people at Elsevier for their assistance and support in bringing this volume to fruition. It has been great pleasure working with all of you. You made my task simple and enjoyable.

Khuloud T. Al-Jamal
King's College London
July 2016

CHAPTER ONE

Recent Trends in Nanotechnology Toward CNS Diseases: Lipid-Based Nanoparticles and Exosomes for Targeted Therapeutic Delivery

A.M. Cardoso[*], J.R. Guedes[*], A.L. Cardoso[*], C. Morais[*,†], P. Cunha[*], A.T. Viegas[*,‡], R. Costa[*], A. Jurado[*,†], M.C. Pedroso de Lima[*,1]

[*]CNC – Center for Neuroscience and Cell Biology, University of Coimbra, Coimbra, Portugal
[†]Department of Life Sciences, University of Coimbra, Coimbra, Portugal
[‡]Inter-University Doctoral Programme in Ageing and Chronic Disease, Faculty of Medicine, University of Coimbra, Coimbra, Portugal
[1]Corresponding author: e-mail address: mdelima@ci.uc.pt

Contents

Abstract

Central nervous system (CNS) diseases constitute a set of challenging pathological conditions concerning diagnosis and therapeutics. For most of these disorders, there is a lack of early diagnosis, biomarkers to allow proper follow-up of disease progression and effective therapeutic strategies to allow a persistent cure. The poor prognosis of most CNS diseases is, therefore, a global concern, especially regarding chronic age-related neurodegenerative disorders, which are already considered problems of public health due to the increasing average of life expectancy.

International Review of Neurobiology, Volume 130
ISSN 0074-7742
http://dx.doi.org/10.1016/bs.irn.2016.05.002

The difficulties associated with the treatment of CNS diseases are owed, at least in part, to very specific characteristics of the brain and spinal cord, when compared to peripheral organs. In this regard, the CNS is physically and chemically protected by the blood–brain barrier (BBB), which, while maintaining essential brain homeostasis, significantly restricts the delivery of most therapeutic agents to the brain parenchyma. On the other hand, regenerative properties of the tissue are lacking, meaning that a CNS insult resulting in neuronal death is a permanent phenomenon.

Approaches for transposing the BBB aiming to treat CNS diseases, relying on specific properties of nanosystems, have been reported for therapeutic delivery to CNS without interfering with the normal function of the brain. In this chapter, we address the latest advances concerning the principles of such approaches, employing lipid-based nanoparticles and cell-produced exosomes as drug and nucleic acid delivery systems, and summarize recent example of applications in the context of neurological diseases. Major achievements obtained in preclinical studies and the trends identified by these studies are emphasized to provide new prospects for further developments in this area, thus enabling us to move from the research realm to the clinical arena.

1. INTRODUCTION

The nervous system is one of the most complex and mysterious systems of all organisms. It is usually divided into peripheral and central nervous system (CNS), the latter being composed of the brain and spinal cord that constitute the operation control center of the body, regulating all life processes from conscious actions to homeostatic functions. The main structural and functional unit of the nervous system is the neuron, a highly polarized cell type that participates in neuronal circuits, allowing the transmission of information across brain regions and the guidance of electrical impulses to peripheral organs in order to control their function. For decades, it was thought that the production of new neurons was confined to the period of embryonic development. However, neuronal stem cells (NSCs) localized in NSC niches, such as the subventricular zone of the lateral ventricles (Lois & Alvarez-Buylla, 1994), were shown to present persistent neurogenesis properties, indicating that neuronal division is a slow but natural process in adulthood (Reynolds, Tetzlaff, & Weiss, 1992). In addition to neurons, several other cell types also found in the nervous system play important roles in the maintenance of the electrical and chemical balances that allow the successful operation of neuronal circuits. The main non-neuronal cells of the brain parenchyma are astrocytes, oligodendrocytes, and microglia. While astrocytes contribute to physical scaffolding and to the restoration of normal concentration of neurotransmitters (the chemical unit of neuronal transmission) in the extracellular space (Khakh & Sofroniew, 2015), oligodendrocytes provide insulation of axons, endeavoring electrical efficacy in neuronal transmission (Chang, Redmond, & Chan, 2016). Microglia, which are the only cells of the CNS with a nonneuronal origin, deriving from primitive hematopoietic precursors, constitute the macrophages of the brain and spinal cord, carrying out functions of immune defense (Guedes, Viegas, Pedroso de Lima, & Cardoso, 2015). In addition to a difference in ontology (Ginhoux et al., 2010), microglia are distinct from other tissue-resident macrophages by their morphology, tissue-derived transcriptome (Gosselin et al., 2014), and specific functions related with neuronal activity (Zhan et al., 2014). The main role of microglia is to serve as the first line of defense against exogenous and endogenous threats to tissue homeostasis. Toward this goal, microglia processes constantly scan the surrounding environment, detecting, through specific receptors localized in their membrane, the appearance of particular pathogenic molecular patterns. This surveillance action is always present, changing only in a disease setting.

Nevertheless, the changes elicited by microglia activation, in the presence of virus, bacteria, and even endogenous proteins, can be so drastic that may lead to severe repercussions in the fragile equilibrium of the nervous tissue. Therefore, the immunosuppressive environment reported to exist in the brain is often erroneously interpreted as the lack of immune players. In fact, the activation of immune cells within the nervous system is tightly controlled, and the immunosuppressive status is maintained by physical and chemical interactions between microglia and neurons (Kierdorf & Prinz, 2013), which allow microglia to remain in a surveilling, nonactivated state. However, microglia-driven neuroinflammation can sometimes occur, simply due to the abolishment of immune repressor production by neurons, even in the absence of an exogenous threat, or, more frequently, triggered by pathogenic invasion of the nervous system, trauma, tumors, or abnormalities in the molecular brain composition causing neurodegeneration (Guedes et al., 2015). Under these conditions, microglia adopts a plethora of reactive phenotypes characterized by morphological alterations from a ramified to an amoeboid shape, cell proliferation, upregulation of activation markers, and production of inflammatory and antiinflammatory mediators, which follow cell migration to sites of injury (Guedes et al., 2015). The consequences of these phenotypic alterations depend on the type and duration of the activation stimulus and can range from tissue protective to highly destructive, even within the same disease setting.

The importance of the CNS in the control of all body processes is patent on its privileged localization and intrinsic protection mechanisms, which allow the maintenance of the brain microenvironment, essential for its normal function, with minimal interference from the outside. The most sophisticated of these mechanisms is ensured by the blood–brain barrier (BBB), a physical physiological barrier between the brain and the blood stream that allows the maintenance of CNS stability. The BBB was first discovered upon the injection of water-soluble dyes in the peripheric circulation, which did not stain the brain tissue, whereas their injection in the subarachnoid space colored the brain but not peripheral tissues (reviewed in Liddelow, 2011). This barrier is localized in the CNS microvessels and is composed of specialized brain endothelial cells (BECs), which are continuously joined by tight junctions and surrounded by a basement membrane, pericytes, and astrocytes (Chow & Gu, 2015).

Specific characteristics of BECs, which are induced and regulated by the neuronal environment, are responsible for maintaining BBB function and integrity: (1) tight junctions at the apical and lateral membranes of the BECs, which prevent the passage of small hydrophilic molecules and ions; (2) minimization of the endothelium vesicular trafficking or transcytosis (the vesicle-dependent process of transcellular passage of material); (3) highly polarized cellular transporters that regulate the influx of nutrients and efflux of toxic materials resulting from the intrinsic brain metabolism; and (4) lack of expression of leukocyte adhesion molecules, such as E-selectin and intracellular adhesion molecule-1 (ICAM1), preventing the entry of immune cells (reviewed in Chow & Gu, 2015). Although all these characteristics limit the passage of molecules with a molecular weight of more than 400–600 kDa, the existence of specialized nutrient transporters, like glucose transporter 1 (GLUT1), allows the specific access of larger compounds to the brain (Simpson, Carruthers, & Vannucci, 2007). Several cellular pathways have been implicated in these processes, including Wnt/β-catenin signaling, which has been demonstrated to be essential for proper angiogenesis and regulation of GLUT1 expression (Daneman et al., 2009; Stenman et al., 2008).

In a therapeutic context, specific characteristics of the BBB can provide unexpected advantages for the improved CNS drug delivery (reviewed in Banks, 2016). The use of drugs that act as analogs of endogenous molecules and that can bind to BBB transporters, being quickly carried into the CNS, is a strategy that benefits from the complex structure of the BBB.

For example, some oligonucleotide analog transporters have been shown to efficiently deliver antisense oligonucleotides against APP to the brain, improving cognitive deficits in Alzheimer's disease mouse models (Farr, Erickson, Niehoff, Banks, & Morley, 2014). A physiological approach for enhancing therapeutic delivery into the brain takes advantage of the transcytosis activity of specific receptors expressed at the BBB, such as transferrin receptors, this mechanism being extensively used to target drug delivery systems to the brain (reviewed in Johnsen & Moos, 2016). Importantly, lipids could also be used to functionalize nanoparticles serving as targeting molecules and also signaling modulators, rather than being simple structural components. In support of this notion, MFSD2A, a major transporter of the omega-3 fatty acid DHA, which is essential for normal brain growth and cognitive function modulating cell signaling (Nguyen et al., 2014), was found to be exclusively and abundantly expressed in BECs (Ben-Zvi et al., 2014). Thus, DHA incorporated in nanoparticles could act as an effective agent to target MFSD2A to BECs, enabling the delivery systems to transpose the BBB. In this context, the unique lipid composition of the brain and the brain requirement for lipid mediators represents an important opportunity for the development of new brain-targeted delivery systems, as well as of multifunctional lipid-based vectors, which, besides allowing the delivery of specific therapeutic agents (including drugs and nucleic acids), can also mediate the delivery of bioactive lipid components.

Nanoparticles for the delivery of therapeutic biomolecules to the CNS have been produced from a diversity of molecules, namely, peptides and polymers, and the reader is referred to excellent and recent reviews on these systems (Patel, Zhou, Piepmeier, & Saltzman, 2012; Zou, Ma, Wang, Yang, & Liu, 2013). This chapter will focus on the latest research employing lipid-based delivery systems for brain targeting, from liposomal vehicles to cell-produced exosomes. Lipids were, in fact, the first nonviral molecules to be explored in the context of CNS-targeted delivery. Their ability to assemble into vesicles composed of hydrophilic and hydrophobic moieties, allow for the aqueous encapsulation of water-soluble therapeutic molecules, and for the membrane incorporation of lipid-soluble drugs. In fact, the wide variety of existing lipids, the numerous possibilities for different combinations, and the potential for chemical modifications with functional ligands, make lipids suitable molecules for generating targeted nanoparticles. Fig. 1 illustrates the versatility of lipid-based delivery systems concerning the

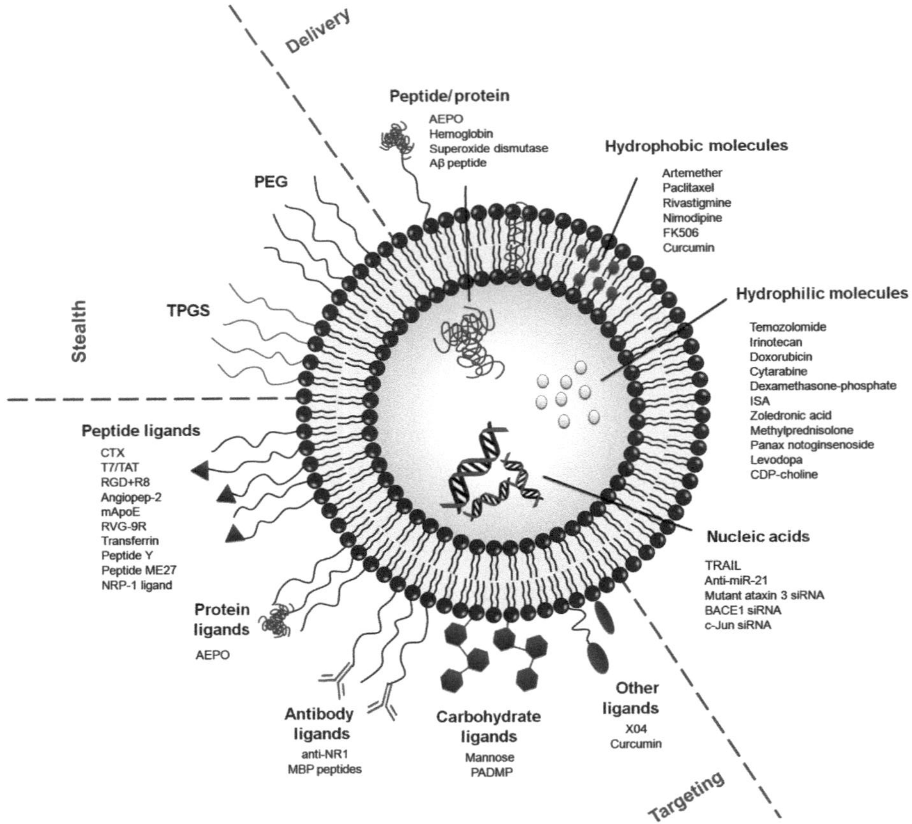

Fig. 1 Multifunctional lipid-based systems used in the treatment of CNS disorders. These vehicles present two compartments, one hydrophobic and one hydrophilic, which confers them the ability to carry a variety of drug molecules. Vector targeting promotes the delivery of therapeutic molecules to the site of interest, which can be achieved through ligand coupling to confer cell specificity, and improved by the incorporation of polymeric compounds, providing them with stealthiness. See text for detail.

variety of therapeutic molecules carried to address CNS diseases, as well as the multiple possibilities of surface modifications to improve cell-targeting ability and circulation lifetime. On the other hand, the recent use of cell-produced exosomes to specific cargo delivery paved the way for exploiting new brain-targeted therapeutic strategies minimizing unwanted side effects.

2. LIPID-BASED NANOPARTICLES FOR NUCLEIC ACID AND DRUG DELIVERY TO THE CNS

CNS diseases comprise a set of pathological states considered to be notably challenging in terms of diagnosis, assessment, monitoring, and treatment. As previously discussed, some characteristics of the brain and spinal cord, namely, the physical and chemical protection offered by the BBB, the immune-privileged environment, and the lack of regenerative properties, render the CNS an extremely isolated system, preventing access of therapeutics in case of disease. CNS diseases are often associated with processes of neurodegeneration that can range from mild effects in neuronal structures, such as changes in dendrite arborization and synaptic remodeling, to more permanent damage, which culminates in death of specific subtypes of neurons. Moreover, neurodegeneration is almost always associated with neuroinflammatory processes, which comprise production of cytokines and neurotrophic factors, as well as the clearance of cellular debris in the case of critical neuronal dysfunction, the extent of the damage depending on the type of insult. CNS disorders can occur as consequence of a combination of environmental and genetic factors, which can occur at different degrees. In fact, ranging from completely exogenous origin, such as is the case of infections, to hereditary monogenic diseases, a large spectrum of conditions can be listed.

In this section, the employment of lipid-based nanoparticles to deliver therapeutic agents to the CNS will be addressed in the context of different CNS diseases with relevant epidemiological impact in society, including brain injuries, mental disorders, neurodegenerative, and autoimmune diseases, and brain tumors. The advantages of applying drug delivery liposomal systems to treat brain diseases, due to their ability to accumulate in the CNS and overcome the serious health risks associated with most drugs used to treat CNS disorders, will also be emphasized.

2.1 Ischemic Stroke

Stroke, an ever-growing burden with an estimated worldwide incidence of 15 million cases each year and a mortality rate of approximately 30%, is the first leading cause of lifelong disability and the third leading cause of death in the developed world (Bendixen & Ocava, 2002; Gorelick & Ruland, 2010; Shafi & Levine, 2010). Accounting for more than 80% of all strokes, acute ischemic stroke is caused by a thrombotic or embolic occlusion of a major cerebral artery (commonly the middle cerebral artery), or its branches, resulting in a transient or permanent cerebral blood flow reduction, and triggering a metabolic cascade of events that, unless promptly reversed, culminate in nonselective neuronal cell death within the affected tissue (Adibhatla & Hatcher, 2007; Durukan & Tatlisumak, 2007). The long-term functional effects of an ischemic episode largely depend on the size and location of the arterial occlusion, as well as on the duration of the reduction of cerebral blood flow (Sims & Muyderman, 2010).

Currently, the only drug approved for neuroprotection following cerebral ischemia is the tissue plasminogen activator (t-PA), whose use has been associated with increased risk of intracerebral hemorrhage (Ishii, Fukuta, et al., 2013). Thus, alternative therapeutic strategies toward acute ischemic stroke are urgent and, in this context, liposome-based drug delivery appears as a promising approach. In fact, accumulation of DiI-labeled liposomes of distearoylphosphatidylcholine (DSPC)/cholesterol (Chol) (2/1 molar ratio) in the ischemic core (80% reduced blood flow) and penumbra regions (20–40% blood flow depression) was demonstrated in an ischemia rat model of middle cerebral artery occlusion (MCAO), even in the absence of blood circulation recovery, suggesting the possibility of delivering liposome-encapsulated drugs into the ischemic brain (Ishii, Fukuta, et al., 2013). Although this accumulation has been partly attributed to BBB disruption in ischemic brain (Ishii, Fukuta, et al., 2013), the formulation composed of DSPC/Chol (prepared at 5/1 molar ratio) was shown to mediate brain delivery of the encapsulated cerebral protective antiinflammatory drug, isopropylidene shikimic acid (ISA), following intravenous (i.v.) administration, even in healthy animals with noncompromised BBB (Qu et al., 2013).

Liposomes prepared from dipalmitoylphosphatidylcholine (DPPC)/dioleoylphosphatidylcholine (DOPC)/Chol have been used to allow a controlled release of nitric oxide (NO), a regulator of cerebral artery tone and neuroprotector, thus providing a means of preventing NO scavenging by hemoglobin and achieving a high local NO concentration (Kim et al.,

2014). Another liposomal formulation, composed of a mixture of phosphatidylcholine (PC), Chol, and phosphatidylethanolamine, was used to deliver angiogenic peptides derived from the vascular endothelial growth factor to rat ischemic brain (Hwang et al., 2015). The authors have shown that the generated small and stable liposomes were able to efficiently encapsulate the peptides and, upon intraarterial delivery (via carotid artery), accumulated in the ischemic hemisphere of the brain. Importantly, the application of this strategy resulted in attenuation of perfusion defect and increased expression of a gene involved in angiogenesis (angiopoietin-2), with the consequent increase of glucose consumption and vascular density, without promoting inflammation (Hwang et al., 2015).

Liposome modifications aiming at improving their circulation time and ability to avoid the mononuclear phagocyte system (MPS) include the use of polyethylene glycol (PEG) to decorate the liposomal particles. This polymeric molecule is highly hydrated and confers stealthiness to the particles, helping them to evade immune system recognition. Due to the hydration layer of the particles, minimal interaction with opsonins and cells of the MPS occurs, as well as increase in the circulation time of the PEGylated particles has been observed (Allen & Hansen, 1991; Avgoustakis, 2003). In fact, PEG is present as an essential component of the majority of drug delivery systems used for brain tumor treatment in in vivo studies, usually as a distearoylphosphatidylethanolamine (DSPE)-PEG conjugate. Similarly to what has been observed with nonmodified liposomes, vesicles of DSPC/Chol functionalized with PEG-2000 accumulated in the ischemic region of permanent MCAO rats, despite the lack of cerebral perfusion, as evaluated by positron emission tomography scan (Fukuta et al., 2014). The preferential accumulation of the PEGylated liposomes in the ischemic region has been attributed to a reduction of integrity of the BBB caused by secretion of matrix metallopeptidase 9 (MMP-9) by neutrophils recruited to the ischemic injury site, which leads to BBB collagen degradation (Gidday et al., 2005; Latour, Kang, Ezzeddine, Chalela, & Warach, 2004; Rosell et al., 2008) and depletion of proteins of tight junctions (Zehendner, Librizzi, de Curtis, Kuhlmann, & Luhmann, 2011). Thus, PEGylated liposomes could be used to deliver drugs in order to protect the ischemic region from damage during recovery from ischemia.

PEGylated liposomes of DPPC/DSPE/Chol and DPPC/DSPE were used to encapsulate the antiinflammatory compounds dexamethasone phosphate and FK506 (Tacrolimus), respectively, in order to counteract the inflammation in ischemic stroke. These liposomal formulations were shown to enhance drug accumulation in the ischemic hemisphere, which was attributed to the enhanced permeability and retention (EPR) effect, and to improve the neurological outcome of ischemic animals, when compared to that of empty liposomes and to the free drug (Ishii, Asai, et al., 2013; Tiebosch et al., 2012).

Liposomes functionalized with PEG were also used to deliver hemoglobin to treat ischemic stroke. Liposomal hemoglobin-mediated oxygen infusion allows oxygenation of the ischemic tissue, without neutrophil infiltration, therefore reducing brain damage. In this regard, hemoglobin encapsulated in liposomes of soybean PC, Chol, and stearic acid functionalized with PEG (Kaneda et al., 2009) showed to be protective of brain ischemia in rats (Kaneda, Ishizuka, Sekiguchi, Morimoto, & Kasukawa, 2014; Kawaguchi et al., 2014; Shimbo et al., 2014) and in non-human primates (Kawaguchi et al., 2013), reducing the volume of the infarct volume and preserving the cortical region from ischemic and reperfusion damage.

Another approach toward the treatment of brain ischemia mediated by lipid-based systems was reported, which lies in drug chemical conjugation with lipids before encapsulation into liposomes. This procedure was employed in the case of the cytoprotector asialo-erythropoietin (AEPO), which fails to accumulate in the brain due to its short circulation half-life. This drawback was surpassed by using liposomal AEPO (Ishii, Asai, Fukuta, et al., 2012), which resulted from AEPO chemical conjugation to DSPE-PEG (DSPE-PEG-AEPO), followed by incubation with liposomes of DSPC, Chol, and DSPE-PEG. Liposome-formulated AEPO accumulated preferentially at the ischemic site, in a time-dependent manner, which was attributed to the increased circulation time of the liposomal formulation and to the EPR effect. AEPO liposomes prevented cerebral cell death in the striatum and improved neurological function (improvement of motor abilities), allowing dose reduction of AEPO required for a therapeutic effect (Ishii, Asai, Fukuta, et al., 2012; Ishi, Asai, Oyama, et al., 2012).

Another strategy for drug delivery is based on the use of mixed polymer–lipid vectors. Formulations with this feature were employed to deliver antioxidants, such as panax notoginsenoside (PNS) and luteolin, used to ameliorate oxidative stress resulting from ischemic brain damage. Thus, PNS, which inhibits lipid peroxidation, was loaded into PEG-poly lactic-*co*-glycolic acid polymeric nanoparticles and the resulting suspension was used to hydrate a lipid film of soybean PC and Chol. The polymer–lipid formulation was able to slowly release PNS, when compared to either the liposomal or the polymeric formulations per se. In addition, treatment with this formulation protected animals from ischemia by reducing infarct volume and brain edema (Zhang et al., 2012). Similarly, luteolin, a polyphenolic compound with antioxidant activity similar to that of vitamin E, has been encapsulated into liposomes of soybean lecithin and Chol, hydrated with a solution containing the nonionic triblock copolymer poloxamer. Repeated intraperitoneal (i.p.) administration of the luteolin-containing liposomes in MCAO rats resulted in decreased symptoms associated with ischemia/reperfusion (regarding limb cooperation and body balance), decreased histological injury (monitored in brain slices), decreased reactive oxygen species (ROS) production, as well as increased glutathione levels and catalase activity. These effects indicate that liposomal luteolin has protective antiischemic activity that can be attributed to the correction of the dysfunctional prooxidant/antioxidant imbalance resulting from ischemic stroke (Zhao et al., 2011). Oxidative stress in ischemia was also tackled using superoxide dismutase (SOD) conjugated liposomes of hydrogenated soybean PC (HSPC)/Chol (Yun et al., 2013). In this study, anti-*N*-methyl-D-aspartate (NMDA) receptor 1 (NR1) antibody was added to liposomes in order to specifically target the nanoparticles to cerebral neurons. Both SOD and NR1 antibodies were covalently attached to HSPC/Chol PEGylated liposomes, and the resulting SOD-coated-NR1-targeted liposomes showed to be more efficient than their nontargeted counterparts, decreasing the infarct volume in an ischemia mouse model to ca. 35% (Yun et al., 2013).

Liposome functionalization with mannose was also reported as a strategy to enhance brain accumulation of therapeutic agents, due to an increase in liposome uptake mediated by the mannose receptors present in brain macrophages and microglia. This approach was used with cytidine diphosphate-choline (CDP-choline) (Ghosh, Das, Mandal, Dungdung, & Sarkar, 2010), a neuroprotective agent able to decrease mitochondrial damage caused by ischemia, whose poor bioavailability is due to its rapid liver metabolization, resulting in the production of metabolites which, in contrast to CDP-choline, are not able to cross the BBB. Liposomal encapsulation of CDP-choline showed the potential to increase the half-life of this drug and to promote its accumulation in a model of brain ischemia. CDP-choline delivered by mannosylated liposomes protected brain cells against mitochondrial damage, membrane lipid peroxidation, ROS production, and the release of cytochrome *c* at a lower concentration than the free drug (Ghosh et al., 2010).

Liposome functionalization with transferrin was used in our laboratory with the purpose of enhancing neuronal delivery, through transferrin receptor-mediated endocytosis (Cardoso et al., 2008). Thus, Cardoso and coworkers developed and characterized dioleoyltrimethylammoniumpropane (DOTAP)/Chol liposomes associated to transferrin, via electrostatic interaction, to target hippocampal neurons, in order to mediate the delivery of siRNAs anti-c-Jun, a transcription factor that plays a major role in triggering neuronal apoptosis (Cardoso et al., 2008). Thereafter, the same authors employed this delivery system in an in vivo model of excitotoxic injury, triggered by intraventricular injection of the glutamate receptor agonist kainate, since excitotoxicity, one of the main features of acute brain injury, results from the excessive release of glutamate by damaged neurons (Cardoso et al., 2010). The authors reported a significant reduction in neuronal death in the CA1 and CA3 hippocampal regions, following specific c-Jun silencing in these regions, as well as a decrease in the expression of inflammatory mediators and a decrease in the number and severity of the seizures that are characteristic of this acute injury model, which can also be valuable in the context of ischemic stroke. In fact, the acute energy deprivation that occurs as a consequence of an ischemic insult leads to an excessive release of the excitatory neurotransmitter glutamate from presynaptic terminals, resulting in the overstimulation of ionotrophic receptors and in an unexpected increase in intracellular calcium levels. This massive increase in intracellular calcium activates a series of signaling pathways, involving kinases, phosphatases, proteases, and the production of free radicals, which culminate in DNA damage, lipid peroxidation, mitochondria dysfunction, and, ultimately, neuronal death (Lo, Dalkara, & Moskowitz, 2003; Lo, Moskowitz, & Jacobs, 2005). Although early neuronal loss in ischemia occurs through necrosis, a substantial number of neurons die, in the days following the acute event, through apoptosis, in which the kinases p38 and Jun and their primary target, the transcription factor c-Jun, are important players.

2.2 Neuropathic Pain

Neuropathic pain is a chronic pain state that usually results from nerve fiber dysfunction or damage, as a consequence of central or peripheric nerve injury. This condition is secondary to a number of diseases, from trauma injury to multiple sclerosis, has a high prevalence worldwide, and is extremely difficult to treat. Although not definitely characterized in terms of cell mechanisms, incorrect pain signaling results from activation of astrocytes and microglia, which causes neuronal sensitization following peripheral nerve injury. Damaged neural fibers promote neuronal release of nociceptor mediators, which contribute to the establishment of mechanical allodynia, the neurological symptom of chronic neuropathic pain. Astrocyte and microglia activation causes a proinflammatory state, characterized by mitogen-activated protein kinases (MAPKs) signaling cascades (Caraglia et al., 2013). Among MAPKs, the extracellular signal-regulated kinase (ERK), which is induced in dorsal horn neurons at the initial stage of nerve damage, and in microglia and astrocytes at a later stage after injury in a sustained fashion, is at least partially responsible for sustaining chronic pain (Zhuang, Gerner, Woolf, & Ji, 2005).

Liposomes composed of egg PC/Chol/DSPE-PEG were used to encapsulate zoledronic acid (ZOL), a drug able to inhibit Ras-dependent ERK, and thus to ameliorate neuropathic pain (Caraglia et al., 2013). This encapsulation strategy was shown to improve drug accumulation and bioavailability within the CNS. Importantly, liposomal ZOL was found to reduce significantly mechanical allodynia in a mouse model of this pathological condition (spared nerve injury), as opposed to empty liposomes or free ZOL. In addition, the number of hypertrophic-activated astrocytes was significantly reduced by two administrations of liposomal ZOL. Such protective phenotype conferred by liposomal ZOL to astrocytes was supported by the increased expression of the antiinflammatory cytokine IL-10 on these cells (Caraglia et al., 2013).

In another study, vesicle-forming amphiphiles or "bolalipids," containing two hydrophilic head groups at each end of a hydrophobic alkyl chain, were synthesized and mixed with Chol and cholesteryl hemisuccinate to produce liposomes able to encapsulate analgesic compounds and to improve their delivery to the CNS (Popov et al., 2013). The approach consisted of generating bolalipids containing an acetylcholine (ACh) headgroup, aiming at fulfilling two goals: (1) to confer a cationic surface to the delivery systems and (2) to provide a cleavable moiety for choline esterases. The therapeutic strategy involved intramuscular administration of pyridostigmine, a choline esterase inhibitor that does not cross the BBB, prior to liposome i.v. injection. Thus, even though the liposomes did not exclusively accumulate in the brain, their cargo was not released in other tissues, due to choline esterase inhibition (Popov et al., 2013). Further improvement of the liposomal vehicle was achieved by surface decoration with chitosan, a polymer that enhances the ability of these liposomal formulations to cross the BBB. Efficient and prolonged analgesia was achieved when these formulations were used to encapsulate and deliver the analgesic peptides leu-enkephalin or kyotorpin into mice brain (Popov et al., 2013).

2.3 Alzheimer's Disease

Alzheimer's disease (AD) is the most common form of dementia worldwide, being characterized by progressive neuronal loss, neuroinflammation, and cognitive impairment (Wyss-Coray & Rogers, 2012). Two hallmark lesions are found in this disease: senile plaques, composed of the Aβ peptide aggregated in fibrils and surrounded by glial cells and intracellular neurofibrillary tangles, containing hyperphosphorylated tau protein (Oddo et al., 2003). Despite major efforts, the etiology of AD remains unclear. However, excessive Aβ production and consequent aggregation appears to be upstream of tau hyperphosphorylation and genetic mutations in proteins involved in Aβ production, namely, amyloid precursor protein (encoded by the APP gene), β-secretase (encoded by the BACE1 gene), and γ-secretase, composed of the subunits presinilin 1 and 2 (encoded by PS1 and PS2 genes), were shown to recapitulate AD pathological markers (reviewed in Bohm et al., 2015).

In the context of AD, liposomes have been used not only to deliver therapeutic compounds but also as delivery agents of vaccines or as "sponges" for pathological molecules, these two strategies having been applied to tackle AD with promising results. In this regard, Aβ peptide emulsified in liposomes composed of DOPC, palmitoyloleoylphosphatidylglycerol, Chol, and sphingosine-1-phosphate was used both as a vaccine and as a therapy in the mutant APP/PS1 mouse model of AD (Carrera et al., 2013). The immunization procedure, performed by i.p. injection of Aβ-liposomes, prior to the onset of Aβ deposition, reduced β-amyloid plaque formation, and inflammation in the hippocampus and cortical areas, attenuating AD neuropathology and impairing the development of the disease. Even after Aβ plaque formation, liposomal Aβ administration reduced inflammation and restricted it to neurodegeneration sites, pointing to the therapeutic potential of this strategy in AD (Carrera et al., 2013).

A therapeutically relevant concept in AD is based on the assumption that brain and peripheral blood AD are in equilibrium with each other. Hence, the removal of Aβ from the periphery is expected to result in an equilibrium shift, with consequent reduction of Aβ in the brain. This strategy is extremely appealing because it does not implicate introducing foreign compounds in the exquisite brain environment (Ordonez-Gutierrez et al.,

2015). Liposomes composed of sphingomyelin and Chol containing either phosphatidic acid (PA) or cardiolipin (5 mol%) were reported to bind Aβ with high affinity and were shown to reduce Aβ-induced toxicity in mouse neuroblastoma cell lines (Bereczki, Re, Masserini, Winblad, & Pei, 2011). Consistently, upon repeated i.p. injection of these vesicles in APP/PS1 mice, Aβ levels in the plasma were significantly reduced (Ordonez-Gutierrez et al., 2015). Although brain levels of Aβ were only mildly affected, this procedure resulted in decreased activation of glycogen synthase kinase 3 (GSK3) in the brain and in the consequent reduction of tau phosphorylation. In addition, despite being anionic, these liposomes did not activate the complement system and did not elicit a proinflammatory response, which suggests that liposomes targeting circulating Aβ may be used as therapeutic agents in AD (Ordonez-Gutierrez et al., 2015).

In order to prevent cholinergic dysfunction resulting from the overexpression of acetylcholinesterase (AChE), which is promoted by the chronic inflammatory state present in the AD brain, reflected by microglia IL-1β production, the cholinesterase inhibitor rivastigmine (RIVA), encapsulated in liposomes composed of egg PC, dihexadecyl phosphate (DCP), and Chol, was administered via i.p. injection in an $AlCl_3$-induced rat model of AD, which shows the same behavior, biochemical, and histological symptoms as AD patients. Liposomal RIVA was found to be less toxic than free RIVA, and the observed decrease of amyloidogenic process was attributed to the decrease of AChE activity in parallel with a downregulation of AChE, IL1B, and BACE1 gene expression with respect to untreated animals (Ismail, Elmeshad, & Salem, 2013). The superior therapeutic effect of liposomal RIVA over that of free RIVA was evidenced by its ability to protect the rat brain by preventing amyloid plaque formation and to confer memory recovery in $AlCl_3$-treated animals. Furthermore, addition of stabilizing agents, such as methyl cellulose and dimethyl-β-cyclodextrin to liposomes of DPPC and Chol containing RIVA, resulted in a prolonged AChE inhibition and in a longer beneficial effect of RIVA in the mouse brain (Muthu, Kulkarni, Xiong, & Feng, 2011).

Multifunctional liposomes have also been designed for AD therapy, aiming to prevent the formation of Aβ aggregates and to promote their dissolution in the brain, as well as to enhance their clearance (Balducci et al., 2014). With this purpose, liposomes of sphingomyelin and Chol were functionalized with PA (to bind Aβ peptide) and with a peptide derived from the receptor-binding domain of ApoE (mApoE) to facilitate BBB crossing (Bana et al., 2014). The use of these two components resulted in disaggregation of Aβ peptide aggregates, presumably due to synergistic interaction of the anionic PA phosphate group and of the cationic amino acids of mApoE peptide with the Aβ residues of opposite charge (Bana et al., 2014). APP/PS1 mice treated with bifunctionalized liposomes showed a significant and sustained reduction in the number and area of Aβ plaques in the cortex and striatum and had their long-term recognition memory restored to levels similar to those presented by wild-type animals. Furthermore, both soluble and insoluble Aβ levels were significantly lowered in the brain of treated animals, when compared with untreated transgenic controls (Balducci et al., 2014). A similar approach, in which curcumin conjugated to phosphatidylserine (PS) was used as the Aβ ligand in liposomes of DPPC and Chol, was tested in cultured cells. Curcumin is a poorly hydrosoluble molecule that labels senile plaques. It was shown that liposomal curcumin, injected stereotactically in the hippocampus and neocortex of APP/PS1 mice, colocalized with Aβ deposits and was able to partly prevent Aβ-induced cell death and to downregulate Aβ expression (Lazar et al., 2013). Methoxy-X04 (4,4′-[(2-methoxy-1,4-phenylene)di-(1E)-2,1-ethenediyl]bisphenol, X04), another Aβ ligand, was also shown to bind Aβ plaques with the additional advantage of being fluorescent, which allows its use in vivo as a localization tool for Aβ aggregates. Liposomal X04, produced as the lipid conjugate DSPE-PEG-X04, maintains the ability of X04 to bind Aβ deposits and, after i.v. administration, colocalizes with parenchymal Aβ plaques and with blood vessels affected by cerebral amyloid angiopathy in APP/PS1 mice (Tanifum et al., 2012).

Taking into consideration the role of β-secretase in Aβ production, thus constituting an interesting target for AD therapy, the ability of liposomes composed of dioleoylphosphatidylglycerol (DOPG) and dioleoylphosphatidylethanolamine (DOPE), functionalized with PEG linked to DOPE or to dipalmitoylphosphatidylethanolamine (DPPE), to deliver anti-BACE1 siRNAs was evaluated (Tagalakis et al., 2014). In addition to PEGylation, which increased the stability of the formulations and the bioavailability of the delivered nucleic acids resulting in enhanced activity in vivo, a cationic peptide (peptide Y or peptide ME27) was added to facilitate the entry of the nanoparticles into brain cells. Interestingly, nanoparticles containing DPPE-PEG showed higher ability to mediate BACE1 gene silencing than those containing DOPE-PEG. Tagalakis and colleagues attributed this effect to shorter alkyl chains (DPPE, C16), which would be associated with greater flexibility and lower gel-to-fluid phase transition temperature than longer alkyl chains (DOPE, C18)

(Tagalakis et al., 2014). However, the alkyl chains of DOPE are unsaturated, presenting a *cis* double bond (C18:1,Δ9*cis*), whereas those of DPPE (C16:0) are entirely saturated, which explains the substantially lower transition temperature of DOPE (−16°C) as compared to that of DPPE (+63°C), ie, in the opposite direction to that reported by the authors. Therefore, a more convincing explanation for the higher efficiency of DPPE-PEG delivery systems to mediate BACE1 gene silencing, as compared to those prepared from DOPE-PEG, could reside, in our opinion, in the higher compaction of DPPE-containing particles. This could account for the authors reported observations of smaller diameter DPPE-containing particles and the higher protection conferred to siRNAs in the presence of serum, thus contributing to the delayed release of nucleic acids, thereby preventing extracellular degradation and allowing more efficient internalization by the target cells.

2.4 Parkinson's Disease

Parkinson's disease (PD) is the second most common neurodegenerative disease, being characterized by motor dysfunctions, such as uncontrollable tremor, bradykinesis, muscular rigidity, and postural instability, caused by progressive loss of dopaminergic neurons in the *substantia nigra pars compacta* (SNpc) and the presence of cytoplasmic inclusions composed of insoluble aggregates of α-synuclein (α-Syn) in association with other proteins, named Lewy bodies (Ruiperez, Darios, & Davletov, 2010). Elevated α-Syn expression in neurons was correlated with neuronal apoptosis resistance, suggesting a key role of this protein in neuroprotection in response to oxidative stress conditions (Quilty et al., 2006). However, extracellular α-Syn was found to oligomerize into neurotoxic aggregates, similar to those produced by Aβ peptide in AD brain. Presumably, α-Syn can form ion-permeable pores in the plasma membrane of neurons, which are thought to be the main drivers of neurotoxicity (Pacheco, Aguayo, & Opazo, 2012).

About 90–95% of PD cases do not appear to have a genetic basis, whereas 5–10% of the cases are linked to inherited mutations and half of early-onset PD is caused by loss-of-function mutations in the *parkin* gene, which encodes an E3 ubiquitin ligase (reviewed in Farooqui & Farooqui, 2011). Despite all these evidences, the etiology of the disease remains unknown.

In the context of PD, liposomes of HSPC/Chol/DSPE-PEG functionalized with chlorotoxin (CTX), a molecule initially isolated from scorpion venom (DeBin & Strichartz, 1991) and described to bind brain gliomas (Lyons, O'Neal, & Sontheimer, 2002) and proliferating vascular endothelial cells (Jacoby et al., 2010), were used to deliver the dopamine precursor levodopa to BECs (Xiang et al., 2012). When compared to control nontargeted liposomes, CTX-functionalized particles mediated increased BEC uptake of an encapsulated fluorescent probe. The efficiency of this strategy was evaluated in a PD animal model established through the parkinsonian neurotoxin 1-methyl-4-phenyl-1,2,3,6-tetrahydro pyridine, which shows a significant decrease in the levels of dopamine and its metabolite, the dihydroxyphenyl acetic acid (DOPAC), in the *S. nigra* and *striata*. Mild recovery of dopamine levels was observed in both brain regions of mice treated with CTX-targeted liposomes containing levodopa. Nevertheless, DOPAC presented a much more expressive increase, especially in the *S. nigra*, in which DOPAC normal levels were almost completely restored. Importantly, Xiang and collaborators observed that movement coordination impairment was significantly attenuated, as assessed by the rotarod endurance test, but only when the targeted formulation was employed. In fact, neither free levodopa nor nontargeted liposome-encapsulated levodopa resulted in improved behavioral performance. Recovery of dopaminergic neurons in the *S. nigra* and *striata* was found to support the functional gain of motor coordination (Xiang et al., 2012).

2.5 Machado–Joseph Disease

Machado–Joseph disease (MJD) is the most common autosomal dominant-inherited form of spinocerebellar ataxia, caused by a mutation on the ataxin-3 gene (ATXN3/MJD1). The genetic cause of this neurodegenerative disorder is the over-repetition of a CAG tract in the coding region of the ATXN3/MJD1 gene, which confers a toxic gain of function of ataxin-3 protein, with consequent neuronal dysfunction and brain degeneration (Taroni & DiDonato, 2004). Currently, no efficient treatment is available for this disease, which remains fatal and incurable.

To address MJD, stable nucleic acid lipid particles (SNALPs) were developed to deliver siRNAs designed to target the mRNA resulting from the transcription of mutant ataxin-3 gene to neuronal cells. These lipid-based nanoparticles were produced in two steps, comprising the encapsulation of the nucleic acids in ceramide-PEGylated liposomes composed of dioleoyldimethylammoniumpropane (DODAP), DSPC, and Chol, followed by postinsertion of micelles composed of DSPE-PEG, previously functionalized with the rabies virus-derived peptide RGV-r9. This peptide was added to confer neuron-targeting ability to SNALP particles

(Conceicao et al., 2016). In a MJD transgenic mouse model expressing mutant ataxin-3, the i.v. administration of neuron-targeting SNALPs carrying siRNAs against mutant ataxin-3 resulted in significant decrease of mutant ataxin-3 expression and aggregate formation, thus reversing MJD-impaired features, such as motor behavior disabilities and cerebellar-associated neuropathology (Conceicao et al., 2016).

2.6 Multiple Sclerosis

Multiple sclerosis (MS) is an inflammatory demyelinating autoimmune disease (Adibhatla & Hatcher, 2007). The immune system of MS patients attacks the myelin sheath of nerve cell fibers in the brain and spinal cord, causing severe disabling. The inflammatory process is activated through T-cells that recognize and attack myelin. This process triggers the activity of other immune cells with the release of cytokines and antibodies.

Eighty-five percent of MS patients suffer from relapsing remitting MS (RRMS), while the remaining 15% have primary progressive MS (PPMS). RRMS is characterized by periods of acute symptomatology and neurologic damage followed by partial recovery that ultimately leads to permanent neurologic impairments (termed secondary progressive disease), while PPMS comprises an unremitting neurologic function deterioration (Wingerchuk & Carter, 2014). Currently, the available treatments are only able to attenuate RRMS-associated attacks, with no approved treatments being effective on PPMS (Wingerchuk & Carter, 2014). The experimental autoimmune encephalomyelitis (EAE) model is the most extensively used murine model for the study of prospective new therapeutic strategies in MS, which, depending on the mice strain and method of induction, can mimic features of different MS forms (McCarthy, Richards, & Miller, 2012).

Due to the strong inflammatory component of MS, glucocorticoids, widely used to treat inflammatory diseases, have been employed to treat this disease. However, these compounds present unfavorable biodistribution properties, thus being good candidates for lipid-based encapsulation. Combination of liposome ability to reach the inflamed brain and to carry antiinflammatory prodrugs, such as methylprednisolone, was explored in an EAE animal model. Methylprednisolone formulated in liposomes of HSPC/Chol/DSPE-PEG demonstrated superior therapeutic efficacy as compared to free drug administration, regarding mortality, rate of disease progression, and severity of disease symptoms. Furthermore, liposome-encapsulated methylprednisolone presented higher therapeutic efficacy than conventional MS drugs (Avnir et al., 2011). In addition to the improved drug bioavailability, the liposomal encapsulation has been associated to an alteration of glucocorticoid mode of action. In fact, free methylprednisolone mainly targets T-lymphocytes, whereas liposomal methylprednisolone preferentially targets macrophages, reducing the expression of proinflammatory cytokines and chemokines (IL-1β, IL-6, IP-10), while upregulating antiinflammatory cytokines and M2-associated genes (IL-10, CD206, CD163, Arginase 1, Ym1, and Fizz1) (Schweingruber et al., 2011). Methylprednisolone-liposomes were further modified to target the brain by postinsertion of glutathione-PEG micelles. This improved formulation showed superior therapeutic efficacy with respect to the nontargeted methylprednisolone in terms of disease burden, when administered at the disease onset (Gaillard et al., 2012). These results correlated with a decrease in T-cell and macrophage/microglia CNS infiltration and astrocyte activation. Consequently, axonal loss and spinal cord demyelinization were less extensive, when compared to that of animals administered with free methylprednisolone, even at 10 times lower dose (Lee et al., 2014).

Another approach to MS treatment involved liposome encapsulation of immunodominant peptides of the myelin basic protein (MBP) (Belogurov et al., 2013). MBP is one of the autoantigens found to be responsible for the axonal loss secondary to demyelination. MBP-derived peptides, corresponding to different fragments of the MBP protein (MBP_{46-62}, $MBP_{124-139}$, and $MBP_{147-170}$), were shown to induce the production of polyclonal antibodies in MS patients. Belogurov and colleagues encapsulated these peptides in mannosylated liposomes of egg PC and dioleylglycerol and observed that the three peptides acted synergistically to suppress the progression of the disease, reducing neurodegeneration and demyelination by decreasing the production of autoantibodies to full-length MBP and by downregulating IL-2 and IFNγ cytokine production. These effects were not observed in MS model animals treated with nonencapsulated peptides or with peptides encapsulated in nonmannosylated liposomes, indicating that both liposome encapsulation and mannosylation are critical for achieving a therapeutic effect (Belogurov et al., 2013).

2.7 CNS Tumors

Gliomas are infiltrative tumors originated from glial cells that account for 30% of all brain tumors and 80% of primary malignant brain tumors

(Omuro & DeAngelis, 2013; Schwartzbaum, Fisher, Aldape, & Wrensch, 2006). Based on their histological morphology, gliomas are subdivided in astrocytomas, oligodendrogliomas, ependymomas, and mixed gliomas, astrocytomas being the most common subtype (Omuro & DeAngelis, 2013). According to the classification of tumors of the CNS of the World Health Organization, in 2007, gliomas are further categorized into four histologic grades based on their malignancy and aggressiveness, ranging from I to IV (Louis et al., 2007). In addition to gliomas, other tumors may originate within the CNS, from different types of cells, namely, meningiomas and pituitary adenomas, which are mostly benign, as well as malignant medulloblastomas, primary CNS lymphomas, and the rarely occurring CNS germ cell tumors (Buckner et al., 2007).

Grade IV astrocytoma, also known as glioblastoma (GBM), is the most frequent and malignant primary brain tumor, accounting for more than 50% of all gliomas (Thakkar et al., 2014). Although being a rare tumor, with a worldwide incidence rate of only 3–5 newly diagnosed cases per 100,000 individuals, GBM has a dramatic outcome, providing a median survival of only 14.6 months postdiagnosis (Preusser et al., 2011). Currently, the diagnosis and grading of gliomas rely on the pathological examination of a biopsy sample, in terms of cellular and nuclear morphology, cell proliferation, vascularization, and a limited amount of biomarkers, which significantly delays medical intervention (Eberlin et al., 2010). Despite advances in standard therapies involving surgery, radiotherapy, and chemotherapy, a curative treatment for GBM patients is still missing.

The design of formulations of anticancer drugs in liposomes, extensively reported in the literature, emerges as a strategy with two main objectives: to increase the lifetime and bioavailability of anticancer drugs at the tumor site, thus enhancing their therapeutic efficacy, and to minimize unwanted effects in nontargeted organs, taking into account the very aggressive character of these drugs. Therefore, significant efforts have been made to achieve specific drug delivery into the tumor. In the case of brain tumors, a selective accumulation of anticancer drugs in the CNS, sparing heart, lungs, and other major organs, has been the main concern of the researchers.

Lipid-based nanoparticles are known to accumulate in CNS tumors due to the EPR effect, thus presenting potential as therapeutic strategies to tackle these life-threatening diseases. In this context, cationic liposomes consisting of a mixture of dimyristoylphosphatidylcholine (DMPC), Chol and 25 mol % DOTAP were shown to be retained in the brain after intracarotid injection, in contrast to neutral liposomes composed of DMPC/Chol or anionic liposomes composed of dimyristoylphosphatidylglycerol (DMPG)/Chol. Due to particular characteristics of brain tumor tissue, such as disruption of the brain tumor barrier, angiogenesis and leaky capillaries, liposomes of DMPC, Chol and DOTAP encapsulating the fluorescent probe Dil, were delivered to glioma and shown to reach the tumor core and persist for long periods in the peritumor brain tissue of glioma-bearing rats (Joshi et al., 2015).

Liposomes of soybean phospholipids and Chol were shown to efficiently encapsulate temozolomide (TMZ), a DNA alkylating agent and the first line treatment of GBM, and to promote its slow release into the plasma of rabbits after i.v. administration, as compared to TMZ delivered in the free form, resulting in longer circulation time, and higher plasma and brain concentrations (Gao, Wang, Liu, Wang, & Huang, 2015; Verreault et al., 2012). On the other hand, increased animal survival (by up to 83%), improved health score (which takes into consideration the percentage of body weight loss, hydration levels, signs of weakness, and levels of activity), and decreased tumor volume were observed in mice bearing an orthotopic GBM when treated with liposomal irinotecan (a topoisomerase inhibitor that prevents DNA unwinding, thus blocking DNA transcription and replication), as compared to animals treated with the free drug (Verreault et al., 2012).

Despite the accumulated experimental evidences of improved biodistribution and decreased toxicity of chemotherapeutic drugs delivered via liposomal vehicles, translation to the clinic is still in process. Some liposomal formulations of chemotherapeutic drugs are in clinical trials, such as Myocet® and Depocyte®, which are liposomal presentations of doxorubicin and cytarabine, respectively. Doxorubicin is able to improve cancer survivorship in pediatric tumor, but exhibits severe side effects, namely, cardiac disease associated with cumulative cardiotoxicity. A study in which Myocet® (doxorubicin encapsulated in egg PC/Chol vesicles) was administered to a cohort of 12 children bearing refractory high-grade glioma (Chastagner et al., 2015) showed that the encapsulation of doxorubicin in liposomes resulted in extended drug half-life, reduced cardiotoxicity, extended lifetime, and antitumor activity similar to that observed with free doxorubicin (Chastagner et al., 2015; Veringa et al., 2013). In another study, cytarabine encapsulated in vesicles of DOPC and dipalmitoylphosphatidylglycerol (DPPG) (Depocyte®) was used to address neoplastic meningitis, which is diagnosed in 1–2% of brain cancer patients (Gaviani et al., 2013). Intrathecal Depocyte® was given to a cohort of patients diagnosed with primary medulloblastoma, germinoma, pylocitic astrocytoma, teratocarcinoma, meningeal melanoma, or ependimoma that progressed to neoplastic meningitis. Patients with neoplastic meningitis presented neoplastic cells in the cerebrospinal fluid (CSF) and nodular or linear meningeal enhancement in brain and spine, as revealed by contrast magnetic resonance imaging (MRI). Depocyte® showed only mild toxicity, which was associated with the route of administration (lumbar puncture), and resulted in clearance of tumor cells from the CSF and restoration of normality in brain and spine MRI images (Gaviani et al., 2013).

As previously referred, a great advantage of employing lipid-based drug delivery systems is to alter drug pharmacokinetics and to increase drug efficiency. This is illustrated in a study showing that liposomes of DSPC/Chol/DSPE-PEG encapsulating doxorubicin were able to modulate tumor vascular permeability of rat glioblastoma, improving drug efficacy with respect to the free drug (Chaudhuri et al., 2012). Results from this study suggested that liposome-encapsulated doxorubicin reduced tumor vasculature, while CD31 labeling of vascular content indicated normal vascular phenotype (from chaotic and leaky tumor-like vessels to organized normal vessels). According to the authors, this process may be explained by the accumulation of liposomal doxorubicin at the edge of the tumor. Liposome accumulation in this region would result from (1) liposome extravasation from the leaky blood vessels in close proximity to the tumor and (2) impaired liposomal diffusion into the tumor due to the large particle hydration radius. Drug release would occur close to tumor vessels, resulting in damage of endothelial cells and vascular collapse (Chaudhuri et al., 2012; Verreault et al., 2011). In consequence of decreased tumor perfusion, drug clearance from the tumor site would also be diminished, resulting in prolonged drug accumulation and increased therapeutic effect (Chaudhuri et al., 2012).

PEGylated vitamin E, D-α-tocopherylpolyethylene glycol 1000 succinate (TPGS), is another PEG conjugate used in drug delivery systems, which was shown to produce nanoparticles with roughly the same size as DSPE-PEG-containing particles. Encapsulation in TPGS-modified liposomes of DPPC/Chol was highly efficient and the in vitro drug release kinetics was slower than that observed for DSPE-PEG-modified liposomes, whereas the extent of glioma cell uptake was similar for liposomes containing either PEG or TPGS conjugates. Interestingly, TPGS-liposomes presented surface charge close to neutrality, as opposed to the highly negative DSPE-PEG-liposomes, which may represent a further advantage of this modification regarding escape from the MPS (Muthu et al., 2011).

The ability of PEGylated liposomes of egg PC and Chol modified with a mannose analog (*P*-aminophenyl-α-D-mannopyranoside, PADMP), which mediates brain entry through glucose transporters overexpressed in the BBB, to cross the BBB was simulated in an in vitro model, in which BECs were plated in the superior compartment of a transwell, and glioma cells were seeded in the bottom compartment (Li et al., 2014). After BEC incubation with the modified liposomes, containing the drugs paclitaxel (inhibitor of microtubule assembly) and artemether (a mitochondria-targeting drug), PADMP-functionalized liposomes were shown to promote much higher cytotoxicity than their nonfunctionalized counterparts, leading to a decrease of glioma cell viability to 20%, in contrast to 50% of cell survival detected for liposomes without PADMP. Despite the promising therapeutic effect of the employed strategy resulting in a significant decrease of tumor size and increased animal survival, biodistribution data pointed to the need of further improvement. In fact, particle accumulation occurred in all major organs, with potential for damaging side effects (Li et al., 2014). Another approach to overcome the BBB is based on the use of peptides. Thus, the above described transwell system was used to evaluate the BBB crossing ability of soybean PC/Chol liposomes, modified with PEG, and containing two peptides, T7, a ligand of the transferrin receptor used to target both BBB and glioma cells, and TAT, a cell-penetrating peptide (CPP) capable of mediating cell internalization and used to facilitate BBB transposition. The presence of the two ligands (T7 and TAT) resulted in an expressive increase in particle uptake by glioma cells, as compared to single-peptide functionalization (T7 or TAT) and to PEGylated particles without peptides. Interestingly, T7/TAT double functionalization produced particles able to penetrate the tumor core with high efficiency, when compared with non-functionalized or functionalized liposomes with either peptide or with PEG. In fact, tumor cell spheroids transfected with these liposomes carrying a fluorescent probe exhibited high fluorescence intensity at different depths, whereas a mild fluorescence was observed in the spheroid core with T7-, TAT-, and PEG-modified liposomes (Zong et al., 2014).

The preparation of delivery systems provided with tumor-specific ligands constitutes an approach to confer tumor targeting and selectivity to carried anticancer drugs. Such ligands are often molecules whose receptors are overexpressed in tumors, such as transferrin (Gao et al., 2013), lactoferrin (Chen et al., 2011), folate (Liu et al., 2013), and angiopep (Sun et al., 2012). Angiopep-2, which targets the low-density lipoprotein receptor-related protein, has a dual targeting effect, since this receptor is overexpressed both in BBB and glioma cells. Angiopep-2 was used to decorate the surface of liposomes composed of dimethylaminoethanecarbamoyl-cholesterol (DC-Chol), DOPE, and DSPE-PEG, encapsulating paclitaxel and complexed with plasmid DNA encoding the human tumor necrosis factor-related apoptosis-inducing ligand (TRAIL). The system provided with the ligand angiopep-2 showed to preferentially accumulate in the brain of mice, sparing other major tissues, whereas liposomes lacking the ligand were also found in the liver and spleen of treated animals (Sun et al., 2012). Brain immunohistochemistry images of mice bearing GBM showed that angiopep-liposomes accumulated preferentially in the tumor and tumor boundaries, whereas liposomes devoid of ligand also accumulate in healthy brain tissue (Sun et al., 2012). In fact, terminal deoxynucleotidyl transferase (TdT) dUTP nick-end labeling (TUNEL) analysis showed that the most extensive TRAIL-induced apoptosis phenomena occurred in the tumor site of animals treated with angiopep-liposomes, correlating with the significantly extended survival of treated animals (Sun et al., 2012).

Besides endogenous ligands, advantage can be taken from other natural molecules to target brain tumor cells, such as the CTX peptide. This peptide binds to a lipid-raft-anchored complex containing a chloride ion channel and matrix metallopeptidase 2 (MMP-2) endopeptidase and has been used to improve the ability of delivery systems to specifically target GBM. This strategy was used in liposomes of HSPC/Chol/DSPE-PEG-CTX loaded with doxorubicin, by Xiang et al. (2011). The authors showed that CTX-functionalized liposomal doxorubicin displayed superior ability to reduce tumor size as compared to nontargeted liposomes. In studies carried out in our laboratory to modulate miR-21, overexpressed in GBM and responsible for promoting tumorigenesis, CTX-targeted SNALPs were used to deliver encapsulated anti-miRNA-21 oligonucleotides, both in vitro and in vivo (Costa et al., 2013, 2015). These studies showed that silencing of miR-21 enhanced significantly the antitumoral action of the tyrosine kinase inhibitor sunitinib in GBM cellular models (Costa et al., 2013). Furthermore, in mice bearing an orthotopic GBM, i.v. administration of anti-miR-21 oligonucleotides formulated in CTX-functionalized SNALPs combined with oral chemotherapy with sunitinib (Costa et al., 2015) resulted in a significant decrease of tumor volume and extended animal survival (Costa et al., 2015).

Synthetic peptides have also been added to the surface of liposomes, in order to improve liposome accumulation in glioma and to favor penetration of the delivery system into the tumor core. This is the case of a cyclical RGD peptide, conjugated with a R8 CPP. The RGD motif acts as a specific ligand for an integrin family ($\alpha_v\beta_3$) known to be overexpressed both in angiogenic endothelial cells and tumor cells, including gliomas. Liposomes of soybean PC, Chol, and DSPE-PEG modified with this chimeric peptide and loaded with paclitaxel were efficiently internalized by BECs, as well as by glioma cells, reaching the deepest layers of tumor spheroids and inducing severe toxicity in tumor cells. Extensive liposome accumulation in rat brain tumors was observed and resulted in a 85% increase in animal survival time, which was significantly superior to that observed in animals treated with free paclitaxel, unmodified liposomes, or liposomes modified with each one of the peptides (R8 or RGD) separately (Liu et al., 2014). Another peptide, RGERPPR, acts as a ligand of neuropilin-1 (NRP-1), a receptor overexpressed in GBM cells and in tumor endothelium. This peptide was called a "tumor penetrating peptide" (TPP) due to its ability to penetrate deep into the tumor core. Thus, liposomes of HSPC, Chol, and DSPE-PEG, further modified with this TPP and encapsulating doxorubicin, showed to be able to efficiently target GBM, accumulating in mice brain tumors, and to elicit a strong therapeutic effect, which was reflected in the extended survival time of treated animals (Yang et al., 2013).

2.8 Other Neurological Disorders

Epilepsy is a striking neurological disorder characterized by disruption of balance between neuronal excitability and inhibition, with subsequent occurrence of recurrent and spontaneous seizures. Thus, epileptic brain dysfunction shares some features with ischemic stroke, namely, regarding the excitotoxic profile that characterizes glutamate release as a consequence of acute brain injury. First line therapy for epilepsy involves treatment with antiepileptic drugs (AEDs). However, there is about one-third of patients whose seizures fail to respond to AEDs (Laxer et al., 2014), and these drugs present many adverse effects (Laxer et al., 2014), suggesting that alternative therapies for this medical condition are urgently needed.

Liposomal nimodipine was evaluated as an anticonvulsant with potential for epileptic seizure inhibition, due to its action as selective antagonist of L-type Ca^{2+} channels (Moreno et al., 2015). Nimodipine is a good candidate for liposomolization due to its low solubility and high first-pass effect in the liver, which need to be overcome in order to obtain a satisfactory therapeutic effect. In a mouse model of pilocarpine-induced seizures, nimodipine encapsulated in liposomes composed of soybean PC/Chol promoted significantly higher anticonvulsant activity than the free drug, suggesting improved bioavailability of the liposomal drug formulation. On the other hand, this formulation showed better therapeutic efficacy than diazepam per se, preventing 100% of seizures and death of mice, without showing any toxicity (Moreno et al., 2015).

Anxiety is another disease characterized by dysfunctional calcium ion levels, caused by increased influx of Ca^{2+} across neuronal membrane (Maigaard, Hageman, Jorgensen, Jorgensen, & Wortwein, 2012). Although anxiety evolved as a physiological response to potentially dangerous situations (Richey, Schmidt, Hofmann, & Timpano, 2010), exaggeration of normal vigilant behaviors can configure a pathological process. Thus, pathological anxiety is a disproportionate reaction to a number of factors, leading to an uncontrollable emotional state. Currently, treatments for anxiety include benzodiazepines and buspirone, associated with severe side effects, such as amnesia, muscle relaxation, sleepiness, and, most worrying, physical dependence. Due to its action as Ca^{2+} channel inhibitor, the antiepileptic drug nimodipine showed potential to treat anxiety without promoting side effects associated with currently used anxiolytic drugs (Yanpallewar, 2004). In fact, nimodipine encapsulated in small and highly homogenous liposomes of soybean PC and Chol and injected intraperitoneally in an animal model of anxiety showed significantly higher anxiolytic activity than the nonencapsulated drug (Moreno et al., 2014).

3. EXOSOMES AS DELIVERY SYSTEMS FOR CNS DISEASES

One of the most important breakthroughs in the field of exosome research was the discovery that these vesicles can transport nucleic acids, such as mRNA and miRNAs, and deliver them to acceptor cells in the same tissue or in a distant organ, where they induce important functional changes. This observation sparked the interest of the pharmaceutical and biotechnology industry and suggested that exosomes could be engineered as a new class of therapeutic vehicles for the delivery of siRNAs and other small nucleic acid molecules in different disease contexts.

Exosomes share many of the biochemical and biophysical properties of liposomes, which paved the way for these structures to be considered useful for the transport, protection, and delivery of nucleic acid cargos. However, exosomes also present several important differences that confer them additional advantages as delivery vectors (van der Meel et al., 2014). Contrarily to most liposome formulations, which present neutral or positive charge, exosomes are exclusively negatively charged and tend to exhibit small sizes (between 50 and 150 nm) with broad size distributions. Nevertheless, since they originate from endogenous membranes, they incorporate a rich set of transmembrane proteins and surface receptors, the so-called self-markers, which contribute to specific cell recognition, thus promoting cell interaction and internalization, while simultaneously protecting them from opsonization and enhancing their circulation time. Several families of membrane proteins, including GTPases, annexins, flotillin, heat-shock proteins, tetraspanins (CD9, CD63, and CD81) and integrins, have been described as being present in exosomes from different cellular origins and are believed to be responsible for receptor–ligand interactions with recipient cells. Another important advantage of exosomes over other nanoparticles, including liposomes, is their lack of accumulation in the liver, which is usually responsible for unwanted side effects and toxicity (Janas, Sapon, Janas, Stowell, & Janas, 2016).

In addition to exosome cell-targeting capacity, exosome-producing cells can also be genetically engineered to pack different types of therapeutic molecules, including mRNA, miRNAs, siRNAs, proteins, and even drugs, within the core or membrane bilayer of exosomes. This versatility has allowed exosomes to be employed as delivery systems for RNAi therapy, immunotherapy, and chemotherapy in the context of brain diseases.

In what concerns their application in gene silencing strategies targeting the brain, Alvarez-Erviti et al. (2011) were the first to demonstrate that systemically injected exosomes were able to deliver antiglyceraldehyde-3-phosphate dehydrogenase (GAPDH) siRNAs to neurons, oligodendrocytes, and microglia in the mouse brain, achieving efficient and specific gene knockdown. These authors genetically engineered self-derived dendritic cells to promote the expression of Lamp2b, an exosomal membrane protein, fused to the rabies 29-mer peptide RVG, derived from the rabies virus glycoprotein. The exosome population obtained from the supernatant of the dendritic cell culture was loaded with siRNAs by electroporation and featured the Lamp2-RVG fusion protein at the membrane level, which allowed the vesicles to cross the BBB and target brain cells expressing the ACh receptor. As a result, specific silencing of the GAPDH mRNA was observed in different mouse brain regions, including the striatum, cortex, and midbrain. The therapeutic potential of this siRNA delivery strategy was demonstrated in AD mouse brain by the observation of a significant reduction of BACE1 mRNA (60%) and protein (62%) in the cortical region, which led to a significant decrease in the release of Aβ1–42, a toxic peptide resulting from BACE1-mediated cleavage of APP (Alvarez-Erviti et al., 2011).

More recently, another study by the same group evaluated the therapeutic potential of RVG-modified exosomes in the context of Parkinson's disease (Cooper et al., 2014). The authors encapsulated anti-α-Syn siRNAs in the exosomes and evaluated their ability to reduce intracellular α-Syn aggregates in a transgenic mouse line expressing the human phosphorylation-mimic S129D α-Syn. This study presented evidence of significant decrease in α-Syn mRNA and protein levels, as well as a reduction in intraneuronal protein aggregates in the *S. nigra*, further highlighting the potential of exosomes to cross the BBB and exert therapeutic effects within the CNS.

A surprising number of studies have been focused on exosomes released by mesenchymal stem cells (MSCs). Indeed, these cells are believed to secrete different tissue remodeling factors, including specific miRNAs, which play important roles during neuroprotection and neurodegeneration. However, recent evidence suggests that most effects mediated by these cells, in physiological and pathological conditions, are not due to the cells per se but are a result of their extracellular vesicles. The protective effects of MSC-derived exosomes vs MSCs themselves were tested by Doeppner et al. (2015) in a model of focal cerebral ischemia. The authors compared long-term neuroprotection, brain remodeling, and neurological impairments in C57/BL6 mice that received a single administration of MSCs, 1 day after stroke, with animals that received three i.v. injections of MSC-derived exosomes, on days 1, 3, and 5 after the ischemic procedure. Their findings clearly show that MSC-derived exosomes do not perform inferiorly to MSCs and are able to promote neuronal recovery, associated with enhanced angiogenesis, as well as to modulate poststroke immune responses, preventing postischemic immunosupression. Another recent work, by Zhang et al. (2015), analyzed the effects of MSC-derived exosomes in trauma brain injury (TBI) with similar results. In this work, two groups of Wistar rats undergoing TBI received, 24 h later, a tail vein injection of exosomes derived from MSCs or equal volume of a saline solution. The authors observed that exosome-treated rats presented improved spatial learning in the Morris water maze test and reduced neurological deficits. Exosome treatment also led to a reduction of inflammation and promoted neurogenesis, leading to an increase in the number of immature neurons in the *dentate gyrus*.

As opposed to their neuroprotective properties, MSC-derived exosomes have also been shown to contribute to tumor resistance and immunosuppression in the brain, by facilitating tumor growth and invasion, stimulating angiogenesis and allowing tumor cells to repress microglia-mediated antitumoral immune responses. Munoz et al. (2013) took advantage of the communication axis between GBM cells and MSCs to deliver anti-miRNA oligonucleotides, aiming at silencing miR-9, a miRNA involved in the expression of the drug efflux transporter P-glycoprotein. The transfer of anti-miR-9 oligonucleotides from MSCs to GBM cells occurred via gap junctions and through released microvesicles, leading to the reversal of P-glycoprotein expression and sensitizing GBM cells toward TMZ, which resulted in increased cell death and caspase activity. Katakowski et al. (2013) followed a similar approach and transfected MSCs with a plasmid encoding miR-146, in order to harvest exosomes enriched in this miRNA. The intratumoral injection of these exosomes significantly reduced the growth of glioma xenografts in a rat model of primary brain tumor. Taken together, the recent studies employing MSC-derived exosomes provide compelling evidence that MSCs are easy to genetically engineer in order to produce exosome populations enriched in specific therapeutic nucleic acids. Moreover, MSC-derived exosomes seem to mediate most of the neuroprotective properties attributed to MSCs and have been reported to be well tolerated in humans, which opens new opportunities for their application in clinical settings associated with acute brain damage and brain cancer.

Another interesting discovery in the field of exosome therapeutics was recently made by Bryniarski et al. (2013). These authors found that exosomes collected from the supernatant of CD8+ suppressor T-cells presenting antigen-specific antibodies at their surface, allowed targeted delivery of miR-150 to the small subset of antigen-presenting cells, contact sensitivity effector T-cells, inducing T-cell tolerance of allergic contact sensitivity in mice by exposure of high doses of reactive hapten. This interesting finding can help to explain the surprising in vivo targeting capacity of these vesicles and opens new possibilities regarding the targeting of therapeutic exosomes to specific cell populations within the brain, such as neurons, oligodendrocytes, astrocytes, and microglia, as long as specific antigens can be identified in each of these target cell populations.

Concerning drug delivery, several studies have addressed the possibility of using drug-loaded exosomes as vehicles to reach the nervous tissue. In 2010, Sun and colleagues first reported the encapsulation of the drug curcumin inside exosomal particles, improving its solubility and bioavailability and decreasing lipopolysaccharide-induced inflammation (Sun et al., 2010).

In 2011, the same authors used a similar formulation to deliver curcumin and JS1124, an activator of stat3 inhibitor, through the intranasal route, to the rodent brain (Zhuang et al., 2011). This system resulted in an improvement of animal survival in a model of EAEs. More recently, another group has evaluated different strategies to load exosomes with the antioxidant catalase (Haney et al., 2015). In this study, permeabilization of exosomes with saponin or sonication followed by extrusion resulted in high loading efficacy and catalase preservation against degradation. The exosomes were delivered through nasal administration to the rodent brain, resulting in significant neuroprotection in an in vivo model of Parkinson's disease. Taken together, these studies suggest that, similarly to liposomes, exosomes can be useful tools to preserve drug activity and increase drug circulation times, resulting in less toxic and more effective therapeutic outcome toward many CNS diseases, including cancer, neurodegenerative disorders, and acute brain injury.

4. CONCLUSIONS

The treatment of CNS pathologies has faced several difficulties: first, the inability to have access to the patient nervous tissue, in order to treat it in situ, avoiding invasive surgical techniques; second, the need to transpose the BBB by systemically administered chemotherapeutics targeting brain diseases; third, the likelihood of drugs to affect healthy nervous cells, with dramatic physiological consequences in the immediate vicinity of the diseased tissue.

The advantage of using nanosized formulations carrying conventional drugs to treat CNS diseases lies in the infinite range of possibilities to manipulate surface and structural properties of these nanosystems in order to: (i) make them competent to target diseased cells, sparing the healthy ones; (ii) carry large loads of different types of molecules, conferring them protection against degradation and preventing early delivery; (iii) provide drugs with long circulation lifetimes and lastly; (iv) allow drug controlled and sustained release at the desired target site. Lipid-based nanocarriers and cell-derived exosomes benefit from the extra advantage of being able to carry hydrophobic and hydrophilic molecules in separate compartments within the same particle, and to encapsulate or complex nucleic acids with therapeutic potential.

In the context of brain diseases, the high lipid content and particular lipid composition of brain cells, as well as the requirement for lipid mediators, make them a favorable target for lipid-containing nanosystems, in which lipids, besides constituting highly versatile carrier components, can play a role as functional groups and bioactive molecules. Clinical trials designed to evaluate the effectiveness of liposomal presentations of the anticancer drugs doxorubicin and cytarabine to tackle brain tumors underline the potential of lipid-based nanoparticles as drug carriers to CNS. On the other hand, exosomes, besides meeting the criteria that make lipid-based nanosystems efficient multifunctional drug carriers, benefit from incorporating a set of membrane proteins, which naturally endow them with specific cell recognition ability, high efficiency in cargo delivery to target cells and the capacity to prevent opsonization and subsequent clearance from circulation.

In summary, the collected literature in this chapter points to significant advances in the research of nanosized lipid-based drug/nucleic acid carriers, supporting the conviction that next generation pharmaceutical bullets will become available in a near future, opening new avenues toward the clinical treatment of a large spectrum of CNS diseases.

ACKNOWLEDGMENTS

This work is funded by FEDER funds through the Operational Programme Competitiveness Factors—COMPETE and national funds by FCT—Foundation for Science and Technology under the strategic project UID/NEU/04539/2013. A.M.C., A.L.C., C.M., and A.T.V. are recipients of fellowships from the FCT with references SFRH/BPD/99613/2014, SFRH/BPD/108312/2015, SFRH/BD/79077/2011, and SFRH/BD/89470/2012, respectively.

REFERENCES

Adibhatla, R. M., & Hatcher, J. F. (2007). Role of lipids in brain injury and diseases. *Future Lipidology*, *2*, 403–422.

Allen, T. M., & Hansen, C. (1991). Pharmacokinetics of stealth versus conventional liposomes: Effect of dose. *Biochimica et Biophysica Acta*, *1068*, 133–141.

Alvarez-Erviti, L., Seow, Y., Yin, H., Betts, C., Lakhal, S., & Wood, M. J. (2011). Delivery of siRNA to the mouse brain by systemic injection of targeted exosomes. *Nature Biotechnology*, *29*, 341–345.

Avgoustakis, K. (2003). Effect of copolymer composition on the physicochemical characteristics, in vitro stability, and biodistribution of PLGA–mPEG nanoparticles. *International Journal of Pharmaceutics*, *259*, 115–127.

Avnir, Y., Turjeman, K., Tulchinsky, D., Sigal, A., Kizelsztein, P., Tzemach, D., et al. (2011). Fabrication principles and their contribution to the superior in vivo therapeutic efficacy of nano-liposomes remote loaded with glucocorticoids. *PloS One*, *6*, e25721.

Balducci, C., Mancini, S., Minniti, S., La Vitola, P., Zotti, M., Sancini, G., et al. (2014). Multifunctional liposomes reduce brain beta-amyloid burden and ameliorate memory impairment in Alzheimer's disease mouse models. *Journal of Neuroscience*, *34*, 14022–14031.

Bana, L., Minniti, S., Salvati, E., Sesana, S., Zambelli, V., Cagnotto, A., et al. (2014). Liposomes bi-functionalized with phosphatidic acid and an ApoE-derived peptide affect Abeta aggregation features and cross the blood-brain-barrier: Implications for therapy of Alzheimer disease. *Nanomedicine*, *10*, 1583–1590.

Banks, W. A. (2016). From blood-brain barrier to blood-brain interface: New opportunities for CNS drug delivery. *Nature Reviews Drug Discovery, 2016*, 1–18.

Belogurov, A. A., Jr., Stepanov, A. V., Smirnov, I. V., Melamed, D., Bacon, A., Mamedov, A. E., et al. (2013). Liposome-encapsulated peptides protect against experimental allergic encephalitis. *FASEB Journal, 27*, 222–231.

Bendixen, B., & Ocava, L. (2002). Evaluation and management of acute ischemic stroke. *Current Cardiology Reports, 4*, 149–157.

Ben-Zvi, A., Lacoste, B., Kur, E., Andreone, B. J., Mayshar, Y., Yan, H., et al. (2014). Mfsd2a is critical for the formation and function of the blood-brain barrier. *Nature, 509*, 507–511.

Bereczki, E., Re, F., Masserini, M. E., Winblad, B., & Pei, J. J. (2011). Liposomes functionalized with acidic lipids rescue Abeta-induced toxicity in murine neuroblastoma cells. *Nanomedicine, 7*, 560–571.

Bohm, C., Chen, F., Sevalle, J., Qamar, S., Dodd, R., Li, Y., et al. (2015). Current and future implications of basic and translational research on amyloid-beta peptide production and removal pathways. *Molecular and Cellular Neuroscience, 66*, 3–11.

Bryniarski, K., Ptak, W., Jayakumar, A., Pullmann, K., Caplan, M. J., Chairoungdua, A., et al. (2013). Antigen-specific, antibody-coated, exosome-like nanovesicles deliver suppressor T-cell microRNA-150 to effector T cells to inhibit contact sensitivity. *Journal of Allergy and Clinical Immunology, 132*, 170–181.

Buckner, J. C., Brown, P. D., O'Neill, B. P., Meyer, F. B., Wetmore, C. J., & Uhm, J. H. (2007). Central nervous system tumors. *Mayo Clinic Proceedings, 82*, 1271–1286.

Caraglia, M., Luongo, L., Salzano, G., Zappavigna, S., Marra, M., Guida, F., et al. (2013). Stealth liposomes encapsulating zoledronic acid: A new opportunity to treat neuropathic pain. *Molecular Pharmaceutics, 10*, 1111–1118.

Cardoso, A. L., Costa, P., de Almeida, L. P., Simoes, S., Plesnila, N., Culmsee, C., et al. (2010). Tf-lipoplex-mediated c-Jun silencing improves neuronal survival following excitotoxic damage in vivo. *Journal of Controlled Release, 142*, 392–403.

Cardoso, A. L., Simoes, S., de Almeida, L. P., Plesnila, N., Pedroso de Lima, M. C., Wagner, E., et al. (2008). Tf-lipoplexes for neuronal siRNA delivery: A promising system to mediate gene silencing in the CNS. *Journal of Controlled Release, 132*, 113–123.

Carrera, I., Etcheverria, I., Li, Y., Fernandez-Novoa, L., Lombardi, V., Vigo, C., et al. (2013). Immunocytochemical characterization of Alzheimer disease hallmarks in APP/PS1 transgenic mice treated with a new anti-amyloid-beta vaccine. *BioMed Research International, 2013*, 709145.

Chang, K. J., Redmond, S. A., & Chan, J. R. (2016). Remodeling myelination: Implications for mechanisms of neural plasticity. *Nature Neuroscience, 19*, 190–197.

Chastagner, P., Devictor, B., Geoerger, B., Aerts, I., Leblond, P., Frappaz, D., et al. (2015). Phase I study of non-pegylated liposomal doxorubicin in children with recurrent/refractory high-grade glioma. *Cancer Chemotherapy and Pharmacology, 76*, 425–432.

Chaudhuri, R. T., Arnold, R. D., Yang, J., Turowski, S. G., Qu, Y., Spernyak, J. A., et al. (2012). Mechanisms of tumor vascular priming by a nanoparticulate doxorubicin formulation. *Pharmaceutical Research, 29*, 3312–3324.

Chen, H., Qin, Y., Zhang, Q., Jiang, W., Tang, L., Liu, J., et al. (2011). Lactoferrin modified doxorubicin-loaded procationic liposomes for the treatment of gliomas. *European Journal of Pharmaceutical Sciences, 44*, 164–173.

Chow, B. W., & Gu, C. (2015). The molecular constituents of the blood-brain barrier. *Trends in Neurosciences, 38*, 598–608.

Conceicao, M., Mendonca, L., Nobrega, C., Gomes, C., Costa, P., Hirai, H., et al. (2016). Intravenous administration of brain-targeted stable nucleic acid lipid particles alleviates Machado-Joseph disease neurological phenotype. *Biomaterials, 82*, 124–137.

Cooper, J. M., Wiklander, P. B., Nordin, J. Z., Al-Shawi, R., Wood, M. J., Vithlani, M., et al. (2014). Systemic exosomal siRNA delivery reduced alpha-synuclein aggregates in brains of transgenic mice. *Movement Disorders*, *29*, 1476–1485.

Costa, P. M., Cardoso, A. L., Custodia, C., Cunha, P., Pereira de Almeida, L., & Pedroso de Lima, M. C. (2015). MiRNA-21 silencing mediated by tumor-targeted nanoparticles combined with sunitinib: A new multimodal gene therapy approach for glioblastoma. *Journal of Controlled Release*, *207*, 31–39.

Costa, P. M., Cardoso, A. L., Nobrega, C., Pereira de Almeida, L. F., Bruce, J. N., Canoll, P., et al. (2013). MicroRNA-21 silencing enhances the cytotoxic effect of the anti-angiogenic drug sunitinib in glioblastoma. *Human Molecular Genetics*, *22*, 904–918.

Daneman, R., Agalliu, D., Zhou, L., Kuhnert, F., Kuo, C. J., & Barres, B. A. (2009). Wnt/β-catenin signaling is required for CNS, but not non-CNS, angiogenesis. *Proceedings of the National Academy of Sciences of the United States of America*, *106*, 641–646.

DeBin, J. A., & Strichartz, G. R. (1991). Chloride channel inhibition by the venon of the scorpion Leiurus quinquestriatus. *Toxicology*, *29*, 1403–1408.

Doeppner, T. R., Herz, J., Gorgens, A., Schlechter, J., Ludwig, A. K., Radtke, S., et al. (2015). Extracellular vesicles improve post-stroke neuroregeneration and prevent post-ischemic immunosuppression. *Stem Cells Translational Medicine*, *4*, 1131–1143.

Durukan, A., & Tatlisumak, T. (2007). Acute ischemic stroke: Overview of major experimental rodent models, pathophysiology, and therapy of focal cerebral ischemia. *Pharmacology, Biochemistry, and Behavior*, *87*, 179–197.

Eberlin, L. S., Dill, A. L., Golby, A. J., Ligon, K. L., Wiseman, J. M., Cooks, R. G., et al. (2010). Discrimination of human astrocytoma subtypes by lipid analysis using desorption electrospray ionization imaging mass spectrometry. *Angewandte Chemie International Edition in English*, *49*, 5953–5956.

Farooqui, T., & Farooqui, A. A. (2011). Lipid-mediated oxidative stress and inflammation in the pathogenesis of Parkinson's disease. *Parkinson's Disease*, *2011*, 247467.

Farr, S. A., Erickson, M. A., Niehoff, M. L., Banks, W. A., & Morley, J. E. (2014). Central and peripheral administration of antisense oligonucleotide targeting amyloid-beta protein precursor improves learning and memory and reduces neuroinflammatory cytokines in Tg2576 (AbetaPPswe) mice. *Journal of Alzheimer's Disease*, *40*, 1005–1016.

Fukuta, T., Ishii, T., Asai, T., Nakamura, G., Takeuchi, Y., Sato, A., et al. (2014). Real-time trafficking of PEGylated liposomes in the rodent focal brain ischemia analyzed by positron emission tomography. *Artificial Organs*, *38*, 662–666.

Gaillard, P. J., Appeldoorn, C. C., Rip, J., Dorland, R., van der Pol, S. M., Kooij, G., et al. (2012). Enhanced brain delivery of liposomal methylprednisolone improved therapeutic efficacy in a model of neuroinflammation. *Journal of Controlled Release*, *164*, 364–369.

Gao, J. Q., Lv, Q., Li, L. M., Tang, X. J., Li, F. Z., Hu, Y. L., et al. (2013). Glioma targeting and blood-brain barrier penetration by dual-targeting doxorubincin liposomes. *Biomaterials*, *34*, 5628–5639.

Gao, J., Wang, Z., Liu, H., Wang, L., & Huang, G. (2015). Liposome encapsulated of temozolomide for the treatment of glioma tumor: Preparation, characterization and evaluation. *Drug Discoveries & Therapeutics*, *9*, 205–212.

Gaviani, P., Corsini, E., Salmaggi, A., Lamperti, E., Botturi, A., Erbetta, A., et al. (2013). Liposomal cytarabine in neoplastic meningitis from primary brain tumors: A single institutional experience. *Journal of the Neurological Sciences*, *34*, 2151–2157.

Ghosh, S., Das, N., Mandal, A. K., Dungdung, S. R., & Sarkar, S. (2010). Mannosylated liposomal cytidine 5' diphosphocholine prevent age related global moderate cerebral ischemia reperfusion induced mitochondrial cytochrome c release in aged rat brain. *Neuroscience*, *171*, 1287–1299.

Gidday, J. M., Gasche, Y. G., Copin, J. C., Shah, A. R., Perez, R. S., Shapiro, S. D., et al. (2005). Leukocyte-derived matrix metalloproteinase-9 mediates blood-brain barrier breakdown and is proinflammatory after transient focal cerebral ischemia. *American Journal of Physiology. Heart and Circulatory Physiology, 289*, H558–H568.

Ginhoux, F., Greter, M., Leboeuf, M., Nandi, S., See, P., Gokhan, S., et al. (2010). Fate mapping analysis reveals that adult microglia derive from primitive macrophages. *Science, 330*, 841–845.

Gorelick, P. B., & Ruland, S. (2010). Cerebral vascular disease. *Disease-A-Month, 56*, 39–100.

Gosselin, D., Link, V. M., Romanoski, C. E., Fonseca, G. J., Eichenfield, D. Z., Spann, N. J., et al. (2014). Environment drives selection and function of enhancers controlling tissue-specific macrophage identities. *Cell, 159*, 1327–1340.

Guedes, J. R., Viegas, A. T., Pedroso de Lima, M. C., & Cardoso, A. L. (2015). Microglia: A double-edged sword for Alzheimer's disease. In E. R. Giffard (Ed.), *Microglia: Physiology, regulation and health implications* (pp. 1–66). New York: Nova Science Publisher.

Haney, M. J., Klyachko, N. L., Zhao, Y., Gupta, R., Plotnikova, E. G., He, Z., et al. (2015). Exosomes as drug delivery vehicles for Parkinson's disease therapy. *Journal of Controlled Release, 207*, 18–30.

Hwang, H., Jeong, H. S., Oh, P. S., Na, K. S., Kwon, J., Kim, J., et al. (2015). Improving cerebral blood flow through liposomal delivery of angiogenic peptides: Potential of 18F-FDG PET imaging in ischemic stroke treatment. *Journal of Nuclear Medicine, 56*, 1106–1111.

Ishii, T., Asai, T., Fukuta, T., Oyama, D., Yasuda, N., Agato, Y., et al. (2012a). A single injection of liposomal asialo-erythropoietin improves motor function deficit caused by cerebral ischemia/reperfusion. *International Journal of Pharmaceutics, 439*, 269–274.

Ishii, T., Asai, T., Oyama, D., Agato, Y., Yasuda, N., Fukuta, T., et al. (2013). Treatment of cerebral ischemia-reperfusion injury with PEGylated liposomes encapsulating FK506. *FASEB Journal, 27*, 1362–1370.

Ishii, T., Asai, T., Oyama, D., Fukuta, T., Yasuda, N., Shimizu, K., et al. (2012b). Amelioration of cerebral ischemia-reperfusion injury based on liposomal drug delivery system with asialo-erythropoietin. *Journal of Controlled Release, 160*, 81–87.

Ishii, T., Fukuta, T., Agato, Y., Oyama, D., Yasuda, N., Shimizu, K., et al. (2013). Nanoparticles accumulate in ischemic core and penumbra region even when cerebral perfusion is reduced. *Biochemical and Biophysical Research Communications, 430*, 1201–1205.

Ismail, M. F., Elmeshad, A. N., & Salem, N. A. (2013). Potential therapeutic effect of nanobased formulation of rivastigmine on rat model of Alzheimer's disease. *International Journal of Nanomedicine, 8*, 393–406.

Jacoby, D. B., Dyskin, E., Yalcin, M., Kesavan, K., Dahlberg, W., Ratliff, J., et al. (2010). Potent pleiotropic anti-angiogenic effects of TM601, a synthetic chlorotoxin peptide. *Anticancer Research, 30*, 39–46.

Janas, A. M., Sapon, K., Janas, T., Stowell, M. H., & Janas, T. (2016). Exosomes and other extracellular vesicles in neural cells and neurodegenerative diseases. *Biochimica et Biophysica Acta, 1858*, 1139–1151.

Johnsen, K. B., & Moos, T. (2016). Revisiting nanoparticle technology for blood-brain barrier transport: Unfolding at the endothelial gate improves the fate of transferrin receptor-targeted liposomes. *Journal of Controlled Release, 222*, 32–46.

Joshi, S., Singh-Moon, R. P., Ellis, J. A., Chaudhuri, D. B., Wang, M., Reif, R., et al. (2015). Cerebral hypoperfusion-assisted intra-arterial deposition of liposomes in normal and glioma-bearing rats. *Neurosurgery, 76*, 92–100.

Kaneda, S., Ishizuka, T., Goto, H., Kimura, T., Inaba, K., & Kasukawa, H. (2009). Liposome-encapsulated hemoglobin, TRM-645: Current status of the development and important issues for clinical application. *Artificial Organs, 33*, 146–152.

Kaneda, S., Ishizuka, T., Sekiguchi, A., Morimoto, K., & Kasukawa, H. (2014). Efficacy of liposome-encapsulated hemoglobin in a rat model of cerebral ischemia. *Artificial Organs*, *38*, 650–655.

Katakowski, M., Buller, B., Zheng, X., Lu, Y., Rogers, T., Osobamiro, O., et al. (2013). Exosomes from marrow stromal cells expressing miR-146b inhibit glioma growth. *Cancer Letters*, *335*, 201–204.

Kawaguchi, A. T., Endo, H., Aikawa, H., Yamano, M., Kawaguchi, Y., Haida, M., et al. (2014). Effects of liposome-encapsulated hemoglobin on learning ability in tokai high-avoider rat after total brain ischemia and reperfusion. *Artificial Organs*, *38*, 667–674.

Kawaguchi, A. T., Haida, M., Ohba, H., Yamano, M., Fukumoto, D., & Tsukada, H. (2013). Liposome-encapsulated hemoglobin ameliorates ischemic stroke in nonhuman primates: Longitudinal observation. *Artificial Organs*, *37*, 904–912.

Khakh, B. S., & Sofroniew, M. V. (2015). Diversity of astrocyte functions and phenotypes in neural circuits. *Nature Neuroscience*, *18*, 942–952.

Kierdorf, K., & Prinz, M. (2013). Factors regulating microglia activation. *Frontiers in Cellular Neuroscience*, 7, 44.

Kim, H., Britton, G. L., Peng, T., Holland, C. K., McPherson, D. D., & Huang, S. L. (2014). Nitric oxide-loaded echogenic liposomes for treatment of vasospasm following subarachnoid hemorrhage. *International Journal of Nanomedicine*, *9*, 155–165.

Latour, L. L., Kang, D. W., Ezzeddine, M. A., Chalela, J. A., & Warach, S. (2004). Early blood-brain barrier disruption in human focal brain ischemia. *Annals of Neurology*, *56*, 468–477.

Laxer, K. D., Trinka, E., Hirsch, L. J., Cendes, F., Langfitt, J., Delanty, N., et al. (2014). The consequences of refractory epilepsy and its treatment. *Epilepsy & Behavior*, *37*, 59–70.

Lazar, A. N., Mourtas, S., Youssef, I., Parizot, C., Dauphin, A., Delatour, B., et al. (2013). Curcumin-conjugated nanoliposomes with high affinity for Abeta deposits: Possible applications to Alzheimer disease. *Nanomedicine*, *9*, 712–721.

Lee, D. H., Rotger, C., Appeldoorn, C. C., Reijerkerk, A., Gladdines, W., Gaillard, P. J., et al. (2014). Glutathione PEGylated liposomal methylprednisolone (2B3-201) attenuates CNS inflammation and degeneration in murine myelin oligodendrocyte glycoprotein induced experimental autoimmune encephalomyelitis. *Journal of Neuroimmunology*, *274*, 96–101.

Li, X. Y., Zhao, Y., Sun, M. G., Shi, J. F., Ju, R. J., Zhang, C. X., et al. (2014). Multifunctional liposomes loaded with paclitaxel and artemether for treatment of invasive brain glioma. *Biomaterials*, *35*, 5591–5604.

Liddelow, S. A. (2011). Fluids and barriers of the CNS: A historical viewpoint. *Fluids and Barriers of the CNS*, *8*, 1–16.

Liu, X., Ma, S., Dai, C., Cai, F., Yao, Y., Yang, Y., et al. (2013). Antiproliferative, antiinvasive, and proapoptotic activity of folate receptor alpha-targeted liposomal doxorubicin in nonfunctional pituitary adenoma cells. *Endocrinology*, *154*, 1414–1423.

Liu, Y., Ran, R., Chen, J., Kuang, Q., Tang, J., Mei, L., et al. (2014). Paclitaxel loaded liposomes decorated with a multifunctional tandem peptide for glioma targeting. *Biomaterials*, *35*, 4835–4847.

Lo, E. H., Dalkara, T., & Moskowitz, M. A. (2003). Mechanisms, challenges and opportunities in stroke. *Nature Reviews. Neuroscience*, *4*, 399–415.

Lo, E. H., Moskowitz, M. A., & Jacobs, T. P. (2005). Exciting, radical, suicidal: How brain cells die after stroke. *Stroke*, *36*, 189–192.

Lois, C., & Alvarez-Buylla, A. (1994). Long-distance neuronal migration in the adult mammalian brain. *Science*, *264*, 1145–1148.

Louis, D. N., Ohgaki, H., Wiestler, O. D., Cavenee, W. K., Burger, P. C., Jouvet, A., et al. (2007). The 2007 WHO classification of tumours of the central nervous system. *Acta Neuropathologica*, *114*, 97–109.

Lyons, S. A., O'Neal, J., & Sontheimer, H. (2002). Chlorotoxin, a scorpion-derived peptide, specifically binds to gliomas and tumors of neuroectodermal origin. *Glia, 39*, 162–173.
Maigaard, K., Hageman, I., Jorgensen, A., Jorgensen, M. B., & Wortwein, G. (2012). Electroconvulsive stimulations prevent chronic stress-induced increases in L-type calcium channel mRNAs in the hippocampus and basolateral amygdala. *Neuroscience Letters, 516*, 24–28.
McCarthy, D. P., Richards, M. H., & Miller, S. D. (2012). Mouse models of multiple sclerosis: Experimental autoimmune encephalomyelitis and Theiler's virus-induced demyelinating disease. *Methods in Molecular Biology, 900*, 381–401.
Moreno, L. C., Cavalcanti, I. M., Satyal, P., Santos-Magalhaes, N. S., Rolim, H. M., & Freitas, R. M. (2015). Acute toxicity and anticonvulsant activity of liposomes containing nimodipine on pilocarpine-induced seizures in mice. *Neuroscience Letters, 585*, 38–42.
Moreno, L. C., da Silva Oliveira, G. Z., Cavalcanti, I. M., Santos-Magalhaes, N. S., Rolim, H. M., & de Freitas, R. M. (2014). Development and evaluation of liposomal formulation containing nimodipine on anxiolytic activity in mice. *Pharmacology, Biochemistry, and Behavior, 116*, 64–68.
Munoz, J. L., Bliss, S. A., Greco, S. J., Ramkissoon, S. H., Ligon, K. L., & Rameshwar, P. (2013). Delivery of functional anti-miR-9 by mesenchymal stem cell-derived exosomes to glioblastoma multiforme cells conferred chemosensitivity. *Molecular Therapy. Nucleic Acids, 2*, e126.
Muthu, M. S., Kulkarni, S. A., Xiong, J., & Feng, S. S. (2011). Vitamin E TPGS coated liposomes enhanced cellular uptake and cytotoxicity of docetaxel in brain cancer cells. *International Journal of Pharmaceutics, 421*, 332–340.
Nguyen, L. N., Ma, D., Shui, G., Wong, P., Cazenave-Gassiot, A., Zhang, X., et al. (2014). Mfsd2a is a transporter for the essential omega-3 fatty acid docosahexaenoic acid. *Nature, 509*, 503–506.
Oddo, S., Caccamo, A., Shepherd, J. D., Murphy, M. P., Golde, T. E., Kayed, R., et al. (2003). Triple-transgenic model of alzheimer's disease with plaques and tangles. *Neuron, 39*, 409–421.
Omuro, A., & DeAngelis, L. M. (2013). Glioblastoma and other malignant gliomas: A clinical review. *Journal of the American Medical Association, 310*, 1842–1850.
Ordonez-Gutierrez, L., Re, F., Bereczki, E., Ioja, E., Gregori, M., Andersen, A. J., et al. (2015). Repeated intraperitoneal injections of liposomes containing phosphatidic acid and cardiolipin reduce amyloid-beta levels in APP/PS1 transgenic mice. *Nanomedicine, 11*, 421–430.
Pacheco, C., Aguayo, L. G., & Opazo, C. (2012). An extracellular mechanism that can explain the neurotoxic effects of alpha-synuclein aggregates in the brain. *Frontiers in Physiology, 3*, 297.
Patel, T., Zhou, J., Piepmeier, J. M., & Saltzman, W. M. (2012). Polymeric nanoparticles for drug delivery to the central nervous system. *Advanced Drug Delivery Reviews, 64*, 701–705.
Popov, M., Abu Hammad, I., Bachar, T., Grinberg, S., Linder, C., Stepensky, D., et al. (2013). Delivery of analgesic peptides to the brain by nano-sized bolaamphiphilic vesicles made of monolayer membranes. *European Journal of Pharmaceutics and Biopharmaceutics, 85*, 381–389.
Preusser, M., de Ribaupierre, S., Wöhrer, A., Erridge, S. C., Hegi, M., Weller, M., et al. (2011). Current concepts and management of glioblastoma. *Annals of Neurology, 70*, 9–21.
Qu, C., Ni, J., Yang, P., Zhang, H., Lin, L., Li, J., et al. (2013). Altered plasma and brain disposition of isopropylidene shikimic acid liposome in rats and the brain protection in cerebral ischemia-reperfusion. *Drug Development and Industrial Pharmacy, 39*, 1291–1295.

Quilty, M. C., King, A. E., Gai, W. P., Pountney, D. L., West, A. K., Vickers, J. C., et al. (2006). Alpha-synuclein is upregulated in neurones in response to chronic oxidative stress and is associated with neuroprotection. *Experimental Neurology*, *199*, 249–256.

Reynolds, B. A., Tetzlaff, W., & Weiss, S. (1992). A multipotent EGF-responsive striatal embryonic progenitor cell produces neurons and astrocytes. *The Journal of Neuroscience*, *12*, 4565–4574.

Richey, J. A., Schmidt, N. B., Hofmann, S. G., & Timpano, K. R. (2010). Temporal and structural dynamics of anxiety sensitivity in predicting fearful responding to a 35% CO2 challenge. *Journal of Anxiety Disorders*, *24*, 423–432.

Rosell, A., Cuadrado, E., Ortega-Aznar, A., Hernandez-Guillamon, M., Lo, E. H., & Montaner, J. (2008). MMP-9-positive neutrophil infiltration is associated to blood-brain barrier breakdown and basal lamina type IV collagen degradation during hemorrhagic transformation after human ischemic stroke. *Stroke*, *39*, 1121–1126.

Ruiperez, V., Darios, F., & Davletov, B. (2010). Alpha-synuclein, lipids and Parkinson's disease. *Progress in Lipid Research*, *49*, 420–428.

Schwartzbaum, J. A., Fisher, J. L., Aldape, K. D., & Wrensch, M. (2006). Epidemiology and molecular pathology of glioma. *Nature Clinical Practice. Neurology*, *2*, 494–503.

Schweingruber, N., Haine, A., Tiede, K., Karabinskaya, A., van den Brandt, J., Wust, S., et al. (2011). Liposomal encapsulation of glucocorticoids alters their mode of action in the treatment of experimental autoimmune encephalomyelitis. *Journal of Immunology*, *187*, 4310–4318.

Shafi, N., & Levine, J. M. (2010). Emergency management of acute ischemic stroke. *Current Atherosclerosis Reports*, *12*, 230–235.

Shimbo, D., Abumiya, T., Shichinohe, H., Nakayama, N., Kazumata, K., & Houkin, K. (2014). Post-ischemic intra-arterial infusion of liposome-encapsulated hemoglobin can reduce ischemia reperfusion injury. *Brain Research*, *1554*, 59–66.

Simpson, I. A., Carruthers, A., & Vannucci, S. J. (2007). Supply and demand in cerebral energy metabolism: The role of nutrient transporters. *Journal of Cerebral Blood Flow & Metabolism*, *27*, 1766–1791.

Sims, N. R., & Muyderman, H. (2010). Mitochondria, oxidative metabolism and cell death in stroke. *Biochimica et Biophysica Acta*, *1802*, 80–91.

Stenman, J. M., Rajagopal, J., Carroll, T. J., Ishibashi, M., McMahon, J., & McMahon, A. P. (2008). Canonical Wnt signaling regulates organ-specific assembly and differentiation of CNS vasculature. *Science*, *21*, 1247–1250.

Sun, X., Pang, Z., Ye, H., Qiu, B., Guo, L., Li, J., et al. (2012). Co-delivery of pEGFP-hTRAIL and paclitaxel to brain glioma mediated by an angiopep-conjugated liposome. *Biomaterials*, *33*, 916–924.

Sun, D., Zhuang, X., Xiang, X., Liu, Y., Zhang, S., Liu, C., et al. (2010). A novel nanoparticle drug delivery system: The anti-inflammatory activity of curcumin is enhanced when encapsulated in exosomes. *Molecular Therapy*, *18*, 1606–1614.

Tagalakis, A. D., Lee do, H. D., Bienemann, A. S., Zhou, H., Munye, M. M., Saraiva, L., et al. (2014). Multifunctional, self-assembling anionic peptide-lipid nanocomplexes for targeted siRNA delivery. *Biomaterials*, *35*, 8406–8415.

Tanifum, E. A., Dasgupta, I., Srivastava, M., Bhavane, R. C., Sun, L., Berridge, J., et al. (2012). Intravenous delivery of targeted liposomes to amyloid-beta pathology in APP/PSEN1 transgenic mice. *PloS One*, *7*, e48515.

Taroni, F., & DiDonato, S. (2004). Pathways to motor incoordination: The inherited ataxias. *Nature Reviews. Neuroscience*, *5*, 641–655.

Thakkar, J. P., Dolecek, T. A., Horbinski, C., Ostrom, Q. T., Lightner, D. D., Barnholtz-Sloan, J. S., et al. (2014). Epidemiologic and molecular prognostic review of glioblastoma. *Cancer Epidemiology, Biomarkers & Prevention*, *23*, 1985–1996.

Tiebosch, I. A., Crielaard, B. J., Bouts, M. J., Zwartbol, R., Salas-Perdomo, A., Lammers, T., et al. (2012). Combined treatment with recombinant tissue plasminogen activator and dexamethasone phosphate-containing liposomes improves neurological outcome and restricts lesion progression after embolic stroke in rats. *Journal of Neurochemistry, 123*(Suppl. 2), 65–74.

van der Meel, R., Fens, M. H., Vader, P., van Solinge, W. W., Eniola-Adefeso, O., & Schiffelers, R. M. (2014). Extracellular vesicles as drug delivery systems: Lessons from the liposome field. *Journal of Controlled Release, 195*, 72–85.

Veringa, S. J., Biesmans, D., van Vuurden, D. G., Jansen, M. H., Wedekind, L. E., Horsman, I., et al. (2013). In vitro drug response and efflux transporters associated with drug resistance in pediatric high grade glioma and diffuse intrinsic pontine glioma. *PloS One, 8*, e61512.

Verreault, M., Strutt, D., Masin, D., Anantha, M., Waterhouse, D., Yapp, D. T., et al. (2012). Irinophore C, a lipid-based nanoparticulate formulation of irinotecan, is more effective than free irinotecan when used to treat an orthotopic glioblastoma model. *Journal of Controlled Release, 158*, 34–43.

Verreault, M., Strutt, D., Masin, D., Anantha, M., Yung, A., Kozlowski, P., et al. (2011). Vascular normalization in orthotopic glioblastoma following intravenous treatment with lipid-based nanoparticulate formulations of irinotecan (Irinophore C), doxorubicin (Caelyx(R)) or vincristine. *BMC Cancer, 11*, 124.

Wingerchuk, D. M., & Carter, J. L. (2014). Multiple sclerosis: Current and emerging disease-modifying therapies and treatment strategies. *Mayo Clinic Proceedings, 89*, 225–240.

Wyss-Coray, T., & Rogers, J. (2012). Inflammation in Alzheimer disease-a brief review of the basic science and clinical literature. *Cold Spring Harbor Perspectives in Medicine, 2*, a006346.

Xiang, Y., Liang, L., Wang, X., Wang, J., Zhang, X., & Zhang, Q. (2011). Chloride channel-mediated brain glioma targeting of chlorotoxin-modified doxorubicine-loaded liposomes. *Journal of Controlled Release, 152*, 402–410.

Xiang, Y., Wu, Q., Liang, L., Wang, X., Wang, J., Zhang, X., et al. (2012). Chlorotoxin-modified stealth liposomes encapsulating levodopa for the targeting delivery against Parkinson's disease in the MPTP-induced mice model. *Journal of Drug Targeting, 20*, 67–75.

Yang, Y., Yan, Z., Wei, D., Zhong, J., Liu, L., Zhang, L., et al. (2013). Tumor-penetrating peptide functionalization enhances the anti-glioblastoma effect of doxorubicin liposomes. *Nanotechnology, 24*, 405101.

Yanpallewar, S. (2004). Nimodipine attenuates biochemical, behavioral and histopathological alterations induced by acute transient and long-term bilateral common carotid occlusion in rats. *Pharmacological Research, 49*, 143–150.

Yun, X., Maximov, V. D., Yu, J., Zhu, H., Vertegel, A. A., & Kindy, M. S. (2013). Nanoparticles for targeted delivery of antioxidant enzymes to the brain after cerebral ischemia and reperfusion injury. *Journal of Cerebral Blood Flow & Metabolism, 33*, 583–592.

Zehendner, C. M., Librizzi, L., de Curtis, M., Kuhlmann, C. R., & Luhmann, H. J. (2011). Caspase-3 contributes to ZO-1 and Cl-5 tight-junction disruption in rapid anoxic neurovascular unit damage. *PloS One, 6*, e16760.

Zhan, Y., Paolicelli, R. C., Sforazzini, F., Weinhard, L., Bolasco, G., Pagani, F., et al. (2014). Deficient neuron-microglia signaling results in impaired functional brain connectivity and social behavior. *Nature Neuroscience, 17*, 400–406.

Zhang, Y., Chopp, M., Meng, Y., Katakowski, M., Xin, H., Mahmood, A., et al. (2015). Effect of exosomes derived from multipluripotent mesenchymal stromal cells on functional recovery and neurovascular plasticity in rats after traumatic brain injury. *Journal of Neurosurgery, 122*, 856–867.

Zhang, J., Han, X., Li, X., Luo, Y., Zhao, H., Yang, M., et al. (2012). Core-shell hybrid liposomal vesicles loaded with panax notoginsenoside: Preparation, characterization and protective effects on global cerebral ischemia/reperfusion injury and acute myocardial ischemia in rats. *International Journal of Nanomedicine*, 7, 4299–4310.

Zhao, G., Zang, S. Y., Jiang, Z. H., Chen, Y. Y., Ji, X. H., Lu, B. F., et al. (2011). Post-ischemic administration of liposome-encapsulated luteolin prevents against ischemia-reperfusion injury in a rat middle cerebral artery occlusion model. *The Journal of Nutritional Biochemistry*, *22*, 929–936.

Zhuang, Z. Y., Gerner, P., Woolf, C. J., & Ji, R. R. (2005). ERK is sequentially activated in neurons, microglia, and astrocytes by spinal nerve ligation and contributes to mechanical allodynia in this neuropathic pain model. *Pain*, *114*, 149–159.

Zhuang, X., Xiang, X., Grizzle, W., Sun, D., Zhang, S., Axtell, R. C., et al. (2011). Treatment of brain inflammatory diseases by delivering exosome encapsulated anti-inflammatory drugs from the nasal region to the brain. *Molecular Therapy*, *19*, 1769–1779.

Zong, T., Mei, L., Gao, H., Shi, K., Chen, J., Wang, Y., et al. (2014). Enhanced glioma targeting and penetration by dual-targeting liposome co-modified with T7 and TAT. *Journal of Pharmaceutical Sciences*, *103*, 3891–3901.

Zou, L.-L., Ma, J.-L., Wang, T., Yang, T.-B., & Liu, C.-B. (2013). Cell-penetrating peptide-mediated therapeutic molecule delivery into the central nervous system. *Current Neuropharmacology*, *11*, 197–208.

CHAPTER TWO

From the Blood to the Central Nervous System: A Nanoparticle's Journey Through the Blood–Brain Barrier by Transcytosis☆

G. Fullstone*,[1], S. Nyberg*,†,[1], X. Tian‡, G. Battaglia*,[1]
*University College London, London, United Kingdom
†Biological Sciences, Sunnybrook Research Institute, Toronto, ON, Canada
‡School of Life Sciences, Anhui University, Hefei, People's Republic of China
[1]Corresponding authors: e-mail address: g.fullstone@ucl.ac.uk; sophie.nyberg@sri.utoronto.ca; g.battaglia@ucl.ac.uk

Contents

Abstract

Designing nanoparticles that effectively enter the central nervous system (CNS) rapidly and without alteration is one of the major challenges in the use of nanotechnology for the brain. In this chapter, we explore the process of transcytosis, a receptor-mediated transport pathway that permits endogenous macromolecules to enter the CNS by crossing the blood–brain barrier. Transcytosis across the blood–brain barrier involves

☆Author contributions: G.F. and S.N. contributed equally. G.F. wrote sections on nanoparticle distribution, binding, wrapping, and release. S.N. wrote sections on endocytosis, trafficking, and exocytosis. G.F., S.N., and X.T. all contributed to the section on nanoparticle transcytosis from the literature. G.F., S.N., and G.B. reviewed the manuscript. Fig. 2 was adapted from Fullstone, Wood, Holcombe, and Battaglia (2015), permission for adaptation and use of this figure is available under the Creative Commons Attribution 4.0 International License (http://creativecommons.org/licenses/by/4.0/legalcode).

International Review of Neurobiology, Volume 130
ISSN 0074-7742
http://dx.doi.org/10.1016/bs.irn.2016.06.001

a number of distinct stages, including receptor binding, endocytosis into a transport vesicle, trafficking of the vesicle to the opposite side of the cell, and finally exocytosis and release of cargo. For each stage, we discuss the current knowledge on biological, physiological, and physical factors that influence nanoparticle transit through that stage of transcytosis, with implications for nanoparticle design. Finally, we look at the current progress in designing nanoparticles that exploit transcytosis for CNS delivery.

1. TRANSCYTOSIS: AN ENTRY PORTAL INTO THE CENTRAL NERVOUS SYSTEM

One of the major challenges in the use of nanotechnology for central nervous system (CNS) applications is the ability to control nanoparticle entry into the CNS parenchyma efficiently and without alteration. A number of entry portals into the CNS exist, including the blood–brain barrier, blood–cerebrospinal fluid barrier, the arachnoid membranes, the olfactory route, and via the peripheral nervous system. Previously, efforts to deliver therapeutics to the CNS have circumvented these barriers by gaining access to the brain via surgery, a highly invasive and risky procedure. Surgery is often used to enable intracerebral (IC) or intracerebroventricular (ICV) injections of a drug, or to place an "extended release" smart material implant. Controlled release over time is essential for these materials, but can be difficult to achieve. IC or ICV injections have produced therapeutic results in some cases (Guerin, Olivi, Weingart, Lawson, & Brem, 2004; Menei et al., 2000). However, a particular concern is that the implantation procedure is not readily repeatable and must therefore carry a large drug dosage. In addition to the inherent risks of surgery and risk of inflammation, a major limiting factor of invasive CNS drug delivery is the limited spread of the drug. For instance, the radial spread of chemotherapeutics from IC implants in rats decreased by 90% just 0.5 mm^2 from the origin (Fung, Shin, Tyler, Brem, & Saltzman, 1996). Consequentially, drugs injected IC or ICV have to utilize extremely high initial concentrations of a drug in order account for this diminishing effect. These high concentrations far exceed physiological values and can result in adverse affects such as astrogliosis (Yamada et al., 1991) and Schwann cell hyperplasia (Winkler et al., 1997) observed in rodents. In 1999, a clinical trial of ICV glial cell line-derived neurotrophic factor infusions into a patient with Parkinson's disease was discontinued due to extensive neurotoxicity (Kordower et al., 1999).

Therefore, controlled entry into the CNS parenchyma by harnessing innate transport across one of the CNS barriers is a major goal for effective and safe long-term therapeutic strategies. The blood–brain barrier is usually cited as the most effective entry route for nanoparticles as it presents the largest surface area between the periphery and the CNS (~20 m^2), is the major access point for most endogenous molecules, and provides the most direct access route to cellular targets in the interstitial fluid (Pardridge, 2005). Recently, a transport route that permits endogenous macromolecules, such as proteins and lipoproteins, to cross the blood–brain barrier endothelial cells has gained wide-scale interest in nanoparticle delivery to the CNS. This transport route, called transcytosis, involves endocytosis into a transport vesicle that is trafficked across the cell to the opposite membrane, where it is fuses and the cargo is released unaltered, as detailed in Fig. 1 (Broadwell, Balin, & Salcman, 1988; Mellman, 1996). A number of cargo molecules are known to induce this pathway by binding specific receptors on the membrane of blood–brain barrier endothelial cells, as detailed in Table 1.

The ability of transcytosis to be induced by a specific ligand or ligands paves the way to control delivery of putative therapeutic agents across the blood–brain barrier. Several antibodies targeting the blood–brain barrier have been engineered with successful delivery in vivo, but require fusion of the drug to the antibody. Alternatively, drugs can be encapsulated into

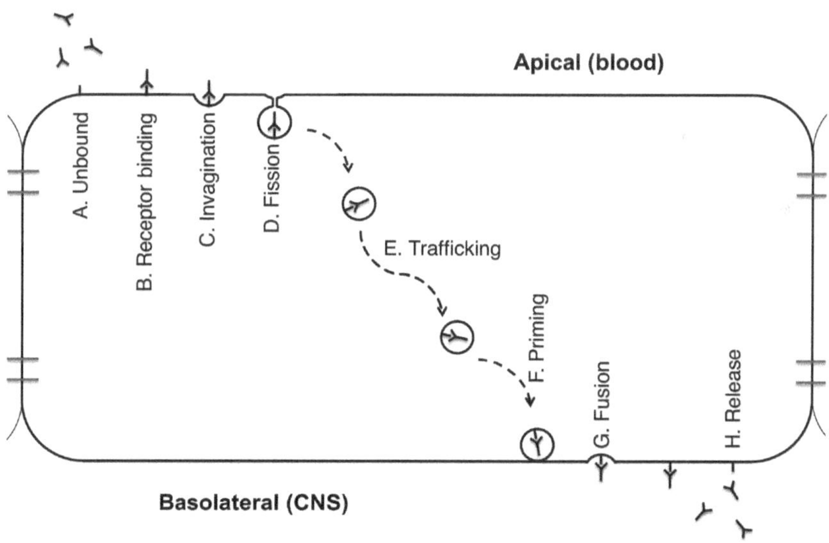

Fig. 1 Schematic of transcytosis. Transcytosis involves the binding of blood-borne unbound ligands (A) to specific transcytosis receptors on the apical membrane (B). This leads to invagination of the cellular membrane around the receptor–ligand complex (C) and fission of this structure to release a membrane-bound transport vesicle (D). The vesicle is trafficked across the cell (E) where it is primed against the basolateral membrane (F), prior to fusion of the vesicle and membrane (G), and subsequent release of the cargo (H).

Table 1 Transcytosis Receptors at the Blood–Brain Barrier and Their Endogenous Substrates

Receptor	Expression	Endogenous Substrates
Transferrin receptor (TfR)	Apical	Transferrin (Tf)
Melanotransferrin receptor (MTfR)	Apical	Melanotransferrin (MTf)
Apolipoprotein E receptor 2 (ApoER2)	Apical	Lipoproteins, ApoE
LDL receptor-related protein 1 and 2 (LRP-1 and LRP-2)	Apical and basolateral	Lipoproteins, amyloid-β, ApoE, lactoferrin
Receptor for advanced glycosylation end products (RAGE)	Apical	Glycosylated proteins, amyloid-β
Insulin receptor	Apical	Insulin
Fcγ-R	Apical	IgG

Banks (2004), Deane, Wu, and Zlokovic (2004), Demeule et al. (2002), Gaillard, Visser, and de Boer (2005), Herz and Marschang (2003), Stern, Du Yan, Fang Yan, and Marie Schmidt (2002), Visser, Voorwinden, Crommelin, Danhof, and de Boer (2004), and Zlokovic et al. (1990).

nanocarriers such as liposomes or polymersomes, which can be chemically modified to display targeting moieties on the surface. Antibodies targeting the blood–brain barrier can carry 1–4 individual drug molecules per antibody, compared to >10,000 drug molecules per individual ~100 nm liposome (Huwyler, Wu, & Pardridge, 1996) or polymersome (Pegoraro et al., 2013). The vesicular route permits delivery of large macromolecules in line with the size of nanoparticles than other transport routes at the blood–brain barrier. In this chapter, we follow the journey of a nanoparticle from the blood to the CNS by transcytosis. At each stage we detail the biological, physiological, and physical factors involved and the subsequent design considerations that should be considered to optimize overall nanoparticle transport across the blood–brain barrier. Finally, we look at examples of nanoparticles exploiting this pathway from the literature.

2. DISTRIBUTION OF NANOPARTICLES IN THE BLOOD

The first challenge in achieving efficient delivery to the CNS is persistence within the blood and effective distribution throughout the microvasculature of the CNS. Persistence within the blood can be modulated by size, shape, and surface chemistry. Generally, nanoparticles of 40–100 nm are favored for in vivo applications as they have a reduced blood clearance through the reticuloendothelial system and through blood extravasation (Alexis, Pridgen, Molnar, & Farokhzad, 2008; Owens & Peppas, 2006).

Moreover, nanoparticles are often equipped with poly[ethylene glycol] (PEG) or similar surface functionalization to reduce the nonspecific interaction between the nanoparticle and blood proteins, such as complement, which can trigger destruction by the immune system through opsonization (Gref et al., 2000; Owens & Peppas, 2006). Moreover, shape has been previously demonstrated to improve the blood circulation times of select nanoparticle compositions (Geng et al., 2007).

In order for nanoparticles in the blood to interact with transcytosis receptors on the vessel wall, they need to be effectively distributed to the edges of capillaries. Capillaries are already optimized for efficient delivery by their high surface area to volume ratio. Furthermore, we have previously modeled how nanoparticles are distributed within a blood capillary (Fullstone, Wood, Holcombe, & Battaglia, 2015). We demonstrated that red blood cells were highly important in creating fluid dynamics that increased dispersion of nanoparticles, as demonstrated in Fig. 2. In the absence of red blood cells (Fig. 2A), the direction of flow is parallel to

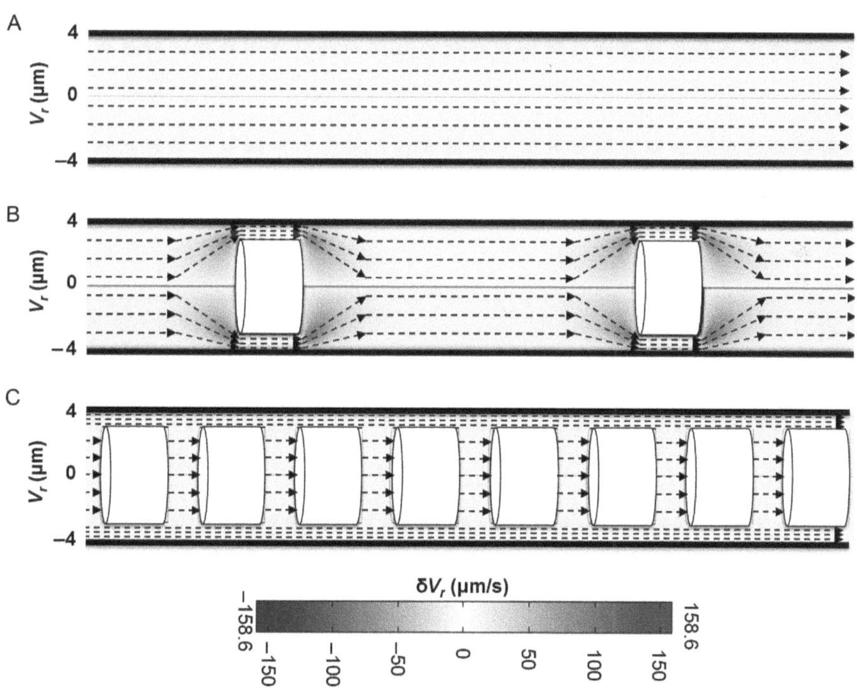

Fig. 2 The dispersal of nanoparticles by blood flow in capillaries. Streamlines of flow observed and heat maps of the radial velocity (V_r) are shown for a capillary with a percentage (v/v) of red blood cells, or hematocrit, of 0% (no red blood cells), 10.7% (physiological in a capillary), and >20%.

the vessel wall, thus moving nanoparticles along the blood vessel without increasing their distribution toward the vessel wall. At a physiological level of red blood cells (Fig. 2B), the blood flows around the blood cells, transiently carrying suspended nanoparticles in closer proximity to the vessel wall where binding can occur. At increased levels of red blood cells (Fig. 2C), blood forms a two-phase flow of a cell phase and cell-free phase. The blood flow in each phase is parallel to the vessel wall decreasing distribution to the vessel wall.

3. BINDING OF A BLOOD-BORNE NANOPARTICLE

After effective dispersion of blood-borne nanoparticles to the vessel walls, the process of transcytosis can occur. The first stage of nanoparticle transcytosis is binding to transcytosis receptors on the apical membrane of the blood–brain barrier. To achieve delivery to the CNS via the blood–brain barrier, carriers need a targeting sequence that not only targets blood–brain barrier receptors but also induces complete transcytosis of the drug. This is achieved by conjugating endogenous ligands (as seen in Table 1) or synthesized ligands, such as antibodies and small peptides (as seen in Table 2), to the surface of nanoparticles.

Table 2 Transcytosis Receptor-Targeting Ligands

Ligand	Receptor Target	Binding Data	References
Ox26 antibody	Transferrin receptor		Ulbrich, Hekmatara, Herbert, and Kreuter (2009)
R17217 antibody	Transferrin receptor		Ulbrich et al. (2009)
8D3 antibody	Transferrin receptor	$k_D = 2.3 \pm 0.3$ nM	Boado, Zhou, Lu, Hui, and Pardridge (2010) and Kissel et al. (1998)
TfR^A, TfR^B, TfR^C, TfR^D	Transferrin receptor	$IC_{50} = 1.7$, 6.9, 65, 111 nM	Yu et al. (2014, 2011)
Angiopep-2	LRP-1		Demeule, Currie, et al. (2008) and Demeule, Régina, et al. (2008)
MAb83-7, MAb83-14	Insulin receptor		Pardridge, Kang, Buciak, and Yang (1995)

Receptor-mediated transcytosis inducing antibodies and small peptides, with receptor targets and binding data (where available).

There are a number of factors to consider when choosing a receptor to target at the blood–brain barrier, especially as cargo entry into the cell is highly regulated at the plasma membrane. Important considerations include receptor expression, ie, whether it is ubiquitous in the body or enriched at the blood–brain barrier; endogenous ligands to the receptor, and whether they would compete with the ligand; efficacy of ligand–receptor interactions; and receptor half-life or rate of recycling to the plasma membrane after endocytosis. The transferrin receptor and insulin receptor, for example, bind only a single endogenous ligand, transferrin, and insulin, respectively. On the other hand, LRP-1 is by far the most diverse receptor, with over 30 different ligands reported such as lipoproteins, proteases, viruses, and toxins (Herz & Strickland, 2001). Interestingly, data suggest that the transcytosis of the LRP-1-targeting peptide Angiopep-2 is only partially inhibited by other LRP-1 ligands (Demeule, Régina, et al., 2008), which may be partially due to the four independent ligand-binding clusters making up the extracellular domain of the receptor.

In order to stably bind to the apical membrane, the strength of adhesion to the blood–brain barrier must be stronger than the dislodging shear forces of blood flow that act upon a stationary bound molecule. The dislodging shear forces consist of a dragging force (Fig. 3A) and a torque (Fig. 3B).

While these forces act upon all substrates, size and shape affect the strength of the dislodging forces acting upon the substrate. Moreover, for nonspherical particles, orientation in respect to the vessel wall is also important. Compared to endogenous substrates, nanoparticles vary greatly in size and shape. Therefore, the strength of binding must account for the dislodging force acting upon a particularly nanoparticle composition. The drag force (F_D) acting upon a spheroid is equal to:

$$F_D = 6\pi a l S_\mu F^S, \quad (1)$$

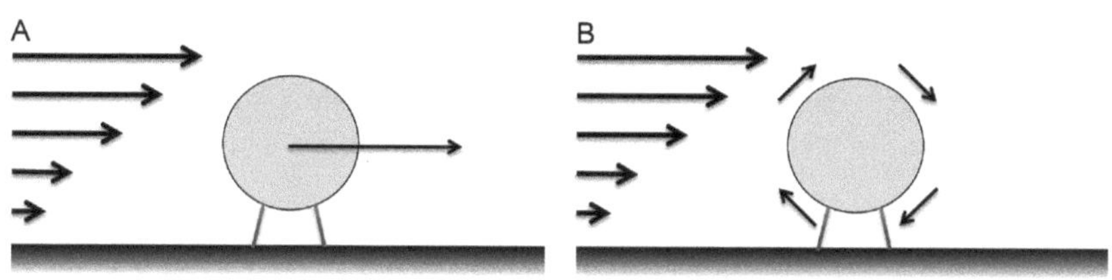

Fig. 3 Dislodging forces acting upon nanoparticles bound to a vessel wall under shear flow. There are two major forces acting on nanoparticles: the drag force (A) in the direction of flow and a rotating force called the torque (B).

where a is the particle radius in the plane parallel to the vessel wall, l is the distance between the cell membrane and nanoparticle center of mass, S_μ is the wall shear stress, and F^S is the force as a function of the particle aspect ratio (γ) (Decuzzi & Ferrari, 2006; Pozrikidis, 1994). γ is the ratio between a and the particle radius in the plane perpendicular to the vessel wall (b), given as:

$$\gamma = \frac{a}{b}, \tag{2}$$

therefore giving:

$$F^S = 1 + \left(1.736 - 0.138\gamma + 0.128\gamma^2 + 0.09\gamma^3\right)e^{-\gamma}. \tag{3}$$

For a spherical nanoparticle ($\gamma = 1$), Eq. (1) simplifies to:

$$F_D = 10\pi r l S_\mu, \tag{4}$$

where r is the particle radius.

The torque force (T_q) is equal to:

$$T_q = 4\pi a^3 S_\mu T^S, \tag{5}$$

where T^S is the torque force as a function of aspect ratio, equal to:

$$T^S = 1 + \left(-20.5 + 46.5\gamma - 35.1\gamma^2 + 8.95\gamma^3\right)e^{-\gamma}. \tag{6}$$

For a spherical nanoparticle, Eq. (4) simplifies to:

$$T_q = 3.78\pi r^3 S_\mu. \tag{7}$$

The effect of these two forces on particles of different sizes, different aspect ratios, and orientation can be seen in Fig. 4. Fig. 4A demonstrates the shape, orientation, and corresponding γ values of various nanoparticle compositions. Fig. 4B and C demonstrates a trend of dislodging forces increasing nonlinearly as particle size increases. Moreover, the drag forces decrease as the aspect ratio increases (Fig. 4B), suggesting that nonspherical particles, orientated to increase surface area with the vessel, may be better at binding. A more complex relationship between aspect ratio and torque is observed (Fig. 4C). Therefore, both size and shape have a considerable effect on the dislodging forces acting on bound nanoparticles. In particular, with larger particles overall binding strength will have to be increased to offset the higher dislodging forces. For example, a spherical nanoparticle of 40 nm will be subjected to >500 times the level of torque and 33 times the level of drag

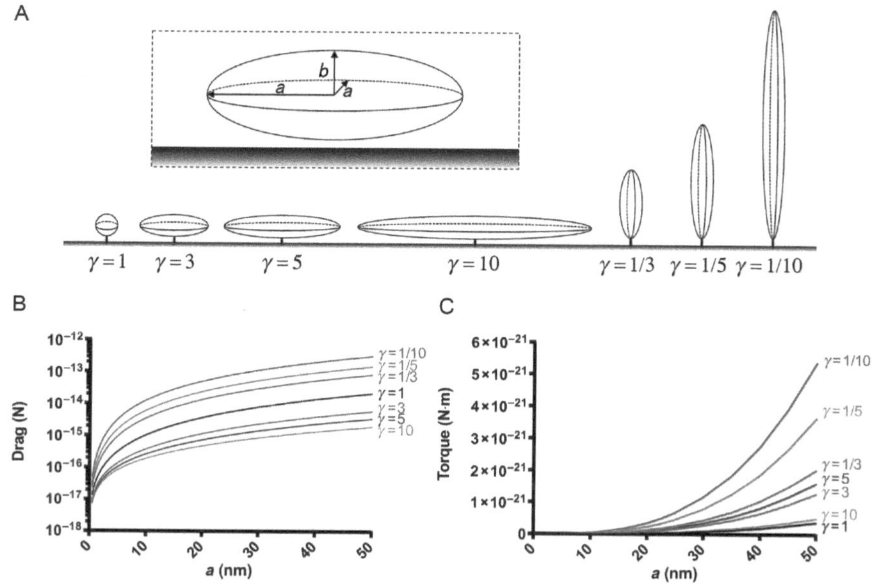

Fig. 4 The effect of nanoparticle size, shape, and orientation on dislodging forces at the vessel wall. Schematic of nanoparticle aspect ratios (γ) with a and b indicated (A). Drag force (B) and torque (C) acting upon nanoparticles of different values of a and γ when $S_\mu = 0.25\,\text{Pa}$, $l = b + \delta_{eq}$.

force as compared to a protein, such as transferrin (~5 nm), undergoing binding. By comparison, a spherical 40 nm nanoparticle has a more modest eight times increase in torque and 1.7 times increase in drag force than lipoprotein binding (~20–30 nm).

A potential benefit of using nanoparticle vectors is their capability to be decorated with multiple ligand molecules on a single nanoparticle. Therefore, the overall binding strength of a nanoparticle is dictated not by the strength of a single interaction but by the cumulative strength of a number of individual interactions. The number of individual interactions is governed by a number of factors, particularly, the density of ligand on the outside of a nanoparticle (Fig. 5A and B), the density of receptor on the cell membrane (Fig. 5C and D), and the contact surface area between the nanoparticle and cell membrane. Receptor density is particularly intriguing as it invites the possibility that nanoparticle binding strength can be controlled by the expression levels of that receptor, a concept discussed later in this chapter.

The contact surface area between a nanoparticle and the cell membrane describes the surface area of a nanoparticle that is within a set length, the receptor–ligand bond length, from the cell membrane (Decuzzi &

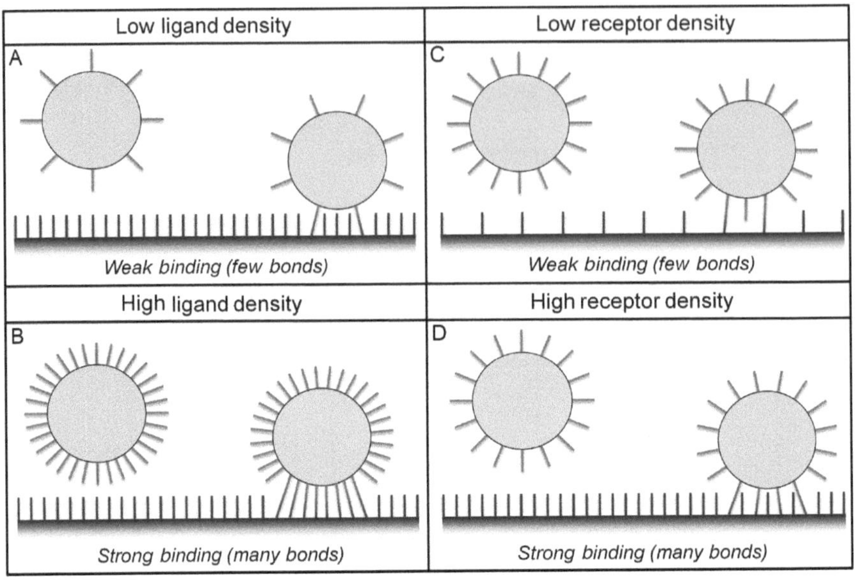

Fig. 5 The effect of ligand density and receptor density on nanoparticle–endothelial interactions. A low ligand density (A) will likely form less nanoparticle–endothelial bonds than a higher ligand density (B). A nanoparticle of fixed ligand density will be less likely to form multiple bonds to a low receptor density (C) than a high receptor density (D).

Ferrari, 2006). This is dictated both by size (Fig. 6A and B), shape (Fig. 6C and D), and orientation of the nanoparticle.

Decuzzi and Ferrari declared that the probability of a nonspherical nanoparticle adhering (P_a) to a vessel wall under shear stress is a balance between bond formation and bond rupture, equal to:

$$P_a = D_L D_R A_C {K_a}^0 e^{(-B_{RL} f / K_B T)}, \tag{8}$$

where D_L is the ligand density, D_R is the receptor density, A_C is the contact surface area, K_a^0 is the receptor–ligand association constant at zero load, B_{RL} is the receptor–ligand bond length, f is the dislodging force *per* receptor–ligand pair, K_B is the Boltzmann constant, and T is the absolute temperature (Decuzzi & Ferrari, 2006). The dislodging force *per* receptor–ligand pair is calculated from the drag and torque acting upon the nanoparticle, as in Eqs. (1)–(7). The total dislodging force (F_{dis}) is given as:

$$F_{dis} = F_d + \frac{T_q}{2}. \tag{9}$$

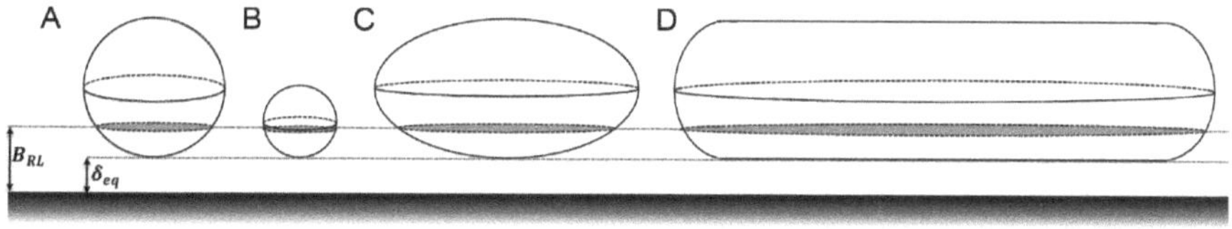

Fig. 6 The effect of nanoparticle size and shape on contact surface area in nanoparticle–endothelial interactions. Nanoparticles, a distance of δ_{eq} from the membrane, have a contact surface area (gray) dictated by the receptor–ligand bond length (B_{RL}), the shape, and size. Larger spherical nanoparticles (A) tend to have a greater contact surface area than smaller particles (B). Spheroidal nanoparticles (C) and tubular nanoparticles (D), with a major axis parallel to the vessel wall, have a greater contact surface area than spherical particles.

The torque particularly stresses bonds on the rear half of the nanoparticle, and therefore the torque is usually halved. The dislodging force *per* receptor bond pair can be estimated by the number of bonds, related to the contact surface area and the receptor density, assuming that the receptor density is the limiting factor:

$$f = \frac{F_{dis}}{D_R A_C}. \tag{10}$$

Eqs. (1)–(10) demonstrate that nanoparticle binding to the blood–brain barrier is a complex process, involving nanoparticle properties (size, shape, and ligand density), biochemical properties (receptor density, receptor–ligand bond length, and receptor–ligand affinity), and physical properties of blood flow (wall shear stress). While most of the biochemical and physical properties are innate, by controlling nanoparticle size, shape, ligand density, and the receptor–ligand affinity, we may be able to control binding for particular scenarios.

One such scenario we can control through such properties is in the design of "superselective" nanoparticles. This tackles a major challenge in general nanoparticle delivery: can we provide nanoparticles with the capability to discriminate between desired targets and other cells/tissues? Ideally, target cells are decorated with a receptor that is unique to those cells. However, this is rarely the case in vivo, therefore new methods to target-specific cells are required. The Frenkel group has extensively studied methods to achieve "superselectivity" in nanoparticle binding. They demonstrated, using Monte Carlo simulations, that nanoparticles could be tuned to bind strongly to cells only when receptor expression is high and not where receptor expression is low (Martinez-Veracoechea & Frenkel, 2011). They

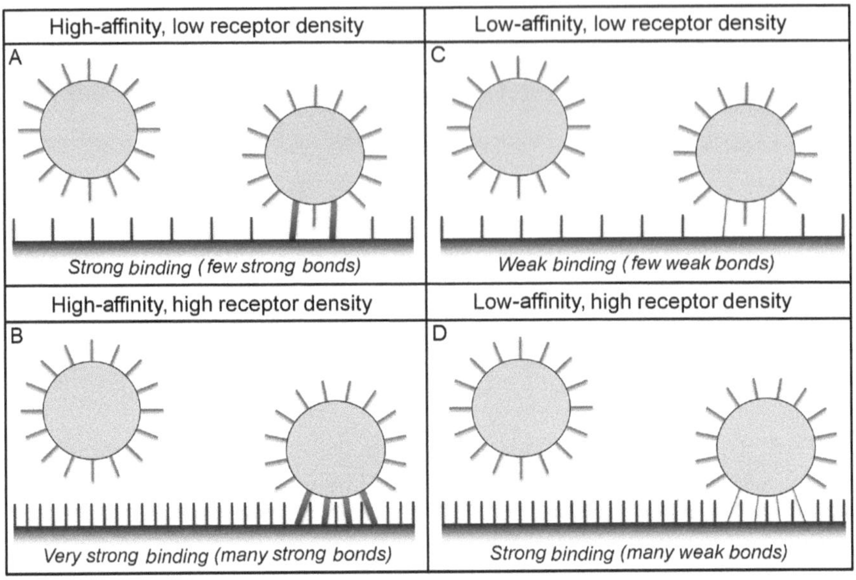

Fig. 7 Superselective binding of nanoparticles to high receptor expression. High-affinity receptor–ligand bonds give strong binding even with few bonds (A). Where there are more bonds, such as when receptor expression is higher (B), the bonding is even stronger due to the cumulative strength of many strong bonds. Low-affinity receptor–ligand bonds give weak binding with few bonds (C). Only by increasing bond number, such as with high receptor expression, is strong binding possible, due to the cumulative effect of many weak bonds and contributions from entropy (D).

demonstrated that if the ligands on a nanoparticle surface have a high affinity for the receptor, then the nanoparticle would bind well to cells with high or low expression of receptor. However, if the ligand is of low affinity, then the nanoparticle will only stably bind if there is a high ligand density and high receptor availability. This is because there is contribution from entropy to nanoparticle binding stability and a cumulative effect from multiple receptor–ligand interactions. Therefore, by increasing the ligand density to excess, the nanoparticles only bind if the corresponding receptor expression is high on the target cells (Fig. 7).

4. WRAPPING

Stably bound nanoparticles undergoing transcytosis must undergo membrane wrapping into a membrane-bound vesicle prior to trafficking. This process is both size and shape dependent, with optimal size ranges obtained from experimental and theoretical studies. Zhang and Li used a theoretical model to study membrane bending in endocytosis and found an optimal radius for membrane wrapping of 25–30 nm, with almost no material endocytosed under 22 nm in radius and very little material endocytosed above 60 nm (Zhang, Li, Lykotrafitis, Bao, & Suresh, 2009). Rejman and colleagues studied this process further, concentrating on size effects upon the predominant entry pathways. They demonstrated that clathrin-mediated endocytosis (CME) is preferred with smaller particles (50–200 nm), whereas larger particles (200–500 nm) increasingly use other mechanisms including caveolae-mediated endocytosis (Rejman, Oberle, Zuhorn, & Hoekstra, 2004). Decuzzi and Ferrari studied the effect of nanoparticle shape, using the same aspect ratios discussed previously in this chapter (Decuzzi & Ferrari, 2008). They demonstrated three endocytosis conditions, dependent on the aspect ratio. Nanoparticles with an aspect ratio of ~1 are completely wrapped and eventually internalized. However, nanoparticles with a very small aspect ratio demonstrated little or no wrapping. Nanoparticles with large aspect ratios demonstrated partial wrapping, a phenomenon termed as "frustrated endocytosis," where endocytosis is considerably slower and less efficient. This would appear to demonstrate that nonspherical nanoparticles, though potentially better at binding to the endothelial wall, might not be transcytosed as readily as their spherical counterparts. Furthermore, a nanoparticle of ~50–60 nm may be more efficient at transcytosis. Interestingly this size falls within the optimal range for better blood half-life, as discussed earlier.

5. TRAFFICKING: TRANSCYTOSIS AT THE BRAIN ENDOTHELIUM

After the extensive plasma membrane remodeling and internalization of nanoparticles as cargo in endocytosis, cargo undergoes intracellular trafficking to the opposing plasma membrane and is released in exocytosis. The cellular machinery involved in transcytosis may differ depending on the phenotype of the polarized monolayer. Transcytosis has frequently been studied in epithelial cell lines such as liver or kidney cell lines MDCK and Caco-2, and the fate of cargo such as immunoglobulin complexes has been well characterized in epithelia (Kaetzel, Robinson, Chintalacharuvu, Vaerman, & Lamm, 1991). However, epithelial cells have a largely different phenotype to endothelial cells. A particular discrepancy is size: epithelial cells span 10 μm across, compared to as little as 0.2 μm across brain endothelial cells (BECs) (Keep & Jones, 1990). Endothelial cell transcytosis has predominantly been studied in pulmonary endothelial cells, which are more permeable and have a higher occurrence of transcytosis than BECs (Brightman & Reese, 1969). Because of the specialized structure and function of BECs, the cellular machinery of BEC transcytosis might differ from that in peripheral endothelial or epithelial transcytosis. Studying transcytosis by the endothelial cells at the blood–brain barrier is complicated, as transportation may be bidirectional between blood and brain. Furthermore, it is unclear which signaling cues endothelial cells receive to determine when to direct the cargo to the opposite side of the plasma membrane and when to target it to intracellular organelles or lysosomal degradation. Much remains to be elucidated about the process of transcytosis and how it might differ in specific tissues. This section aims to outline the general processes of endocytosis, intracellular trafficking, and exocytosis, followed by a review of findings of how these processes might differ in transcytosis. Furthermore, these trafficking events are compared between peripheral and BEC phenotypes.

5.1 Endocytosis

Endocytosis is essential to cells for uptake of nutrients, but is also used for regulation of signaling at the plasma membrane (Shibata et al., 2000) and remodeling of the plasma membrane lipid composition (Bu, Maksymovitch, Geuze, & Schwartz, 1994). In endocytosis, part of the plasma membrane folds inwards together with lipids and proteins and pinches off into discrete vesicles inside the cell together with the ligand and receptor as cargo. The protein and lipid composition of these intracellular vesicles varies, and the different pathways of endocytosis have been well characterized in model organisms such as yeast as well as different kinds of mammalian cells. Some pathways involve uptake of molecules of the micrometer scale, such as phagocytosis and micropinocytosis. These processes are absent at the blood–brain barrier and will therefore not be discussed herein (Brightman & Reese, 1969).

The major endocytosis pathway is CME. In CME, ligand–receptor interaction triggers the formation of pits in the plasma membrane adjacent to the binding site. The pits are surrounded by the protein clathrin, which self-assemble into triskelia or "cage-like" structures together with clathrin supporting adaptor proteins. Many types of clathrin adaptor proteins have been identified and maintain specific subcellular localizations, and only adaptor protein 2 (AP2) associates with clathrin at the plasma membrane to mediate endocytosis. Clathrin and AP2 surround the pits, forming clathrin-coated vesicles (CCVs) as they fold inwards (Brodsky, Chen, Knuehl, Towler, & Wakeham, 2001). The GTPase dynamin assembles around the elongated phospholipid neck of the invaginated plasma membrane and severs it in an energy-dependent manner into a discrete vesicle (Tuma & Hubbard, 2003). Additionally, the cytoskeleton helps in vesicular biogenesis and actin is essential for the invagination process of endocytosis. CCVs can be up to 200 nm in diameter depending on cargo size and have a lifetime of around 20–120 s (Canton & Battaglia, 2012; Ehrlich et al., 2004). The most-studied clathrin-independent mechanism is caveolae-mediated endocytosis. Caveolae are plasma membrane resident 50–70 nm diameter flask-shaped pits coated with the protein caveolin-1. They are present in many cell types and often abundant in vascular endothelial cells, with the exception of at the blood–brain barrier where they are sparse (Frank, Woodman, Park, & Lisanti, 2003). Upon stimulation they can undergo endocytosis in a dynamin-dependent manner (Henley, Krueger, Oswald, & McNiven, 1998; Oh, McIntosh, & Schnitzer, 1998).

Endothelial cells often have a number of intracellular vesicles with openings facing the capillary lumen at any given time. Many morphological studies performed in peripheral tissues such as the myocardium or pulmonary microvasculature have identified these vesicles as caveolae (Predescu, Predescu, & Malik, 2007). However, evidence supporting a role for caveolae in BEC transcytosis is lacking, not least because very few studies examining vesicular transcytosis have been performed in the brain endothelium. Although caveolar proteins are expressed by BECs, these cells are known to contain by far the lowest vesicular population with <100 vesicles per μm^3 (Frank et al., 2003; Henley et al., 1998). Perhaps the most important evidence against a role for caveolae in blood–brain barrier transcytosis is that mice devoid of caveolar coat protein caveolin-1 are viable, suggesting that removing these structures does not disturb the essential process of import and export of compounds across the blood–brain barrier (Drab et al., 2001).

The vast majority of endocytosis occurs through clathrin-mediated vesicles, and evidence exists for an important role of these structures in transcytosis. A recent study using total internal reflection fluorescence (TIRF) microscopy and siRNA gene knockdown of CCVs in adipose microvasculature endothelial cells showed that transcytosis of insulin was mediated by CCVs (Azizi et al., 2015). In BECs, transcytosis of iron–transferrin complexes occurs via CCV internalization in vivo (Roberts, Sandra, Siek,

Lucas, & Fine, 1992). However, there is a lack of studies following the complete steps of endothelial cell transcytosis, from apical to basolateral or vice versa.

5.2 Intracellular Trafficking

Internalized cargo can have a multitude of fates: targeting to an intracellular organelle such as the Golgi apparatus, degradation in lysosomes, or recycling to the plasma membrane via recycling endosomes. The internalized ligand often comes with a specific sequence of amino acid as the cargo sorting cue. For example, nuclear localization signals often consist of a stretch of basic amino acids, whereas proteins resident at the endoplasmic reticulum have a KDEL or KKXX amino acid sequence. Some ligands are targeted for lysosomal degradation, such as the LDL receptor after cholesterol has been released into the cytoplasm. This endolysosomal pathway follows an acidic pH gradient conducive to denaturing proteins, starting at a physiological pH at the plasma membrane and ending at approximately pH 5 in lysosomes. Cargo can also be recycled from endosomes to the plasma membrane, common for receptors that have undergone endocytosis. A simplified summary of these pathways is outlined in Fig. 8.

The intracellular trafficking events in endothelial cell transcytosis are not well known. At the blood–brain barrier, it is unclear whether cargo is

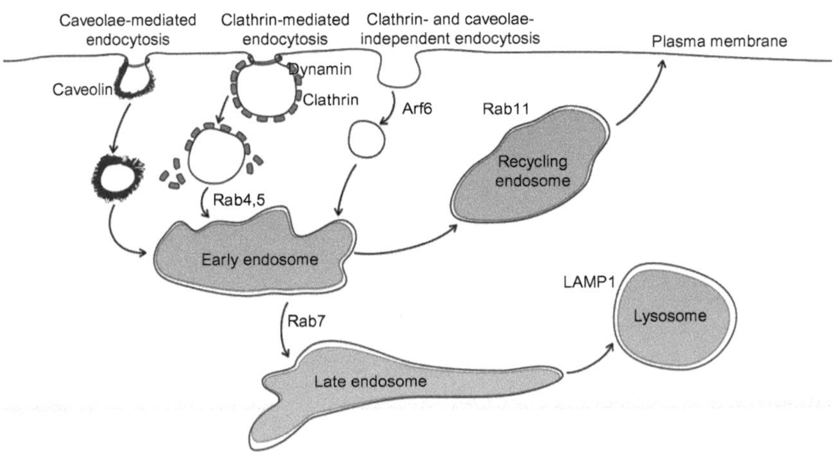

Fig. 8 The endosomal sorting pathway. Some Rabs implicated in endocytosis include Rab5 in early endosomes, Rab4 and 11 in "recycling" endosomes returning to the plasma membrane, and Rab7 in late endosomes.

trafficked through endosomes at all. Experiments tracking cargo with tracer dyes in CME was present in early endosomes within 1–5 min of incubation, late endosomes after 10–15 min, and lysosomes after 30 min (Pfeffer, 2001). Internalization by caveolae can take as long as 1–2 h (Pelkmans, Kartenbeck, & Helenius, 2001; Schnitzer, Oh, Pinney, & Allard, 1994). In both cases the timing is inconsistent with the rapid transcytosis rates observed in vivo, which has been observed using live imaging to occur in <60 s in the lung endothelium (Oh et al., 2007). Furthermore, the putative mechanism of the relatively frequently studied Fe–Tf transcytosis across the brain endothelium illustrates another discrepancy. It has been hypothesized that after internalization in CCVs, Tf shuttles iron across to the opposing plasma membrane, and then recycles in an unbound state (Moos & Morgan, 2000). Normally in the case of ligand–receptor interactions, dissociation occurs as a result of encountering the acidic pH of endosomes. This raises the possibility that Fe–Tf undergoes transcytosis at the brain endothelium through a pathway independent of acidifying endosomes.

The most energetically favorable transport pathway for cargo involves traveling the shortest distance possible. Brain capillary endothelial cells are approximately 40% thinner than peripheral endothelial cells (Coomber & Stewart, 1985), with the nonnuclear area as thin as 0.2–0.5 μm. As a result, cargo requires shorter transportation distances from one membrane leaflet to the other, which facilitates rapid transcytosis. Interestingly, there is ultrastructural evidence for a network of endothelial intracellular vesicles that have merged into "tubes" or pores. Transendothelial channels can be defined as chains of fused intracellular vesicles that are simultaneously open on the endothelial cell apical and basolateral sides. Bundgaard and colleagues performed electron microscopy (EM) of frog mesenteric capillaries to investigate transcytosis, where the addition of electron dense tannic acid seemed to fill out intracellular clusters of 2–5 vesicles that had fused (Bundgaard, Frøkjaer-Jensen, & Crone, 1979). Repeating these studies in hagfish brain endothelium, they found through 3D reconstructions of serial EM sections that many of the vesicles traditionally observed in 2D electron micrographs appeared to be part of a larger branching network when reconstructed in the *z* plane. These branching networks of intracellular vesicles were also observed in the EM work of Simionescu and colleagues, investigating caveolar transcytosis of gold-labeled albumin in the mouse lung, heart, and diaphragm (Simionescu, Popov, & Sima, 2008). The branching networks of vesicles seem more frequent at the blood–brain barrier; a study used 3D reconstructions of serial ~20 nm electron micrographs to compare the number of transendothelial channels in microvasculature from different origins, and found an average of 8 channels per μm^2 in mouse BECs, compared to 4 μm^{-2} in the choroid plexus. These channels were located at the thinnest area of the cell and were approximately the size of a single vesicle rather than a network of fused vesicles, although the latter were also present (Coomber & Stewart, 1985). A caveat of the channel hypothesis is that evidence is predominantly EM based. EM of cell or tissue samples requires the fixation and processing with harsh lipid removal techniques, which undoubtedly affect membrane structures including intracellular vesicles. Further evidence for transendothelial channels should ideally be obtained through alternative methods that do not require processing with fixatives.

In summary, knowledge of the intracellular transportation stages of transcytosis is lacking. It is possible that BEC transcytosis is a system completely independent of the endosomal pathway, and instead potentially consists of tubulation of intracellular vesicles into a fused network to facilitate rapid transcytosis. However, more evidence must be obtained regarding the existence of these tubes and their involvement in transcytosis.

5.3 Exocytosis

Exocytosis is the fusion of an intracellular vesicle with the plasma membrane. Constitutive exocytosis occurs in all cell types as a means for cells to recycle lipids and proteins to the plasma membrane and is very slow. In specialized cells such as neurons, there is additionally a highly regulated form of exocytosis that synchronously releases many vesicles of intracellular contents into the extracellular environment on a millisecond scale. Here, the final steps of fusion between the vesicular and target plasma membrane are under strict temporal control by specialized proteins that also provide the energy for fast fusion. The protein machinery regulating exocytosis has been identified mostly through studying neuron transmission of synaptic vesicles. Using affinity chromatography, Rothman and colleagues initially identified three main components essential for membrane fusion in synaptic vesicle exocytosis: *N*-ethylmaleimide soluble factor (NSF), synaptosomal-associated proteins (SNAPs), and SNAREs (SNAP receptors) (Sollner et al., 1993). He subsequently coined the SNARE hypothesis, which states that membrane fusion is dependent on the association of SNARE proteins on the vesicular and target membranes. Experimental evidence in support for this hypothesis is strong: knockout of SNARE proteins vesicle-associated membrane protein and SNAP-25 in mice is lethal (Schoch et al., 2001). Furthermore, exocytosis and synaptic transmission are inhibited by botulinum toxins and clostridium neurotoxins, which target SNAPs or SNAREs for proteolysis (Xu, Binz, Niemann, & Neher, 1998).

The essential step of membrane fusion involves the association of SNARE proteins expressed on the intracellular vesicle (v-SNARE) and the opposing target membrane (t-SNARE). All SNAREs have the same SNARE motif of 60–70 amino acids, which in their monomeric state lacks structure. When the vesicular and target membranes are brought in close apposition, the appropriate combination of 4 SNAREs on the vesicle and target membranes intertwines into stable core complexes of 4 α helices (sometimes referred to as a SNARE-pin). As this transmembrane *trans*-SNARE complex assembles or "zips up," it docks the vesicle to the membrane. When both bilayers of lipids have merged, a pore opens throughout the membranes to form a channel. This pore collapses eventually, releasing cargo and trapping all proteins in the target membrane. After fusion has occurred, the membrane fusion ATPase NSF disassembles the SNARE complexes via SNAP, its adaptor protein. This "recycling" of *cis*-SNARE complexes is essential for membrane fusion to occur again.

Although SNARE complexes are critical for membrane fusion, they are insufficient to mediate fusion in a timely manner. SNARE regulatory proteins assist with the fusion process and are especially significant in quick regulated exocytosis such as neurotransmission. A range of both positive and negative regulators of secretion has been identified and fine-tunes the exocytosis process. Knockout studies have identified the protein family of Sec1/Munc18 (SM) as regulators essential to every fusion step in neuronal exocytosis (Dulubova et al., 2007). SM proteins bind to the syntaxin family of t-SNAREs in the plasma membrane and can adopt open or closed conformations to negatively regulate secretion. TIRF microscopy of fluorescently labeled cargo in Munc18 knockout adrenal chromaffin cells showed that Munc18 is required for stable vesicle docking, causing on average a vesicular membrane residency of >10 s (Toonen et al., 2006). The regulatory role of SM proteins appears to be conserved in cell types other than neurons. For example, the isoform Munc18c is expressed in human microvascular endothelial cells, and exocytosis was inhibited upon treating cells with siRNA to syntaxin 4 which directly interacts with Munc18c (Fu, Naren, Gao, Ahmmed, & Malik, 2005).

Although SNAREs have been well characterized in neurons, the roles of SNAREs and their regulatory proteins in transcytosis are far less understood. There have been several reports of inhibition of myocardial vasculature transcytosis in vivo through the use of NSF inhibitor *N*-ethylmaleimide (NEM) (Predescu, Horvat, Predescu, & Palade, 1994). The addition of recombinant NSF rescued NEM-inhibited transcytosis in hepatic cell-free systems (Sztul, Colombo, Stahl, & Samanta, 1993). However, implicating certain SNAREs essential in BEC transcytosis is proving complicated. More than 100 SNARE proteins have been identified in mammalian cells (Hong, 2005). Only certain combinations out of the hundreds available fuse, meaning that SNAREs provide specificity to membrane fusion (Südhof & Rothman, 2009). This provides a targeting mechanism that may indeed be conserved in transcytosis. However, combinations of SNAREs specific to transcytosis have not been identified. In polarized cells there may be some asymmetrical distribution of SNAREs in the apical and basolateral leaflets, as reviewed by Tuma and Hubbard (2003). Also, coimmunoprecipitation of lung microvasculature identified a few molecular mediators of transcytosis in a complex containing syntaxin, cellubrevin, SNAP-23, Rab5, NSF, and α- and γ-SNAP (Predescu, Predescu, & Palade, 2001). However, further work needs to be done in order to elucidate the role of SNAREs and their associated proteins in transcytosis. Interestingly, the SNARE components identified in lung microvasculature by Predescu and coworkers were associated with lipids in supramolecular protein–lipid complexes, highlighting the role of lipids as protein segregating platforms in advanced membrane signaling and remodeling events such as exocytosis. A schematic of the different stages of transcytosis, including the potential roles of dynamin and SNAREs, is given in Fig. 9.

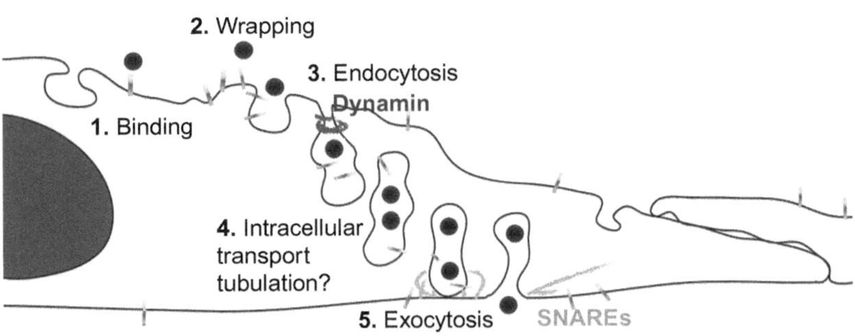

Fig. 9 Transcytosis at the brain endothelium. Schematic of transcytosis at the brain endothelium, including known and likely roles of proteins in each stage.

6. RELEASE AND RECEPTOR RECYCLING

Recently, a number of studies have suggested that the receptor–ligand affinity in binding is also important in the efficiency of unbinding and release after transcytosis. Yu and colleagues demonstrated that by reducing the affinity of antitransferrin receptor antibodies to the transferrin receptor transcytosis across the blood–brain barrier was actually increased when blood antibody concentrations were high (Gadkar et al., 2016; Yu et al., 2014, 2011). Moreover, the high-affinity antibodies triggered eventual trafficking to the lysosome within BECs in vitro and in vivo (Bien-Ly et al., 2014). It was hypothesized that this occurred because binding and unbinding are the forward and reverse of the same reaction. Therefore, high affinity results in excellent binding, but poor unbinding. In the absence of effective release, the receptor–ligand complex was redirected to the lysosome. Simultaneously, low-affinity binding results in excellent unbinding but poor initial binding. In using an intermediate affinity antibody, transcytosis efficiency increases, as there is both good binding and unbinding.

Similar principles may also be of importance in nanoparticle transcytosis and this remains an area for further research. However, as discussed earlier, nanoparticle binding strength is more complex as a single nanoparticle binding event can involve multiple receptor–ligand interactions. Niewoehner and colleagues studied the effect of multiple receptor–ligand interactions by investigating transcytosis efficiency of monovalent and bivalent antitransferrin antibodies. They demonstrated that the monovalent antibody was more effective at transcytosis, as the bivalent antibody was sorted into a lysosomal compartment (Niewoehner et al., 2014). Wiley and colleagues investigated this effect in nanoparticles, focusing on the effect of size and ligand density on transferrin-functionalized nanoparticle transcytosis (Wiley, Webster, Gale, & Davis, 2013). They demonstrated that the dissociation constant (K_D) of the nanoparticles with BECs decreased with increased size and increased ligand density, suggesting a stronger affinity for the BECs. Furthermore, the most effective two compositions trialed had similar, intermediate, values of K_D, 0.71 and 0.89 nM, despite having sizes of $\sim$50 and $\sim$80 nm and ligand densities of 4×10^{15} and 10^{15} m^{-2}, respectively. Compositions with a higher affinity were associated with the blood vessels, suggesting that they lacked the ability to effectively unbind. These two studies, along with the work of Yu and colleagues, demonstrate that reducing the overall binding strength of nanoparticles may be integral to efficient transcytosis.

7. NANOPARTICLE TRANSCYTOSIS: THE STORY SO FAR

In recent years, many different groups have attempted to harness transcytosis at the blood–brain barrier to deliver nanoparticles into the CNS. In this section we look at some examples of these. The most common receptor targeted is the transferrin receptor. A number of groups have used the antitransferrin monoclonal antibody OX26 that has previously been shown to increase brain penetrance of methotrexate conjugated directly to the antibody (Friden et al., 1991; Pardridge, Buciak, & Friden, 1991). Pardridge and colleagues used OX26 conjugated to 121 nm PEG–poly[lactic acid] nanoparticles, at a ligand density of 0.0015 nm^{-2}, to increase CNS delivery (Olivier, Huertas, Lee, Calon, & Pardridge, 2002). However, recent work has concluded that OX26-nanoparticles actually colocalized with brain capillary endothelial cells, rather than entering the parenchyma itself (Gosk, Vermehren, Storm, & Moos, 2004; Moos & Morgan, 2001). Further work by the Zhu group found that an optimal OX26 surface density of ~0.001 nm^{-2} improved delivery of PEG–poly[ε-caprolactone] polymersomes to the CNS, suggesting multiple OX26-transferrin receptor binding events affected transcytosis efficiency (Pang et al., 2008).

R17217 is a mouse-specific monoclonal antibody to the transferrin receptor that has also been used extensively in transcytosis-mediated delivery of nanoparticles to the CNS. The Kreuter group used loperamide-loaded human serum albumin nanoparticles functionalized with the OX26 or R17217 transferrin receptor antibodies, of ~166–168 nm, at ligand densities of 0.003 and 0.002 nm^{-2} to achieve antinociceptive effect in rats (Ulbrich et al., 2009).

The monoclonal antitransferrin receptor antibody 8D3 has also been used to aid targeting of transcytosis across the blood–brain barrier. It provides the additional benefit, compared to OX26 and R17217, of being cross-species specific, demonstrating species reactivity with the human, mouse, rat, hamster, and monkey transferrin receptor. Fornaguera and colleagues used 8D3 to functionalized poly[lactic-*co*-glycolic acid] (PLGA)-based nanoparticles and achieve analgesic effect when loaded with loperamide (Fornaguera, Dols-Perez, et al., 2015). Interestingly, in a separate study they found that functionalization of PLGA nanoparticles increased the activation of complement, demonstrating that functionalization may have unwanted side effects on immune system activation and consequently blood clearance (Fornaguera, Caldero, et al., 2015). Cabezón and coworkers used the 8D3 antibody to functionalize gold nanoparticles. Upon administration to mice, they demonstrated that the nanoparticles were uptake by clathrin-dependent mechanism. Furthermore, the majority of the nanoparticles ended up in the late endosome. A small minority did however cross to the abluminal membrane, however, there it failed to detach (Cabezón et al., 2015). This suggests that the high affinity of 8D3, with a k_D of ~2.3 nM (Boado et al., 2010) to the mouse transferrin receptor, in line with the high-affinity antibody tested by Yu and colleagues, would make 8D3 not particularly suitable for effective transcytosis.

Peptides can be used to target nanocarriers to the CNS, including Angiopeps and G23. The G23 peptide appears to bind the ganglioside GM1 and results in a GM1-enriched caveolae facilitated transcytotic entry portal for such functionalized nanoparticles (Stojanov et al., 2012). Rhodamine-labeled G23 polymersomes revealed the potential of G23 to mediate the blood–brain barrier passage in vivo, by analyzing fluorescence in thin brain slices after 24 h following intravenous injection. Compared to nontargeted polymersomes, significant accumulation was found with G23 polymersomes in the cortex, striatum (forebrain), midbrain, pons, and cerebellum sections. Furthermore, they were not colocalized with endothelial capillaries but found within the brain parenchyma itself.

More recently, targeting LRP-1-mediated transcytosis has become an increasingly popular mechanism for CNS delivery across the blood–brain barrier. Demeule and colleagues identified a family of peptides, called Angiopeps, based on the Kunitz protease inhibitor domain of LRP receptor family substrates. In particular, Angiopep-2 has been shown to stimulate highly efficient transcytosis across the blood–brain barrier by LRP-1-mediated transcytosis (Demeule, Currie, et al., 2008; Demeule, Régina, et al., 2008). The high expression of LRP-1 on glial cells has led to the application of this peptide to target gliomas with antitumor agents. Functionalization of PEG–poly[ε-caprolactone] nanoparticles, PEG-modified poly[amidoamine] dendrimers, and PEG-modified multiwalled carbon nanotubes with Angiopep-2 increased the delivery and antitumor activity of drugs including doxorubicin, docetaxel, and the TRAIL (tumor necrosis factor-related apoptosis-inducing ligand) gene construct against gliomas (Gao et al., 2014; Ren et al., 2012). In addition, Battaglia and colleagues demonstrated the potential of utilizing Angiopep-2 functionalized pH-sensitive polymersomes to deliver a model IgG antibody to the mouse

CNS via LRP-1-mediated transcytosis. The authors used an in vitro three-dimensional model of the blood–brain barrier to perform colocalization experiments of Angiopep-2-polymersomes against endocytosis markers including Rab5, Rab7, Rab11, and LAMP1. Surprisingly, polymersomes were not associated with any of these membrane-trafficking markers during transcytosis. Additionally, fluorescein isothiocyanate (FITC)-IgG-loaded pH-sensitive polymersomes remained intact without releasing the cargo during transcytosis, suggesting that transcytosis occurs independently of acidifying endosomal compartments. The delivery of a model antibody cargo into the parenchyma provides new therapeutic and diagnostic opportunities (Tian et al., 2015). These studies demonstrate that Angiopep-2 can successfully improve delivery of a diverse range of nanoparticles across the blood–brain barrier for potential therapeutic intervention.

Kreuter and colleagues used poly[butyl cyanoacrylate] coated with polysorbate 80 to deliver the analgesics dalargin or loperamide to the CNS of mice. They found a greater analgesic response when administered as a complex (Alyautdin et al., 1997; Kreuter, Alyautdin, Kharkevich, & Ivanov, 1995). They later found the mechanism of entry to be dependent on adsorption of the endogenous apolipoproteins B and E to the complex, likely followed by transcytosis (Alyautdin et al., 1997; Kreuter et al., 2002). A similar study by Schröder and Sabel validated that this formulation could successfully increase CNS penetrance of dalargin (Schröder & Sabel, 1996). Gao and Jiang used the same formulation to test the effect of size on CNS delivery of methotrexate. They concluded that delivery of methotrexate was more efficient for nanoparticles of 70 nm than with 170, 220, and 345 nm nanoparticles (Gao & Jiang, 2006). This is consistent with an ideal size of between 40 and 100 nm for better blood residency and close to 50–60 nm for efficient endocytosis.

In the work of Wiley and colleagues, ligand density and size were considered together when evaluating the efficiency of gold nanoparticle transcytosis (Wiley et al., 2013). They used gold nanoparticles of different sizes and conjugated them to varying levels of transferrin. They demonstrated both a size and ligand density affect on the overall binding of the nanoparticles to Neuro2A cells. The overall k_D of their nanoparticles decreased with increasing size and increasing ligand density, as could be expected from the (Eqs. 1–10). The ligand densities of their formulations varied from 0.001 to 0.007 nm^{-2}. The k_D of monoferric transferrin has previously been estimated to be 10^4 nM, significantly lower than the antibodies dissociation constants given in Table 2 (Sawyer & Krantz, 1986).

8. SUMMARY

Drug delivery to the CNS is restricted by the barriers separating the brain and spinal cord from the rest of the body. The surface area combined with extensive blood supply makes the blood–brain barrier the ideal barrier to target for drug delivery. However, the extremely restrictive permeability combined with the metabolic barrier presents significant challenges. Particularly, BEC transcytosis is an attractive target. However, the cellular mechanisms involved at the blood–brain barrier are poorly understood. Further work is required to give a comprehensive overview of BEC transcytosis, from ligand–receptor binding to membrane trafficking and exocytosis. Nevertheless, a variety of approaches to target transcytosis at the blood–brain barrier with nanocarriers have been successful. These nanocarriers possess several common trends essential to successful drug delivery into the CNS. The first requirement is a targeting mechanism for receptors mediating transcytosis at the blood–brain barrier, for which current approaches include antibodies or peptides attached to the nanocarrier or naked drug. Furthermore, the size of the nanocarrier must be within the right range to induce transcytosis. Additionally, transcytosis must be completed, ie, result in shuttling of the nanocarrier to the parenchyma. Knowledge is lacking on what cues the brain endothelia use to direct a ligand to remain within the endothelia and what cues are used to complete transcytosis. Finally, the cargo must be able to dissociate from its carrier during transcytosis, for release into the parenchyma. Yu and colleagues highlighted the importance of this requirement in 2011 by showing that antibodies with too high affinity for their cargo shuttled back and forth into the parenchyma without releasing the cargo, producing no therapeutic effect. This area is an emerging hypothesis in general CNS delivery and remains relatively untouched in nanoparticle delivery to the CNS. Therefore, future work should aim to better understand these factors and explore methods to obtain maximum transcytosis efficiency at the blood–brain barrier, therefore improving the potential of nanoparticle-mediated delivery for the CNS.

REFERENCES

Alexis, F., Pridgen, E., Molnar, L. K., & Farokhzad, O. C. (2008). Factors affecting the clearance and biodistribution of polymeric nanoparticles. *Molecular Pharmaceutics*, *5*(4), 505–515. http://dx.doi.org/10.1021/mp800051m.

Alyautdin, R., Petrov, V., Langer, K., Berthold, A., Kharkevich, D., & Kreuter, J. (1997). Delivery of loperamide across the blood-brain barrier with polysorbate 80-coated polybutylcyanoacrylate nanoparticles. *Pharmaceutical Research*, *14*(3), 325–328. http://dx.doi.org/10.1023/a:1012098005098.

Azizi, P. M., Zyla, R. E., Guan, S., Wang, C., Liu, J., Bolz, S.-S., ... Lee, W. L. (2015). Clathrin-dependent entry and vesicle-mediated exocytosis define insulin transcytosis across microvascular endothelial cells. *Molecular Biology of the Cell*, *26*(4), 740–750. http://dx.doi.org/10.1091/mbc.E14-08-1307.

Banks, W. A. (2004). The source of cerebral insulin. *European Journal of Pharmacology*, *490*(1–3), 5–12. http://dx.doi.org/10.1016/j.ejphar.2004.02.040.

Bien-Ly, N., Yu, Y. J., Bumbaca, D., Elstrott, J., Boswell, C. A., Zhang, Y., ... Watts, R. J. (2014). Transferrin receptor (TfR) trafficking determines brain uptake of TfR antibody affinity variants. *The Journal of Experimental Medicine*, *211*(2), 233–244. http://dx.doi.org/10.1084/jem.20131660.

Boado, R. J., Zhou, Q.-H., Lu, J. Z., Hui, E. K.-W., & Pardridge, W. M. (2010). Pharmacokinetics and brain uptake of a genetically engineered bi-functional fusion antibody targeting the mouse transferrin receptor. *Molecular Pharmaceutics*, 7(1), 237–244. http://dx.doi.org/10.1021/mp900235k.

Brightman, M. W., & Reese, T. S. (1969). Junctions between intimately apposed cell membranes in the vertebrate brain. *The Journal of Cell Biology*, *40*(3), 648–677. http://dx.doi.org/10.1083/jcb.40.3.648.

Broadwell, R. D., Balin, B. J., & Salcman, M. (1988). Transcytotic pathway for blood-borne protein through the blood-brain barrier. *Proceedings of the National Academy of Sciences of the United States of America*, *85*(2), 632–636.

Brodsky, F. M., Chen, C.-Y., Knuehl, C., Towler, M. C., & Wakeham, D. E. (2001). Biological basket weaving: Formation and function of clathrin-coated vesicles. *Annual Review of Cell and Developmental Biology*, *17*(1), 517–568. http://dx.doi.org/10.1146/annurev.cellbio.17.1.517.

Bu, G., Maksymovitch, E. A., Geuze, H., & Schwartz, A. L. (1994). Subcellular localization and endocytic function of low density lipoprotein receptor-related protein in human glioblastoma cells. *Journal of Biological Chemistry*, *269*(47), 29874–29882.

Bundgaard, M., Frøkjaer-Jensen, J., & Crone, C. (1979). Endothelial plasmalemmal vesicles as elements in a system of branching invaginations from the cell surface. *Proceedings of the National Academy of Sciences of the United States of America*, *76*(12), 6439–6442.

Cabezón, I., Manich, G., Martín-Venegas, R., Camins, A., Pelegrí, C., & Vilaplana, J. (2015). Trafficking of gold nanoparticles coated with the 8D3 anti-transferrin receptor antibody at the mouse blood–brain barrier. *Molecular Pharmaceutics*, *12*(11), 4137–4145. http://dx.doi.org/10.1021/acs.molpharmaceut.5b00597.

Canton, I., & Battaglia, G. (2012). Endocytosis at the nanoscale. *Chemical Society Reviews*, *41*(7), 2718–2739. http://dx.doi.org/10.1039/c2cs15309b.

Coomber, B. L., & Stewart, P. A. (1985). Morphometric analysis of CNS microvascular endothelium. *Microvascular Research*, *30*(1), 99–115. http://dx.doi.org/10.1016/0026-2862(85)90042-1.

Deane, R., Wu, Z., & Zlokovic, B. V. (2004). RAGE (Yin) versus LRP (Yang) balance regulates Alzheimer amyloid β-peptide clearance through transport across the blood–brain barrier. *Stroke*, *35*(11 Suppl. 1), 2628–2631. http://dx.doi.org/10.1161/01.STR.0000143452.85382.d1.

Decuzzi, P., & Ferrari, M. (2006). The adhesive strength of non-spherical particles mediated by specific interactions. *Biomaterials*, *27*(30), 5307–5314. http://dx.doi.org/10.1016/j.biomaterials.2006.05.024.

Decuzzi, P., & Ferrari, M. (2008). The receptor-mediated endocytosis of nonspherical particles. *Biophysical Journal*, *94*(10), 3790–3797. http://dx.doi.org/10.1529/biophysj.107.120238.

Demeule, M., Currie, J.-C., Bertrand, Y., Ché, C., Nguyen, T., Régina, A., … Béliveau, R. (2008). Involvement of the low-density lipoprotein receptor-related protein in the transcytosis of the brain delivery vector angiopep-2. *Journal of Neurochemistry, 106*(4), 1534–1544. http://dx.doi.org/10.1111/j.1471-4159.2008.05492.x.
Demeule, M., Poirier, J., Jodoin, J., Bertrand, Y., Desrosiers, R. R., Dagenais, C., … Béliveau, R. (2002). High transcytosis of melanotransferrin (P97) across the blood–brain barrier. *Journal of Neurochemistry, 83*(4), 924–933. http://dx.doi.org/10.1046/j.1471-4159.2002.01201.x.
Demeule, M., Régina, A., Ché, C., Poirier, J., Nguyen, T., Gabathuler, R., … Béliveau, R. (2008). Identification and design of peptides as a new drug delivery system for the brain. *Journal of Pharmacology and Experimental Therapeutics, 324*(3), 1064–1072. http://dx.doi.org/10.1124/jpet.107.131318.
Drab, M., Verkade, P., Elger, M., Kasper, M., Lohn, M., Lauterbach, B., … Kurzchalia, T. V. (2001). Loss of caveolae, vascular dysfunction, and pulmonary defects in caveolin-1 gene-disrupted mice. *Science, 293*(5539), 2449–2452. http://dx.doi.org/10.1126/science.1062688.
Dulubova, I., Khvotchev, M., Liu, S., Huryeva, I., Südhof, T. C., & Rizo, J. (2007). Munc18-1 binds directly to the neuronal SNARE complex. *Proceedings of the National Academy of Sciences of the United States of America, 104*(8), 2697–2702. http://dx.doi.org/10.1073/pnas.0611318104.
Ehrlich, M., Boll, W., van Oijen, A., Hariharan, R., Chandran, K., Nibert, M. L., & Kirchhausen, T. (2004). Endocytosis by random initiation and stabilization of clathrin-coated pits. *Cell, 118*(5), 591–605. http://dx.doi.org/10.1016/j.cell.2004.08.017.
Fornaguera, C., Caldero, G., Mitjans, M., Vinardell, M. P., Solans, C., & Vauthier, C. (2015). Interactions of PLGA nanoparticles with blood components: Protein adsorption, coagulation, activation of the complement system and hemolysis studies. *Nanoscale,* 7(14), 6045–6058. http://dx.doi.org/10.1039/c5nr00733j.
Fornaguera, C., Dols-Perez, A., Calderó, G., García-Celma, M. J., Camarasa, J., & Solans, C. (2015). PLGA nanoparticles prepared by nano-emulsion templating using low-energy methods as efficient nanocarriers for drug delivery across the blood–brain barrier. *Journal of Controlled Release, 211*, 134–143. http://dx.doi.org/10.1016/j.jconrel.2015.06.002.
Frank, P. G., Woodman, S. E., Park, D. S., & Lisanti, M. P. (2003). Caveolin, caveolae, and endothelial cell function. *Arteriosclerosis, Thrombosis, and Vascular Biology, 23*(7), 1161–1168. http://dx.doi.org/10.1161/01.atv.0000070546.16946.3a.
Friden, P. M., Walus, L. R., Musso, G. F., Taylor, M. A., Malfroy, B., & Starzyk, R. M. (1991). Anti-transferrin receptor antibody and antibody-drug conjugates cross the blood-brain barrier. *Proceedings of the National Academy of Sciences of the United States of America, 88*(11), 4771–4775.
Fu, J., Naren, A. P., Gao, X., Ahmmed, G. U., & Malik, A. B. (2005). Protease-activated receptor-1 activation of endothelial cells induces protein kinase Cα-dependent phosphorylation of syntaxin 4 and Munc18c: Role in signaling P-selectin expression. *Journal of Biological Chemistry, 280*(5), 3178–3184. http://dx.doi.org/10.1074/jbc.M410044200.
Fullstone, G., Wood, J., Holcombe, M., & Battaglia, G. (2015). Modelling the transport of nanoparticles under blood flow using an agent-based approach. *Scientific Reports, 5*, 10649. http://dx.doi.org/10.1038/srep10649. http://www.nature.com/srep/2015/150610/srep10649/abs/srep10649.html. supplementary-information.
Fung, L. K., Shin, M., Tyler, B., Brem, H., & Saltzman, W. M. (1996). Chemotherapeutic drugs released from polymers: Distribution of 1,3-bis(2-chloroethyl)-l-nitrosourea in the rat brain. *Pharmaceutical Research, 13*, 671–682.
Gadkar, K., Yadav, D. B., Zuchero, J. Y., Couch, J. A., Kanodia, J., Kenrick, M. K., … Ramanujan, S. (2016). Mathematical PKPD and safety model of bispecific TfR/BACE1

antibodies for the optimization of antibody uptake in brain. *European Journal of Pharmaceutics and Biopharmaceutics*, *101*, 53–61. http://dx.doi.org/10.1016/j.ejpb.2016.01.009.
Gaillard, P. J., Visser, C. C., & de Boer, A. G. (2005). Targeted delivery across the blood-brain barrier. *Expert Opinion on Drug Delivery*, *2*(2), 299–309. http://dx.doi.org/10.1517/17425247.2.2.299.
Gao, K., & Jiang, X. (2006). Influence of particle size on transport of methotrexate across blood brain barrier by polysorbate 80-coated polybutylcyanoacrylate nanoparticles. *International Journal of Pharmaceutics*, *310*(1–2), 213–219. http://dx.doi.org/10.1016/j.ijpharm.2005.11.040.
Gao, H., Zhang, S., Cao, S., Yang, Z., Pang, Z., & Jiang, X. (2014). Angiopep-2 and activatable cell-penetrating peptide dual-functionalized nanoparticles for systemic glioma-targeting delivery. *Molecular Pharmaceutics*, *11*(8), 2755–2763. http://dx.doi.org/10.1021/mp500113p.
Geng, Y. A. N., Dalhaimer, P., Cai, S., Tsai, R., Tewari, M., Minko, T., & Discher, D. E. (2007). Shape effects of filaments versus spherical particles in flow and drug delivery. *Nature Nanotechnology*, *2*(4), 249–255. http://dx.doi.org/10.1038/nnano.2007.70.
Gosk, S., Vermehren, C., Storm, G., & Moos, T. (2004). Targeting anti-transferrin receptor antibody (OX26) and OX26-conjugated liposomes to brain capillary endothelial cells using in situ perfusion. *Journal of Cerebral Blood Flow and Metabolism*, *24*(11), 1193–1204.
Gref, R., Lück, M., Quellec, P., Marchand, M., Dellacherie, E., Harnisch, S., … Müller, R. H. (2000). 'Stealth' corona-core nanoparticles surface modified by polyethylene glycol (PEG): Influences of the corona (PEG chain length and surface density) and of the core composition on phagocytic uptake and plasma protein adsorption. *Colloids and Surfaces. B, Biointerfaces*, *18*(3–4), 301–313. http://dx.doi.org/10.1016/S0927-7765(99)00156-3.
Guerin, C., Olivi, A., Weingart, J. D., Lawson, H. C., & Brem, H. (2004). Recent advances in brain tumor therapy: Local intracerebral drug delivery by polymers. *Investigational New Drugs*, *22*, 27–37.
Henley, J. R., Krueger, E. W. A., Oswald, B. J., & McNiven, M. A. (1998). Dynamin-mediated Internalization of Caveolae. *The Journal of Cell Biology*, *141*(1), 85–99. http://dx.doi.org/10.1083/jcb.141.1.85.
Herz, J., & Marschang, P. (2003). Coaxing the LDL receptor family into the fold. *Cell*, *112*(3), 289–292. http://dx.doi.org/10.1016/S0092-8674(03)00073-4.
Herz, J., & Strickland, D. K. (2001). LRP: A multifunctional scavenger and signaling receptor. *Journal of Clinical Investigation*, *108*(6), 779–784.
Hong, W. (2005). SNAREs and traffic. *Biochimica et Biophysica Acta*, *1744*(2), 120–144. http://dx.doi.org/10.1016/j.bbamcr.2005.03.014.
Huwyler, J., Wu, D., & Pardridge, W. M. (1996). Brain drug delivery of small molecules using immunoliposomes. *Proceedings of the National Academy of Sciences of the United States of America*, *93*(24), 14164–14169.
Kaetzel, C. S., Robinson, J. K., Chintalacharuvu, K. R., Vaerman, J. P., & Lamm, M. E. (1991). The polymeric immunoglobulin receptor (secretory component) mediates transport of immune complexes across epithelial cells: A local defense function for IgA. *Proceedings of the National Academy of Sciences of the United States of America*, *88*(19), 8796–8800.
Keep, R., & Jones, H. (1990). A morphometric study on the development of the lateral ventricle choroid plexus, choroid plexus capillaries and ventricular ependyma in the rat. *Developmental Brain Research*, *56*(1), 47–53.
Kissel, K., Hamm, S., Schulz, M., Vecchi, A., Garlanda, C., & Engelhardt, B. (1998). Immunohistochemical localization of the murine transferrin receptor (TfR) on blood–tissue barriers using a novel anti-TfR monoclonal antibody. *Histochemistry and Cell Biology*, *110*, 63–72.

Kordower, J. H., Palfi, S., Chen, E.-Y., Ma, S. Y., Sendera, T., Cochran, E. J., … Comella, C. D. (1999). Clinicopathological findings following intraventricular glial-derived neurotrophic factor treatment in a patient with Parkinson's disease. *Annals of Neurology*, *46*(3), 419–424. http://dx.doi.org/10.1002/1531-8249(199909)46:3<419::aid-ana21>3.0.co;2-q.

Kreuter, J., Alyautdin, R. N., Kharkevich, D. A., & Ivanov, A. A. (1995). Passage of peptides through the blood-brain barrier with colloidal polymer particles (nanoparticles). *Brain Research*, *674*(1), 171–174. http://dx.doi.org/10.1016/0006-8993(95)00023-J.

Kreuter, J., Shamenkov, D., Petrov, V., Ramge, P., Cychutek, K., Koch-Brandt, C., & Alyautdin, R. (2002). Apolipoprotein-mediated transport of nanoparticle-bound drugs across the blood-brain barrier. *Journal of Drug Targeting*, *10*(4), 317–325. http://dx.doi.org/10.1080/10611860290031877.

Martinez-Veracoechea, F. J., & Frenkel, D. (2011). Designing super selectivity in multivalent nano-particle binding. *Proceedings of the National Academy of Sciences of the United States of America*, *108*(27), 10963–10968. http://dx.doi.org/10.1073/pnas.1105351108.

Mellman, I. (1996). Endocytosis and molecular sorting. *Annual Review of Cell and Developmental Biology*, *12*(1), 575–625. http://dx.doi.org/10.1146/annurev.cellbio.12.1.575.

Menei, P., Pean, J. M., Nerrière-Daguin, V., Jollivet, C., Brachet, P., & Benoit, J. P. (2000). Intracerebral implantation of NGF-releasing biodegradable microspheres protects striatum against excitotoxic damage. *Experimental Neurology*, *161*(1), 259–272. http://dx.doi.org/10.1006/exnr.1999.7253.

Moos, T., & Morgan, E. H. (2000). Transferrin and transferrin receptor function in brain barrier systems. *Cellular and Molecular Neurobiology*, *20*, 77–95.

Moos, T., & Morgan, E. H. (2001). Restricted transport of anti-transferrin receptor antibody (OX26) through the blood–brain barrier in the rat. *Journal of Neurochemistry*, *79*(1), 119–129. http://dx.doi.org/10.1046/j.1471-4159.2001.00541.x.

Niewoehner, J., Bohrmann, B., Collin, L., Urich, E., Sade, H., Maier, P., … Freskgård, P.-O. (2014). Increased brain penetration and potency of a therapeutic antibody using a monovalent molecular shuttle. *Neuron*, *81*(1), 49–60. http://dx.doi.org/10.1016/j.neuron.2013.10.061.

Oh, P., Borgstrom, P., Witkiewicz, H., Li, Y., Borgstrom, B. J., Chrastina, A., … Schnitzer, J. E. (2007). Live dynamic imaging of caveolae pumping targeted antibody rapidly and specifically across endothelium in the lung. *Nature Biotechnology*, *25*(3), 327–337. http://www.nature.com/nbt/journal/v25/n3/suppinfo/nbt1292_S1.html.

Oh, P., McIntosh, D. P., & Schnitzer, J. E. (1998). Dynamin at the neck of caveolae mediates their budding to form transport vesicles by GTP-driven fission from the plasma membrane of endothelium. *The Journal of Cell Biology*, *141*(1), 101–114. http://dx.doi.org/10.1083/jcb.141.1.101.

Olivier, J.-C., Huertas, R., Lee, H. J., Calon, F., & Pardridge, W. M. (2002). Synthesis of pegylated immunonanoparticles. *Pharmaceutical Research*, *19*(8), 1137–1143. http://dx.doi.org/10.1023/a:1019842024814.

Owens, D. E., & Peppas, N. A. (2006). Opsonization, biodistribution, and pharmacokinetics of polymeric nanoparticles. *International Journal of Pharmaceutics*, *307*(1), 93–102. http://dx.doi.org/10.1016/j.ijpharm.2005.10.010.

Pang, Z., Lu, W., Gao, H., Hu, K., Chen, J., Zhang, C., … Zhu, C. (2008). Preparation and brain delivery property of biodegradable polymersomes conjugated with OX26. *Journal of Controlled Release*, *128*(2), 120–127. http://dx.doi.org/10.1016/j.jconrel.2008.03.007.

Pardridge, W. M. (2005). The blood-brain barrier: Bottleneck in brain drug development. *NeuroRx: The Journal of the American Society for Experimental NeuroTherapeutics*, *2*(1), 3–14. http://dx.doi.org/10.1602/neurorx.2.1.3.

Pardridge, W. M., Buciak, J. L., & Friden, P. M. (1991). Selective transport of an anti-transferrin receptor antibody through the blood-brain barrier in vivo. *Journal of Pharmacology and Experimental Therapeutics*, *259*(1), 66–70.

Pardridge, W. M., Kang, Y.-S., Buciak, J. L., & Yang, J. (1995). Human insulin receptor monoclonal antibody undergoes high affinity binding to human brain capillaries in vitro and rapid transcytosis through the blood–brain barrier in vivo in the primate. *Pharmaceutical Research*, *12*, 807–816.

Pegoraro, C., Cecchin, D., Gracia, L. S., Warren, N., Madsen, J., Armes, S. P., … Battaglia, G. (2013). Enhanced drug delivery to melanoma cells using PMPC-PDPA polymersomes. *Cancer Letters*, *334*(2), 328–337. http://dx.doi.org/10.1016/j.canlet.2013.02.007.

Pelkmans, L., Kartenbeck, J., & Helenius, A. (2001). Caveolar endocytosis of simian virus 40 reveals a new two-step vesicular-transport pathway to the ER. *Nature Cell Biology*, *3*(5), 473–483. http://www.nature.com/ncb/journal/v3/n5/suppinfo/ncb0501_473_S1.html.

Pfeffer, S. R. (2001). Rab GTPases: Specifying and deciphering organelle identity and function. *Trends in Cell Biology*, *11*(12), 487–491. http://dx.doi.org/10.1016/s0962-8924(01)02147-x.

Pozrikidis, C. (1994). The motion of particles in the Hele-Shaw cell. *Journal of Fluid Mechanics*, *261*, 199–222.

Predescu, D., Horvat, R., Predescu, S., & Palade, G. E. (1994). Transcytosis in the continuous endothelium of the myocardial microvasculature is inhibited by N-ethylmaleimide. *Proceedings of the National Academy of Sciences of the United States of America*, *91*(8), 3014–3018.

Predescu, S. A., Predescu, D. N., & Malik, A. B. (2007). Molecular determinants of endothelial transcytosis and their role in endothelial permeability. *American Journal of Physiology—Lung Cellular and Molecular Physiology*, *293*(4), L823–L842. http://dx.doi.org/10.1152/ajplung.00436.2006.

Predescu, S. A., Predescu, D. N., & Palade, G. E. (2001). Endothelial transcytotic machinery involves supramolecular protein–lipid complexes. *Molecular Biology of the Cell*, *12*(4), 1019–1033.

Rejman, J., Oberle, V., Zuhorn, I. S., & Hoekstra, D. (2004). Size-dependent internalization of particles via the pathways of clathrin- and caveolae-mediated endocytosis. *Biochemical Journal*, *377*(Pt. 1), 159–169. http://dx.doi.org/10.1042/bj20031253.

Ren, J., Shen, S., Wang, D., Xi, Z., Guo, L., Pang, Z., … Jiang, X. (2012). The targeted delivery of anticancer drugs to brain glioma by PEGylated oxidized multi-walled carbon nanotubes modified with angiopep-2. *Biomaterials*, *33*(11), 3324–3333. http://dx.doi.org/10.1016/j.biomaterials.2012.01.025.

Roberts, R., Sandra, A., Siek, G. C., Lucas, J. J., & Fine, R. E. (1992). Studies of the mechanism of iron transport across the blood-brain barrier. *Annals of Neurology*, *32*(S1), S43–S50. http://dx.doi.org/10.1002/ana.410320709.

Sawyer, S. T., & Krantz, S. B. (1986). Transferrin receptor number, synthesis, and endocytosis during erythropoietin-induced maturation of Friend virus-infected erythroid cells. *Journal of Biological Chemistry*, *261*(20), 9187–9195.

Schnitzer, J. E., Oh, P., Pinney, E., & Allard, J. (1994). Filipin-sensitive caveolae-mediated transport in endothelium: Reduced transcytosis, scavenger endocytosis, and capillary permeability of select macromolecules. *The Journal of Cell Biology*, *127*(5), 1217–1232. http://dx.doi.org/10.1083/jcb.127.5.1217.

Schoch, S., Deák, F., Königstorfer, A., Mozhayeva, M., Sara, Y., Südhof, T. C., & Kavalali, E. T. (2001). SNARE function analyzed in synaptobrevin/VAMP knockout mice. *Science*, *294*(5544), 1117–1122. http://dx.doi.org/10.1126/science.1064335.

Schröder, U., & Sabel, B. A. (1996). Nanoparticles, a drug carrier system to pass the blood-brain barrier, permit central analgesic effects of i.v. dalargin injections. *Brain Research*, *710*(1–2), 121–124. http://dx.doi.org/10.1016/0006-8993(95)01375-X.

Shibata, M., Yamada, S., Kumar, S. R., Calero, M., Bading, J., Frangione, B., … Zlokovic, B. V. (2000). Clearance of Alzheimer's amyloid-β(1-40) peptide from brain by LDL receptor-related protein-1 at the blood-brain barrier. *Journal of Clinical Investigation, 106*(12), 1489–1499.

Simionescu, M., Popov, D., & Sima, A. (2008). Endothelial transcytosis in health and disease. *Cell and Tissue Research, 335*, 27–40.

Sollner, T., Whiteheart, S. W., Brunner, M., Erdjument-Bromage, H., Geromanos, S., Tempst, P., & Rothman, J. E. (1993). SNAP receptors implicated in vesicle targeting and fusion. *Nature, 362*(6418), 318–324.

Stern, D., Du Yan, S., Fang Yan, S., & Marie Schmidt, A. (2002). Receptor for advanced glycation endproducts: A multiligand receptor magnifying cell stress in diverse pathologic settings. *Advanced Drug Delivery Reviews, 54*(12), 1615–1625. http://dx.doi.org/10.1016/S0169-409X(02)00160-6.

Stojanov, K., Georgieva, J. V., Brinkhuis, R. P., van Hest, J. C., Rutjes, F. P., Dierckx, R. A. J. O., … Zuhorn, I. S. (2012). In vivo biodistribution of prion- and GM1-targeted polymersomes following intravenous administration in mice. *Molecular Pharmaceutics, 9*(6), 1620–1627. http://dx.doi.org/10.1021/mp200621v.

Südhof, T. C., & Rothman, J. E. (2009). Membrane fusion: Grappling with SNARE and SM proteins. *Science (New York, N.Y.), 323*(5913), 474–477. http://dx.doi.org/10.1126/science.1161748.

Sztul, E., Colombo, M., Stahl, P., & Samanta, R. (1993). Control of protein traffic between distinct plasma membrane domains. Requirement for a novel 108,000 protein in the fusion of transcytotic vesicles with the apical plasma membrane. *Journal of Biological Chemistry, 268*(3), 1876–1885.

Tian, X., Nyberg, S., Sharp, P. S., Madsen, J., Daneshpour, N., Armes, S. P., … Battaglia, G. (2015). LRP-1-mediated intracellular antibody delivery to the central nervous system. *Scientific Reports, 5*, 11990. http://dx.doi.org/10.1038/srep11990. http://www.nature.com/articles/srep11990. supplementary-information.

Toonen, R. F., Kochubey, O., de Wit, H., Gulyas-Kovacs, A., Konijnenburg, B., Sørensen, J. B., … Verhage, M. (2006). Dissecting docking and tethering of secretory vesicles at the target membrane. *The EMBO Journal, 25*(16), 3725–3737. http://dx.doi.org/10.1038/sj.emboj.7601256.

Tuma, P. L., & Hubbard, A. L. (2003). Transcytosis: Crossing cellular barriers. *Physiological Reviews, 83*(3), 871–932. http://dx.doi.org/10.1152/physrev.00001.2003.

Ulbrich, K., Hekmatara, T., Herbert, E., & Kreuter, J. (2009). Transferrin- and transferrin-receptor-antibody-modified nanoparticles enable drug delivery across the blood–brain barrier (BBB). *European Journal of Pharmaceutics and Biopharmaceutics, 71*(2), 251–256. http://dx.doi.org/10.1016/j.ejpb.2008.08.021.

Visser, C. C., Voorwinden, L. H., Crommelin, D. J. A., Danhof, M., & de Boer, A. G. (2004). Characterization and modulation of the transferrin receptor on brain capillary endothelial cells. *Pharmaceutical Research, 21*(5), 761–769. http://dx.doi.org/10.1023/B:PHAM.0000026425.69874.8e.

Wiley, D. T., Webster, P., Gale, A., & Davis, M. E. (2013). Transcytosis and brain uptake of transferrin-containing nanoparticles by tuning avidity to transferrin receptor. *Proceedings of the National Academy of Sciences of the United States of America, 110*(21), 8662–8667. http://dx.doi.org/10.1073/pnas.1307152110.

Winkler, J., Ramirez, G. A., Kuhn, H. G., Peterson, D. A., Day-Lollini, P. A., Stewart, G. R., … Thal, L. J. (1997). Reversible Schwann cell hyperplasia and sprouting of sensory and sympathetic neurites after intraventricular administration of nerve growth factor. *Annals of Neurology, 41*(1), 82–93. http://dx.doi.org/10.1002/ana.410410114.

Xu, T., Binz, T., Niemann, H., & Neher, E. (1998). Multiple kinetic components of exocytosis distinguished by neurotoxin sensitivity. *Nature Neuroscience, 1*(3), 192–200. http://dx.doi.org/10.1038/642.

Yamada, K., Kinoshita, A., Kohmura, E., Sakaguchi, T., Taguchi, J., Kataoka, K., & Hayakawa, T. (1991). Basic fibroblast growth factor prevents thalamic degeneration after cortical infarction. *Journal of Cerebral Blood Flow & Metabolism, 11*(3), 472–478. http://dx.doi.org/10.1038/jcbfm.1991.90.

Yu, Y. J., Atwal, J. K., Zhang, Y., Tong, R. K., Wildsmith, K. R., Tan, C., … Watts, R. J. (2014). Therapeutic bispecific antibodies cross the blood-brain barrier in nonhuman primates. *Science Translational Medicine, 6*(261), 261ra154. http://dx.doi.org/10.1126/scitranslmed.3009835.

Yu, Y. J., Zhang, Y., Kenrick, M., Hoyte, K., Luk, W., Lu, Y., … Dennis, M. S. (2011). Boosting brain uptake of a therapeutic antibody by reducing its affinity for a transcytosis target. *Science Translational Medicine, 3*(84), 84ra44. http://dx.doi.org/10.1126/scitranslmed.3002230.

Zhang, S., Li, J., Lykotrafitis, G., Bao, G., & Suresh, S. (2009). Size-dependent endocytosis of nanoparticles. *Advanced Materials (Deerfield Beach, Fla.), 21*, 419–424. http://dx.doi.org/10.1002/adma.200801393.

Zlokovic, B. V., Skundric, D. S., Segal, M. B., Lipovac, M. N., Mackic, J. B., & Davson, H. (1990). A saturable mechanism for transport of immunoglobulin G across the blood-brain barrier of the guinea pig. *Experimental Neurology, 107*(3), 263–270. http://dx.doi.org/10.1016/0014-4886(90)90144-H.

CHAPTER THREE

Application of Nanomedicine to the CNS Diseases

D. Carradori*, A. Gaudin†, D. Brambilla‡, K. Andrieux§,1

*Advanced Drug Delivery and Biomaterials, Louvain Drug Research Institute, Université catholique de Louvain, Bruxelles, Belgium
†Yale University, New Haven, CT, United States
‡Institute of Pharmaceutical Sciences, ETH Zurich, Zurich, Switzerland
§Université Paris Descartes, Université Paris-Sorbonne, UTCBS, UMR CNRS 8258, UE1022 INSERM, Paris, France
[1]Corresponding author: e-mail address: karine.andrieux@parisdescartes.fr

Contents

Abstract

Drug delivery to the brain is a challenge because of the many mechanisms that protect the brain from the entry of foreign substances. Numerous molecules which could be active against brain disorders are not clinically useful due to the presence of the blood–brain barrier. Nanoparticles can be used to deliver these drugs to the brain. Encapsulation within colloidal systems can allow the passage of nontransportable drugs

International Review of Neurobiology, Volume 130
ISSN 0074-7742
http://dx.doi.org/10.1016/bs.irn.2016.06.002

across this barrier by masking their physicochemical properties. It should be noted that the status of the blood–brain barrier is different depending on the brain disease. In fact, in some pathological situations such as tumors or inflammatory disorders, its permeability is increased allowing very easy translocation of carriers. This chapter gathers the promising results obtained by using nanoparticles as drug delivery systems with the aim to improve the therapy of some CNS diseases such as brain tumor, Alzheimer's disease, and stroke. The data show that several approaches can be investigated: (1) carrying drug through a permeabilized barrier, (2) crossing the barrier thanks to receptor-mediated transcytosis pathway in order to deliver drug into the brain parenchyma, and also (3) targeting and treating the endothelial cells themselves to preserve locally the brain tissue. The examples given in this chapter contribute to demonstrate that delivering drugs into the brain is one of the most promising applications of nanotechnology in clinical neuroscience.

1. INTRODUCTION

Nanomedicine can be defined as the use of nanoscale or nanostructured materials in medicine. Because of their structure, nanoparticles (NPs) have some unique medical effects, for example, the ability to cross the biological barriers or the passive targeting of tissues. This application of nanotechnology to health care holds great promise for revolutionizing medical treatments and therapies in areas such as imaging, faster diagnosis and drug delivery, as well as the development of new medical products. Drug delivery appears as the pivotal role of NPs, which are developed in order to modify and improve the bioavailability, biodistribution, and pharmacokinetics of drugs. It has been shown that NPs have the ability to carry various therapeutic agents including low-molecular-weight compounds and also DNA, RNA, proteins, and peptides. The drug is combined with NPs that can protect it from metabolism and allow more control over its release. NPs could be also used for the delivery of two or more drugs simultaneously for combination therapy. The drug when combined with NPs is specifically carried in order to be delivered only to the diseased site, thereby reducing side effects (Couvreur & Vauthier, 2006). Specific design of the NP surface is usually necessary to achieve this goal. To avoid uptake by mononuclear phagocytes, PEG (polyethylene glycol) is grafted on the surface of the NPs increasing their half-life in the blood circulation as a result of "stealth" properties. This aim could be also achieved by coating the NP surface with ligand able to selectively interact with a cell-specific marker. Finally, their very small dimensions and chemical properties allow some kinds of NPs to pass through barriers in the organism.

A clear example is the blood–brain barrier (BBB) which represents an important pharmacological challenge due to the difficulties in passing through it. This barrier is composed of endothelial cells closely packed on the brain capillary's surface presenting a stronger restriction of molecule passage than in capillaries elsewhere in the body. They are together with astrocytes and other cells which ensure their barrier role. Delivering drugs across the BBB is one of the most promising applications of nanotechnology in clinical neuroscience (Garcia-Garcia, Andrieux, Gil, & Couvreur, 2005). First, there is evidence that NPs can enter the brain when a tumor or an inflammatory response is present. Another strategy is to direct NPs to a particular region of capillary endothelium where receptors are expressed and mediate transcytosis across the endothelium. This targeting approach can be achieved either with antibodies or with the natural ligand or an analog which is linked at the surface of the NPs and allows to promote translocation across the BBB.

This chapter gathers results obtained with NPs of different generations (PEGylated or not, functionalized with ligands or not) which have been designed with the goal to improve the treatment of some CNS diseases such as brain tumor, Alzheimer's disease, and stroke. These data are discussed as a function of the BBB status (permeabilised or not) encountered in these specific pathological situations.

2. NPs FOR BRAIN CANCER

2.1 Physiopathology

2.1.1 Brain Tumors and Glioblastoma

2.1.1.1 Brain Tumors

Central nervous system (CNS) can be affected by tumors arising from different types of cells (Kieran, Walker, Frappaz, & Prados, 2010). CNS tumors can occur at any age of the patient and show dissimilarities in terms of epidemiology, symptomatology, development, and treatment depending on their pathogenesis (Jacques & Cormac, 2012). The brain is the most affected area of the CNS counting 80–90 % of all CNS tumors (Zülch, 2012). Cancerous as well as noncancerous brain tumors originate from the brain itself (primary tumors) or spread in it as metastases (secondary tumors) (Huttner, 2012; Nayak, Lee, & Wen, 2012). The World Health Organization (WHO) established a classification, based on four levels, to describe their malignancy (Fuller & Scheithauer, 2007): grade I (no infiltrative and low proliferative), grade II (infiltrative and low proliferative), grade III (infiltrative and proliferative), and grade IV (widespread infiltrative and high craniospinal disseminative). Brain tumors represent a challenge in medicine due to the extremely sensitive equilibrium between the preservation of the cerebral functions and the eradication of the tumor which limits most of the therapeutic approaches.

2.1.1.2 Glioblastoma

Glioblastoma (GB) is the most frequent primary brain tumor (Fig. 1) with an incidence of 3.19/100,000 persons at a median age of 64 years (Ostrom et al., 2014; Thakkar et al., 2014). Patient life expectancy is unfortunately between 8 and 14 months after the diagnosis (Stupp, Roila, & ESMO Guidelines Working Group, 2009). The reason of such short prognosis comes from GB peculiarities (Xu, Gao, Sun, & Duan, 2015): multidrug resistance, due to the dormant glioma cells; limited surgical resection, prevented by the "crab claw-like" invasion attitude of the GB; critical importance of the residual glioma cells, able to easily develop in new primary tumors; and surgery-dependent malignancy (the resection can increase the regenerative proliferation of the tumor). Moreover, the BBB and the blood–brain tumor barrier (BBTB) represent a physical limitation for all of the conventional drugs used in other anticancer treatments (Oberoi et al., 2016; Van Tellingen et al., 2015). Nowadays the therapeutic approach consists in surgery, chemotherapy, and radiotherapy or, in most of the cases, in a combination of them (Venur, Peereboom, & Ahluwalia, 2015).

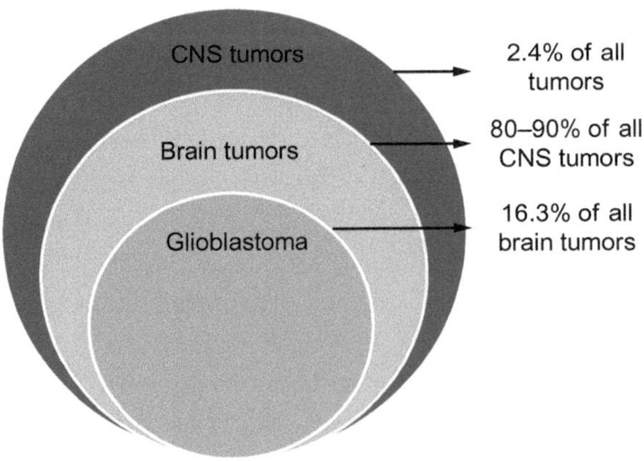

Fig. 1 Schematic representation of epidemiological data for glioblastoma. *Adaptation of Fig. 1 from Aparicio-Blanco, J., & Torres-Suárez, A. I. (2015). Glioblastoma multiforme and lipid nanocapsules: A review.* Journal of Biomedical Nanotechnology, 11*(8), 1283–1311 with permission of the journal.*

2.2 Standard Treatments

2.2.1 Current GB Therapies

The therapy is attributed to the patient after the collection of all of the data produced by the neurological exam (logic, vision, hearing, coordination, and memory evaluation) and the scansion of the brain (by computed tomography or magnetic resonance imaging).

2.2.1.1 Surgery

First a stereotaxic biopsy is performed to establish the type of the tumor and its grade, and then the tumor is totally or partially resected by the surgeon in relation also to the health status of the patient. The surgery is not conclusive and it is always associated to chemotherapy and radiotherapy (Park et al., 2010).

2.2.1.2 Chemotherapy

Nitrosureas, in combination with radiation, represent the most frequent therapeutic approach in the treatment of GB. These molecules are able to cross the BBB and to alkylate cancerous cell DNA, inducing GB cells to apoptosis. Despite any defined chemotherapeutic schedule exists, temozolomide is usually the fist-line chemotherapy while lomustine (and bevaciuzam, an angiogenetic inhibitor) the second one (Chen, Xu, Lu, Chen, & Wu, 2013; Field, Jordan, Wen, Rosenthal, & Reardon, 2015). The Food and Drug Administration (FDA) approved the utilization of carmustin-based polymeric wafers (eg, Gliadel) which are placed in the cerebral cavity after the resection of the tumor (Dixit, Hingorani, Achawal, & Scott, 2011).

2.2.1.3 Radiotherapy

Radiation therapy always follows surgery, increasing survival from 3–4 to 7–12 months (Ryu et al., 2014). Doses between 5000 to 6000 cGy are administered and recently associated with temozolomide over 30 days (Yang, Zhou, & Lin, 2014).

2.2.2 Therapeutic Limitations

The infiltrative nature of GB, together with its heterogeneity, makes standard treatments ineffective (Lima et al., 2012). Surgery is a highly invasive technique which cannot eradicate the totality of the tumor due to the lack of microscopic accuracy in the resection (Koga & Saito, 2012). Chemotherapy and radiotherapy have an effect only on the periphery of the tumor, while the core remains not sensitive and still able to proliferate (Mrugala, 2013). Moreover, the role of GB stem cells is emerging as crucial in the resistance as well as in the recurrence of the tumor (Sundar, Hsieh, Manjila, Lathia, & Sloan, 2014).

The critic localization of the tumor and its cytological characteristics has prompted research to explore nonconventional therapeutic approaches to face this nonconventional convergence of factors.

2.3 NP-Based GB Therapy

Nanomedicine consists in the application of nanotechnology in medicine to overcome the limits of conventional pharmaceutical products and increase their therapeutic effect (Dawidczyk, Russell, & Searson, 2014). Anticancer drug-loaded NPs already showed a more favorable pharmacokinetic profile (half-life, controlled release) and better tumor accumulation compared to their conventional pharmaceutical forms (Wang, Langer, & Farokhzad, 2012). Moreover, NP-based system allows to multidrug coencapsulation and codelivery, limiting toxicity and drug resistance (Malam, Lim, & Seifalian, 2011).

2.3.1 Advantages

2.3.1.1 Suitable Drugs

The development of promising active molecules against GB is frequently arrested by their unfavorable physicochemical and rheological profiles (Rabanel, Aoun, Elkin, Mokhtar, & Hildgen, 2012). Poor solubility and high fragility are the major obstacles. NP-based systems can increase the number of suitable drugs for the treatment of brain tumors. Drug loading in lipid–NPs makes most of these active molecules suitable for anti-GB therapy (Lacoeuille, Garcion, Benoit, & Lamprecht, 2007): NP provides physic protection to the compound as well as its transportation also in noncompatible streams.

2.3.1.2 Pharmacokinetic Modulation

Drug release is crucial for the efficacy of the treatment and for the best patient compliance. The major objective of the pharmacokinetic modulation is to perform the lowest number of the least invasive administration, preserving the therapeutic effect. The fine design of the NP composition (eg, pH sensitive or thermosensitive materials) and architecture (eg, monolayer or multilayer liposome, nanosphere, or nanocapsule), together with a deep knowledge of the physiological factors involved, can highly influence the time and the place of the drug release (Brigger, Dubernet, & Couvreur, 2012; Mura, Nicolas, & Couvreur, 2013).

2.3.1.3 Targeting

Nanoparticulate systems offer much more possibilities to increase the interaction of drug tumor than the drug alone. GB targeting by NPs can be passive or active (Bertrand, Wu, Xu, Kamaly, & Farokhzad, 2014). The size of the NPs can let them accumulating in specific tissue, cellular or subcellular areas. The ultrasmall size is also crucial for the enhanced permeability and retention (EPR) effect (Acharya & Sahoo, 2011): NPs can passively extravasate into the tumor thanks to the high tumor vessel permeability. Moreover, the functionalization of the shell with proteinaceous biomolecules (antibodies, peptides, and aptamers) can increase NP uptake in cancerous cells by active targeting (eg, by receptor-mediated endocytosis; Li, Deng, Peng, & Wang, 2014). One of the most studied for GB stem cell targeting is CD133 (Hong, Jang, Lee, & Nam, 2015). The high specificity drug tumor reduces side effects and increases the efficacy of the treatment.

2.3.1.4 Crossing Barrier Facilitation

The presence of both BBB and BBTB makes harder the design of anti-GB nanoparticulate systems compared to other pathologic situations. While BBB's role is mainly to protect the brain from harmful substances, BBTB contributes to GB growth allowing nutrient delivering through existing, or newly formed, blood vessels around and inside the tumor. The aim of anti-GB nanoparticulate systems is to let active drugs cross BBB and BBTB and reach the tumor. The most critical part is brain capillary endothelial cells (BCECs) internalization (Nag, 2011). The discovery of overexpressed, or specific, BCEC receptors inspired the production of corresponding ligands to functionalize NP surface and increase the internalization of the drug-loaded NPs. Monoclonal antibodies (against human insulin receptor (Ulbrich, Knobloch, & Kreuter, 2011), glutathione receptor (Raval, Mistry, Acharya, & Acharya, 2015) or transferrin receptor (Porru et al., 2014), and apolipoproteins (Kreuter et al., 2007)) are the most used compounds to increase the passage through BBB and BBTB. Physicochemical properties such as size, charge, and intrinsic composition can also have a specific role in the interaction of NP–BBTB, modulating the accumulation in tumoral and peritumoral areas (Kim, Harford, Pirollo, & Chang, 2015).

2.3.1.5 Theranostic Medicine

NP-based systems are able to carry both imaging as well as therapeutic compounds (Chiarelli, Kievit, Zhang, & Ellenbogen, 2015). Such capacity confers to the systems a double function: diagnosis and therapy. Moreover, the live monitoring of the therapeutic effect would allow personalized GB therapies patient by patient.

2.3.1.6 Gene Therapy

GB cells can be reprogrammed in healthy cells by gene delivery (Kane et al., 2015). The incorporation of nucleic acids by tumor cells allows damaged gene replacement, cancer protein knockdown, and angiogenesis modulation. NP-based systems provide a suitable tool to deliver nucleic acids overcoming the limitations due to the fragility and high hydrophilicity (low penetration through plasmatic membranes) of these molecules (Bolhassani et al., 2014; De Jesus & Zuhorn, 2015).

2.3.2 Preclinical Phase and Clinical Trials

In the last decades the rise of nanomedicine allows the revaluation of old anti-GB drugs as well as the fine engineering of new systems which opened to improved GB diagnoses and treatments. While a wide variety of NPs are in preclinical phase, some anti-GB NP-based therapeutics are already under clinical trials. Moreover, as GB stem cell hypothesis lately suggests the critic role of these cells in the development and maintaining of the tumor, anti-GB stem cell NP-based strategies are emerging (Kaluzova, Bouras, Machaidze, & Hadjipanayis, 2015; Wang et al., 2014). The increasing number of the publications concerning anti-GB NP-based therapeutics is supported by the exciting results obtained in preclinical phase and clinical trials (Fig. 2).

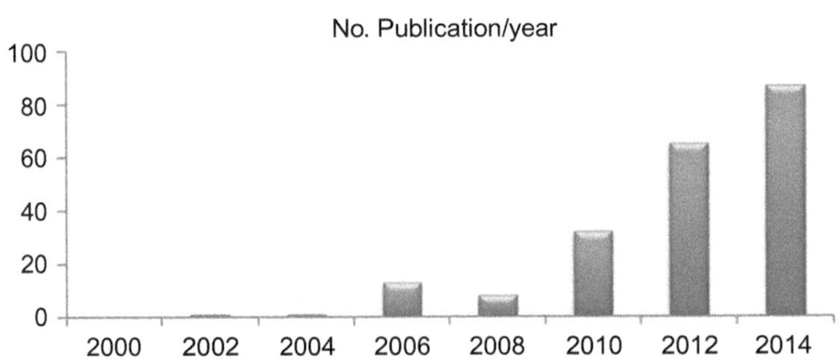

Fig. 2 Evolution of the number of publications related to the application of nanoparticle-based system in the treatment of GB. Data collected performing a bibliographic research on PubMed: "nanoparticle(s)" + "glioblastoma." The articles were first selected according to their relevance to the subject and then grouped according to the year of the publication.

2.3.2.1 Preclinical Phase

The large availability of biocompatible materials, functionalizing elements, NP architectures, and route of administrations allows the development of a wide number of NP-based strategies: BBB crossing, side effect decreasing, tumor targeting, drug resistance decreasing, pharmacokinetic modulation, and immunostimulation (Table 1). Systems are tested in vitro as well as in vivo. The most used in vitro models are represented by 9L (rat), F98 (rat), U87MG (human), and GL261 (mouse) cell lines (Barth, 1998), while intracranic or subcutaneous GB mouse and rat models are used for the in vivo efficacy evaluation (Huszthy et al., 2012; Jacobs, Valdes, Hickey, & De Leo, 2011).

2.3.2.2 Clinical Trials

Nowadays two clinic trials concern anti-GB NP-based therapeutics, NCT02340156 and NCT00734682. The first corresponds to the phase II clinical trial of temozolomide-loaded immunoliposomes functionalized with the human wild-type p53 DNA plasmidic sequence SGT-53. Considering p53 gene abnormalities responsible of chemoresistance, this system is expected to reduce temozolomide resistance while the physicochemical properties of the nanoconstruct will improve BBB penetration and tumor targeting. The second corresponds to the phase I clinical trial of the topoisomerase I inhibitor CPT-11/irinotecan-loaded liposomes. Investigators expect reduced drug toxicity, due to drug encapsulation, with resulting increased patient survival.

2.3.3 Limits of NP-Based GB Therapies

Despite promising results in preclinical phase, clinical translation is limited and slow. Besides conventional obstacles in drug development (design, manufacturing process, biologic efficacy, and industrial scale-up), anti-GB NP-based therapeutics must face additional issues carried by the domain of nanomedicine (Table 2). On the one hand, opsonization and activation of the reticular–endothelial system (RES) lets NPs accumulating in organs other than brain (eg, liver) increasing their toxicity and nullifying the therapeutic effect (Moghimi, Hunter, & Murray, 2001; Suk, Xu, Kim, Hanes, & Ensign, 2015). On the other hand, the lack of worldwide regulatory procedures in nanomedicine strongly discourages investors to base the development of nanotherapeutics in the treatment of GB.

Table 1 Nanoparticle-Based Systems Currently Tested in Preclinical Phases

Nanoparticle-Based System	Administration Route	Aim of the Design	Loaded Drug(s)	Results
LNC (Huynh et al., 2011)	IV	DSPE-mPEG$_{2000}$ (stealth profile and PK modulation)	FcdiOH	Tumor mass progression significantly reduced
LNC (Lamprecht & Benoit, 2006)	SC	Pgp inhibitor (efflux pump inhibition)	PTX	Tumor mass progression significantly reduced
LNC (Basile et al., 2012)	SC	PEG$_{2000}$-LAA (passive targeting)	PTX	Strong inhibition of tumor growth
LNC (Balzeau et al., 2013)	CED	NFL (GB targeting and cytotoxic activity)	PTX	Tumor progression significantly reduced
LNC (Huynh et al., 2012)	IC	DSPE-mPEG$_{2000}$ (stealth profile and PK modulation) HBD (crossing BBB by BBB permeabilization)	FediOH	Increased survival
Nanocomplex (Kim et al., 2014)	IV	P53 gene (GB stem cell targeting)	TMZ	Increased survival
LNC (Messaoudi et al., 2015)	CED	Anti-EGFR and anti-Galectin (drug resistance decreasing)	TMZ siRNA	Increased survival
LNC (Lollo et al., 2015)	CED	Chitosan (positive charge) CpG (immunoresponse stimulation)	PTX	Strong inhibition of tumor growth
PEG-PLGA (Kang et al., 2015)	IV	CRT peptide (crossing BBB by TfR interaction)	PTX	Increased survival

Table 1 Nanoparticle-Based Systems Currently Tested in Preclinical Phases—cont'd

Nanoparticle-Based System	Administration Route	Aim of the Design	Loaded Drug(s)	Results
PLGA (Acharya & Sahoo, 2011)	IV	Drug encapsulation to improve drug solubility	PTX	Increased survival
PEG-PLA (Zhang et al., 2013)	IV	Peptide-22 (crossing BBB by LDLR interaction)	PTX	Increased survival
LNC (Cohen et al., 2015)	CED	Hyaluronan (GB cell targeting by CD44R interaction)	siPLK1	Increased survival
PEG-PTMC (Jiang et al., 2014)	IV	DGlu (crossing BBB by GLUTR interaction)	PTX	Strong inhibition of tumor growth

LNC, lipid nanocapsule; *IV*, intravenous; *PEG*, polyethylene glycol; *DSPE*, distearylphosphatidylethanolamine; *FcdiOH*, ferrociphenol; *PK*, pharmacokinetic; *SC*, subcutaneous; *Pgp*, permeability glycoprotein; *PTX*, paclitaxel; *LAA*, alkyl-lipoamino acid; *CED*, convection-enhanced delivery; *NFL*, neurofilament light; *IC*, intracranic; *HBD*, hydrogen bond donors; *TMZ*, temozolomide; *CpG*, cytosine phosphate guanine; *PLGA*, poly(lactic-*co*-glycolic acid); *CRT*, calreticulin; *LDLR*, low-density lipoprotein receptor; *CD44R*, hyaluronic acid receptor; *siPLK1*, si Polo-like kinase 1; *PTMC*, polytrimethylene carbonate; *Dglu*, D-Glucose; *GLUTR*, glucose transporter receptor.

Table 2 Schematic Representation of the Limits in Nanoparticle-Based Therapy Development

Objective(s)	Obstacles
Optimal formulation	Physicochemical compatibility drug system
	Manufacturing process
	Drug delivery and drug release design
Optimal biodistribution and pharmacokinetic	Biocompatibility
	Blood stream
	Reticular–endothelial system
	Physiological barriers
	Cellular and subcellular interactions
Optimal therapeutic schedule	Route of administration
	Dosage
	Patient compliance
Marketing	Industrial scale-up
	Regulatory authorities

2.4 Novel NP-Based Approaches for GB Treatment

Preexistent anti-GB therapies were recently improved by their association to nanomedicine: eg, GB ablation by hyperthermia, when carried by magnetite-, hematite-, or maghemite-based NPs exposed to altering magnetic fields, reduces healthy tissue damage compared to the classic approach (Verma, Lal, & Van Noorden, 2014).

External devices are used to increase NP penetration through BBB: eg, transcranial-focused ultrasound and microbubble injection are able to temporarily enhance the permeability of the barrier and therefore NPs accumulation in GB cells (Åslund et al., 2015).

Moreover, NPs are currently studied not only as carriers but also as objects to carry. Lately, systems charged in NPs such as hydrogels (Bastiancich et al., 2016) and marrow-isolated adult multilineage inducible (MIAMI) cells (Clavreul et al., 2015) are under development in the treatment of GB.

2.5 Conclusion on the Application of Nanomedicine to GB Therapy

The major challenge in anticancer treatments is the achievement of the therapeutic drug concentration in the tumor. This is much more difficult to reach in brain tumors than in others due to the presence of BBB and BBTB. Nanomedicine provides tools to overcome such limit. A deep knowledge of the processes (eg, receptors, physicochemical affinity, and specific channels) at the physiological barriers allows to functionalize NPs with ligands able to let the system crossing BBB by specific targeting. Moreover, NP design offers many more elements to modulate such as drug pharmacokinetic and pharmacodynamic than conventional formulations.

Despite the slow clinical translation, NP-based therapeutics remain one of the most promising strategies in the treatment of GB thanks to their versatility.

3. NPs FOR ALZHEIMER'S DISEASE THERAPY

3.1 Physiopathology

3.1.1 Neurodegenerative Diseases and Alzheimer's Disease

3.1.1.1 Neurodegenerative Diseases

Neurodegenerative disorders (NDDs) have always affected humans in the past but their higher incidence, related to a general increasing age of the population (Rizzi, Rosset, & Roriz-Cruz, 2014), is currently producing unsustainable costs for societies (Dorsey, George, Leff, & Willis, 2013; social and economic). No resolute therapies exist and NDD diagnoses are made years after the beginning of the pathological process: symptoms are detectable by neurologists only at an advanced stage of the disease (Paulsen et al., 2013). Genetic as well as environmental factors can be ascribed to the pathogenesis of NDD producing progressive impairment and degeneration of neurons (Cannon & Greenamyre, 2011; Gapp, Woldemichael, Bohacek, & Mansuy, 2014). Since there is a natural loss of immune response capacity during the aging process, the development of pathologies is more frequent and severe in aged people. Clinical manifestations are very different depending on the disease while the neuropathology frequently shows a common characteristic: neural accumulation of aggregated or misfolded proteins (Ciechanover & Kwon, 2015) (eg, α-synuclein in Parkinson's disease (Dehay et al., 2015) or β-amyloid (Aβ) in Alzheimer's disease (Knowles, Vendruscolo, & Dobson, 2014)). Early identification of specific NDD-related biomarkers would mean early treatment for patients but until resolute therapies would not be developed, no relevant benefit will be produced in any case. Few therapeutics succeed in preclinical phases and fewer reached the market (Schneider et al., 2014; Wingerchuk & Carter, 2014). The major obstacles in the development of anti-NDD drugs are represented by the BBB, which acts as a physical and enzymatical barrier against more than 98% of the small active molecules, and the multifactorial pathogenesis of these diseases, which makes hard the design of therapeutics (Rowinska-Zyrek, Salerno, & Kozlowski, 2015).

3.1.1.2 Alzheimer's Disease

Alzheimer's disease (AD) is first described by Alois Alzheimer 100 years ago. AD is a devastating NDD which affects 80% of dementia-diagnosed people (Kumar & Singh, 2015). 2010 registered more than 25 million AD cases and, according to the previsions, this number will double each 20 years reaching 100 million cases in 2050 (Reitz & Mayeux, 2014). AD is neuropathologically characterized by two major lesions: extracellular deposition of amyloid plaques mainly composed of β-amyloid (Aβ) peptides and intracellular neurofibrillar tangles composed of hyperphosphorylated Tau proteins (Bloom, 2014). Both of them lead to neuronal and synaptic deterioration. Progressive loss of cognitive function is one of the most serious consequences of this disease (Maresova, Mohelska, Dolejs, & Kuca, 2015). Despite great progresses in research, AD pathogenesis is still unknown.

3.2 Conventional Treatments

Several hypotheses have been proposed over the years and each one of them drove to the development of specific classes of therapeutics.

3.2.1 *Current Therapy and Research Strategies*

3.2.1.1 Therapeutics Based on "Cholinergic Hypothesis"

The oldest hypothesis developed to explain AD pathogenesis proposed a reduced production of acetylcholine (Francis, Palmer, Snape, & Wilcock, 1999). Actually, three of the four medicines currently used in the treatment of AD are acetylcholinesterase inhibitors (donepezil, galantamine, and rivastigmine; Deardorff, Feen, & Grossberg, 2015). The aim of these drugs is to increase acetylcholine concentration by the inhibition of its metabolism. Memantine, the fourth, is an NMDAr antagonist which contrasts glutaminergic overstimulation (Matsunaga, Kishi, & Iwata, 2015). Only drugs modulating cholinergic system have shown interesting, but limited, clinical benefits: those molecules show only transient effects without preventing the progression of the disease (O'Regan, Lanctôt, Mazereeuw, & Herrmann, 2015).

3.2.1.2 Therapeutics Based on "Tau Cascade Hypothesis"

Neurofibrillary tangles (NT) are pathological insoluble aggregates of hyperphosphorylated tau proteins, which are microtubule-associated proteins, observed within the neurons of AD patients. NT have been considered trigger of AD cascade since 1980s (Bakota & Brandt, 2016; Grundke-Iqbal et al., 1986). The precise mechanism of NT formation is not completely understood, and it is still controversial whether NT are a primary causative factor in the disease or play a more peripheral role (Frost, Götz, & Feany, 2015). Intermediate aggregates of abnormal tau molecules are cytotoxic and impair cognition (Mohamed, Herrou, Plouffe, Piperno, & Leclerc, 2013). Despite tau aggregation is driven by Aβ accumulation, elevated levels of phosphotau amino acids, such as T181 and T231, could constitute an important element to predict incipient AD in patients with mild cognitive impairment (Mattsson et al., 2009). In this context, microtubule stabilizers (eg, taxanes (Kinghorn, Altmann, Höfle, Müller, & Prantz, 2009)), Tau-phosphorylation inhibitors (eg, NAP (Gozes, 2010)), and Tau-aggregation inhibitors (eg, methylthioninium chloride (Wischik et al., 2015)) are under development to treat AD.

3.2.1.3 Therapeutics Based on "Amyloid Cascade Hypothesis"

The discovery that cerebral plaques were composed of Aβ induced Hardy and colleagues to develop the "amyloid cascade hypothesis" (Hardy & Allsop, 1991). Aβ plaques originate from an imbalance between Aβ peptide (39–42 amino acids) production and clearance which would represent the initiating factor of AD. In the cerebrospinal fluid and serum of AD patients, its concentration is strongly increased (Toledo, Shaw, & Trojanowski, 2013). It derives from proteolytic cleavages of amyloid precursor protein (APP) which is an integral membrane phosphoprotein type I expressed in many tissues, especially in the synapses of neurons (Goldgaber, Lerman, McBride, Saffiotti, & Gajdusek, 1986). Its primary function is still unknown. It is usually cleaved by proteases of the secretase family: α-, β-, and γ-secretases, and can follow two different exclusive pathways: amyloidogenic and nonamyloidogenic. β-secretase, better known as BACE (beta-site APP-cleaving enzyme) or memapsin-2, is involved in the amyloidogenic pathway (Van der Kant & Goldstein, 2015). BACE is able to cleave APP forming a 99-amino-acid fragment which is cleaved by γ-secretase forming β-amyloid peptides with 40/42 amino acids called $A\beta_{1-40}$ and $A\beta_{1-42}$ (Muresan & Muresan, 2015). The $A\beta_{1-42}$ peptide is the most important component of senile plaques on the extracellular leaflet of neurons in brains of AD patients. The central role of the long form $A\beta_{1-42}$ in AD has been established through several histopathological, genetic, and biochemical studies. Among the observations that have been made, deposition of $A\beta_{1-42}$ in the brain is an early and invariant feature of all forms of AD. Then this hypothesis became the most widely accepted theory in AD research despite the constant criticism (Drachman, 2014). Aβ peptide aggregates were identified as the neurotoxic agents responsible of neurodegeneration (Haass & Selkoe, 2007). Aβ spontaneously self-aggregates into multiple, coexisting physical forms. Its ability to aggregate is strongly influenced by factors such as pH, metallic ions, temperature, and ionic strength. The monomer (4514.15 Da; 1–1.5 nm) is the most unstable species in aqueous media. Monomers tend to aggregate to larger species. Oligomers (2–6 peptides) coalesce into intermediate assemblies and lead to fibrils which arrange themselves into β-pleated sheets to form the insoluble fibers of advanced amyloid plaques (Henry, Querfurth, & LaFerla, 2010). Plaques are the final state of the Aβ aggregation process in vivo and are large extracellular Aβ deposits. The presence of plaques does not correlate with neurotoxicity, as had been initially suggested. In contrast, it is believed that they can contribute to the removal and inactivation of smaller, neurotoxic Aβ species. Soluble oligomers and intermediate amyloids are the most neurotoxic forms of Aβ. Oligomers seem to be 40-fold more toxic than monomers and 10-fold more than fibrils because their small dimension allows them to insert into membrane and strongly interact with cell surface. Soluble spherical aggregates of $A\beta_{1-40}$ and $A\beta_{1-42}$ produced in vitro are both toxic, but the latter are much more so, by a factor of about 100, and their toxicity may involve tau protein kinase I/glycogen synthase kinase-3 activation. The severity of the cognitive defect in AD correlates with levels of oligomers in the brain, not the total Aβ burden. Therapeutics coming from "amyloid cascade hypothesis" can be grouped in Aβ peptide synthesis inhibitors (eg, inhibitors of β and γ secretase (Mikulca et al., 2014)), Aβ peptide aggregation inhibitors (eg, curcumin and derivatives (Ahmed & Gilani, 2014)), and Aβ peptide elimination promotors (eg, immunotherapy (Levites et al., 2015)). The aim of such approaches is to decrease Aβ cerebral concentration obtaining the inhibition of the neurodegenerative process as well as the recovery of the cognitive functions.

3.2.2 Limitations of AD's Therapies

Although the disease is studied for over a century, approved anti-AD therapies are sadly palliative. None of them combines neuron regeneration and inhibition of neurodegenerative processes. BBB remains the main barrier for small molecules tested in preclinic, limiting the evaluation of their pharmacological potential in clinic. Moreover, the multifactorial pathogenesis of AD makes difficult the design of multivalent therapeutics via ordinary pharmaceutical approaches.

3.3 NP-Based AD Therapies

The contribution of nanomedicine in the fight against such devastating disease was carried on two fronts. NP-based systems were employed to improve the physicochemical (solubility or ionization) and consequently the biopharmaceutical (bioavailability, biodistribution, toxicity, pharmacokinetic, and pharmacodynamic) characteristics of known active molecules as well as to support and to develop new therapeutic strategies.

3.3.1 Nanotechnological Improvement and Innovation

3.3.1.1 Nanotechnological Improvement

Piperine is an alkaloid which showed neuroprotective action (Yang, Chen, Liu, & Qu, 2015). The development of the drug as anti-AD therapeutic was initially hindered by its high hydrophobicity and rapid metabolization. Piperine-loaded chitosan NPs were used to overcome these limits: the drug was intranasally administered optimizing its biological activity (Elnaggar, Etman, Abdelmonsif, & Abdallah, 2015). Moreover, the nanoencapsulated form of rivastigmine tartrate showed faster memory recovery in AD animal models than the free drug: L-lactide-depsipeptide polymeric NPs improved the pharmacokinetics of the therapeutic (Pagar, Sardar, & Vavia, 2014). As another example, solid lipid nanocapsules are currently under investigation to improve the solubility and the gastrointestinal absorption of lipoyl-memantine (Laserra et al., 2015).

3.3.1.2 Nanotechnological Innovation

Drug-loaded NPs can be functionalized with ligands able to let them cross BBB and/or to reach their target in the brain (Fig. 3). The most used ligands are ApoE3 (Wagner et al., 2012), which selectively interacts with LDLr (mediating nanovector transcytosis through BBB epithelial cells) and antibodies directed against Aβ peptides (Lannfelt, Relkin, & Siemers, 2014).

However, the development of early diagnosis methods was definitely the most promising innovation introduced in AD research by nanomedicine. The detection and quantification of the most important AD biomarkers, Tau proteins and Aβ peptides, was almost impossible to perform with conventional methods (eg, ELISA) due to the extremely low concentration of these compounds in physiological fluids (serum and cerebrospinal fluid; Henriksen et al., 2014). NP-based procedures were successfully employed to overcome such limitations by the amplification of AD biomarkers signal achieving a clear detection. Gold NPs functionalized with tau-specific monoclonal antibody were tested for Tau protein quantification by immune-polymerase chain reaction (Stegurová et al., 2014) or by Rayleigh scattering assay (Neely et al., 2009). Compared to an ELISA method, the sensibility of the NP-based methods was highly superior in both of cases.

3.3.2 Preclinical Phase and Clinical Trials

The anti-AD potential of a NP-based system is mainly established in relation to:

- the capacity to cross BBB,
- reduce Tau protein/Aβ peptide cerebral concentrations, and
- induce memory recovery.

BBB crossing is quantified by fluorescence or radioactivity: NP-based systems are incubated with simple unidimensional (eg, human brain endothelial

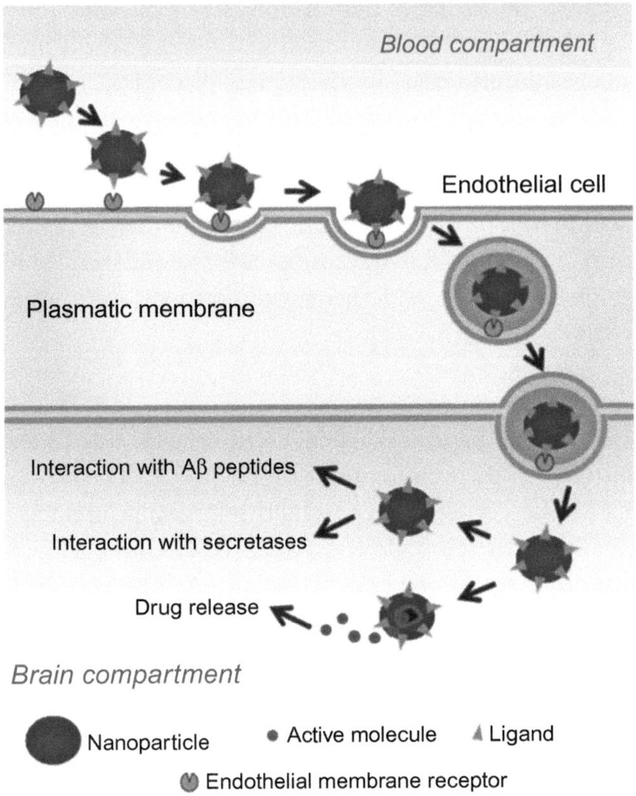

Fig. 3 Nanoparticle-based strategies in the treatment of AD. Functionalization with specific ligands can allow NP crossing through BBB by endothelial cell transcytosis. According to its design a NP is able to interact with Aβ peptides (eg, increasing their clearance) or α-,β-, and γ-secretases (eg, inhibiting their enzymatic action) or releasing active molecules (eg, neurotransmitters).

cells, hCMEC/D3 (Czupalla, Liebner, & Devraj, 2014)) to complex tridimensional (eg, endothelial cells/pericytes/astrocytes (Wang, Khafagy, Khanafer, Takayama, & El-Sayed, 2016)) in vitro BBB models. Similar evaluations can be performed also in vivo by biodistribution studies.

The evaluation of the action on Tau proteins and Aβ peptides is carried out by surface plasmon resonance (Le Droumaguet et al., 2012; affinity, Fig. 4) or by optical density (polymerization/depolymerization) in vitro and by immunohistochemistry (Balducci et al., 2014) or ELISA ex vivo.

Memory recovery assay is commonly performed on AD animal model (Cavanaugh, Pippin, & Barnard, 2014) which shows different physiopathological signs depending on the genetic modification, and quantified by NOR test (Antunes & Biala, 2012).

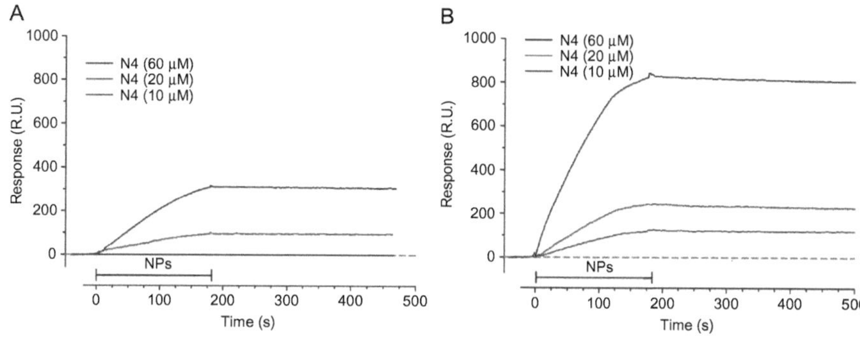

Fig. 4 Nanoparticle-based strategies in the treatment of AD. Functionalization with antibody directed against Aβ peptide can allow NP interacting with circulating blood and brain Aβ peptide (eg, increasing their clearance and avoiding their aggregation). SPR sensorgrams (resonance units, RU, vs time) of anti-$A\beta_{1-42}$ mAb-functionalized (N4) NPs at various concentrations (10, 20, 60 μM), flowed for 3 min (bars) onto a $A\beta_{1-42}$ monomer (A) or $A\beta_{1-42}$ fibrils (B) immobilized on the sensor chip. *Reprinted from Le Droumaguet, B., Nicolas, J., Brambilla, D., Mura, S., Maksimenko, A., De Kimpe, L., et al. (2012). Versatile and efficient targeting using a single nanoparticulate platform: Application to cancer and Alzheimer's disease. ACS Nano, 6(7), 5866–5879 with permission of the journal.*

NP-based systems for the treatment of AD focus on brain delivery of active drugs (growth factors, curcuminoids, acetylcholine esterase inhibitors, hormones, chelating ligands) and immunotherapy (antibodies directed toward Tau proteins and Aβ peptides). Current NP-based therapeutics are reported in Table 3.

According to the European and United States online clinical trial registers (clinicaltrialsregister.eu and clinicaltrial.gov, respectively), no NP-based systems for AD therapy are under any clinical trial in 2016.

3.4 Conclusions on the Application of Nanomedicine to Alzheimer Disease Therapy

The most important limit of currently approved anti-AD therapies is that they have only symptomatic and monovalent effect. None of them acts on the pathogenesis (or presumed pathogeneses) of the disease. Moreover, the late diagnosis of AD hinders full benefits from the therapeutic schedule.

Although encouraging results have been obtained in preclinical phase, no anti-AD NP-based system is currently tested in clinic. The lack in the translation could be ascribed to the relatively novelty of this approach. Nevertheless, nanoparticulate systems offer the possibility to regulate multiple AD pathological dysfunctions at the same time. Furthermore, nanotechnology showed its important contribution in the development of AD diagnostic

Table 3 Summary of Nanoparticle-Based Systems for AD

Strategy	Nanoparticle-Based System	Material	Design	Effect(s)
Drug delivery	NPs (Elnaggar et al., 2015)	Chitosan	Piperine encapsulation	Suitable for administration
	NPs (Pagar et al., 2014)	L-lactide-depsipeptide	Rivastigmine encapsulation	PK improvement
	SLN (Laserra et al., 2015)	Lipids	Memantine encapsulation	Bioavaiability improvement
	NPs (Eslami, Nikkhah, Hashemianzadeh, & Sajadi, 2016)	Chitosan	Tacrine encapsulation	Side effect reduction
	NPs (Joshi, Chavhan, & Sawant, 2010)	PLGA	Rivastigmine encapsulation	PK improvement
	Liposome (Corace et al., 2014)	Lipids	Tacrine encapsulation	Side effect reduction
	NPs (Tiwari et al., 2013)	PLGA	Curcumin coating	Bioavaiability improvement
	NPs (Mittal, Carswell, Brett, Currie, & Kumar, 2011)	PLGA	Estradiol encapsulation	Bioavaiability improvement
AD biomarkers quantification and AD lesions visualization	Magnetic NPs (Chauhan, Narang, & Jain, 2015)	Graphene/platinum	AChE–ChO coating	Acetylcholine quantification
	Magnetic NPs (Cheng et al., 2015)	Iron oxide	Curcumin coating	Aβ plaque MRI visualization

	Magnetic NPs (Yu et al., 2015)	Gold	Gelsolin coating	Aβ peptides quantification
	Magnetic NPs (Zhang et al., 2015)	Iron oxide	DDNP coating	Aβ plaque MRI visualization
	Magnetic NPs (Neely et al., 2009; Stegurová et al., 2014)	Gold	Monoclonal anti-Tau coating	Tau proteins quantification
Multivalent effect	Liposome (Mancini et al., 2016)	Lipids	mApoE–PA coating	BBB crossing and Aβ aggr. modulation
	NPs (Brambilla et al., 2012)	PACA-PLA	PEGylation	BBB crossing and Aβ aggr. modulation
	NPs (Zhang et al., 2014)	PLA	TGN and QSH peptides coating	BBB crossing and Aβ aggr. modulation
	NPs (Liu et al., 2013)	PLA	NAPVSIPQ peptide encapsulation	BBB crossing and neuroprotection
	NPs (Mathew et al., 2012)	PLGA	Curcumin and Tet-1 coating	BBB crossing and Aβ aggr. modulation
	Dendrimers (Liu et al., 2016)	Poly-L-lysines	RVG29 and D-peptide coating	BBB crossing, Aβ and Tau proteins aggr. modulation
	Dendrimers (Wasiak et al., 2012)	Aminothiophosphate	Cationic phosphorus containing	BBB crossing, Aβ and Tau proteins aggr. modulation

NPs, nanoparticles; *PK*, pharmacokinetic; *SLN*, solid–lipid nanocapsules; *PLGA*, poly(lactic-*co*-glycolic acid); *AChE–ChO*, acetylcholinesterase–choline oxidase; *MRI*, magnetic resonance imaging; *mApoE–PA*, ApoE peptide–phosphatidic acid; *PACA*, polyalkylcyanoacrylate.

methods. The combination of the multivalence with early diagnosis makes nanomedicine one of the most promising approaches in the treatment of AD.

4. NANOTECHNOLOGY FOR THE TREATMENT OF STROKE

4.1 Physiopathology of Stroke

With an event occurring every 40 s and a victim every 4 min, stroke represents the fourth leading cause of death in developed countries and a major health concern (Go et al., 2014). Among 4,000,000 people experiencing a new or recurrent stroke each year worldwide, 19% will die in short terms after the event, and, without treatment, more than 30% will remain moderately or severely disabled after 3 months. Being the main cause of long-term disability in most industrialized countries, stroke also represents an economic burden, as its total cost in the United States is estimated at $36.5 billion per year, while the indirect costs from lost productivity and other factors are estimated at $15 million per year. Although these statistics are staggering, stroke is highly preventable: modifiable cardiovascular risk factors account for much of the proportional risk of stroke, and it is estimated that up to 80% of strokes may be prevented (Gorelick & Ruland, 2010). Among all stroke cases, more than 87% result from a blood clot in a major artery, and are called ischemic strokes (Go et al., 2014), inducing a transient or permanent reduction of cerebral blood flow in the territory of a major brain artery. It is now widely accepted that following this event, the ischemic brain injury results from a complex sequence of pathophysiological cascade that evolves over time and space (Dirnagl, Iadecola, & Moskowitz, 1999).

4.2 Current Stroke Therapies

Thanks to prospective randomized clinical trials and other studies performed over the past decade, the general approach to the management of acute stroke has evolved from nihilism to active intervention. Previous management of stroke was mainly based on natural recovery capabilities of the brain, as patients that have experienced a stroke often exhibit continued functional recovery for many years following their initial injury. Natural recoveries include varying degrees of behavioral compensation and cerebral plasticity (Murphy & Corbett, 2009), neurogenesis (Kernie & Parent, 2010), and to a less clear extent, angiogenesis (Krupinski, Kaluza, Kumar, Kumar, & Wang, 1994). Nevertheless, post-stroke capacities are unlikely to be identical to pre-stroke ones owing to the loss of neurons that present specific functions. This leads to chronic disabilities in which severity is directly linked to the volume of the infarct (Lovblad et al., 1997) and infarct localization, raising the urgent need for therapeutic intervention.

4.2.1 Thrombolysis

The current gold standard treatment for stroke patients is thrombolysis, via the use of recombinant tissue plasminogen activator (rt-PA). The aim of lysing the occluding thrombus or embolus by disruption of fibrin, usually by enhancing the endogenous formation of plasmin from plasminogen, is to reduce the volume of cerebral tissue irreversibly damaged (Pereira, Martin, & Warburton, 2001). However, rt-PA is currently given to less than 5% of stroke patients (Wardlaw, Lindley, & Lewis, 2002), showing a lag between positive clinical trials results and routine medical practice. This attitude stems, at least partly, from concern over a 10-fold increased risk of intracerebral hemorrhage in patients receiving the thrombolytic agent (Goldstein, 2007), a neurotoxic potential within the parenchyma if the molecule is not confined to the vasculature (Yepes, Roussel, Ali, & Vivien, 2009), leading to a very restricted time window of administration of 4.5 h (Hacke et al., 2008). Therefore, research currently focuses on different therapeutic approaches for acute ischemic stroke such as the use of ultrasounds (Mahon et al., 2003), possibly with concomitant administration of gaseous microbubbles or lipid microspheres (Alexandrov et al., 2008) to enhance thrombolytics activity, or the use of devices designed specifically for the cerebral vasculature allowing endovascular angioplasty, stenting, or mechanical clot retrieval.

4.2.2 Neuroprotective Therapy

Neuroprotective therapy is another explored treatment, which aims at structurally preserve the intact tissue until adequate blood supply can be reestablished, either through spontaneous or therapeutic recanalization. The elucidation of biochemical and molecular events and the identification of intracellular mediators of the ischemic injury allowed the development of pathophysiological rational for extended classes of neuroprotectants (Ginsberg, 2008). Most of these strategies tend to: (i) decrease neuro-excitotoxicity using calcium channels blockers, glutamate antagonists, or GABA agonists, (ii) to decrease oxidative stress with antioxidants and NO downregulatory molecules, (iii) to decrease the inflammatory response,

(iv) to avoid the no-reflow phenomenon by inhibiting leukocytes and neutrophiles adhesion and activation, and (v) to inhibit apoptosis pathways with caspase-3 inhibitors and modulation of the Bcl-2 family. Other neuroprotective strategies include neural stem cells therapy, exploiting the ability of these cells to differentiate and regenerate, and possibly to restore loss of brain functions (Cheng, Al-Khoury, & Zivin, 2004). Finally, preclinical studies have shown the protective effects of therapeutic hypothermia in models of focal ischemia, and early-phase feasibility studies recently showed its safety and efficacy in patients with acute ischemic stroke (Wu & Grotta, 2013). However, amongt 400 neuroprotective drugs sent to clinical trials, none ever demonstrated real pharmacological efficacy in man. These neuroprotective strategies consisting in a single target therapy suffer from the fact that modulation of one pathway alone may not be sufficient to tip the balance of molecular activity in an extremely complex injury (Zaleska et al., 2009). Thus, the use of multitargeted pleiotropic molecules or a combination of multiple drugs exerting different actions and able to modulate several pathways in all cells of the neurovascular unit may be an interesting strategy for the treatment of acute ischemic injury (Lo, Dalkara, & Moskowitz, 2003; Zhang et al., 2012).

4.3 NP-Based Stroke Therapy

4.3.1 Advantages

Given the repeated failures of classical drug formulations in clinical trials, the use of nanotechnologies for the treatment of stroke has raised interest in the past decades (Nair, Dileep, & Rajanikant, 2012), and has been widely investigated for the delivery of active drugs. Indeed, nanotechnologies represent highly promising opportunities, as they open the way of enhanced drug delivery to the brain, thanks to prolonged circulation time and increased brain penetration, and offer the possibility for controlled release of one or several drugs into the brain parenchyma. Different types of nanomedicine including liposomes, polymeric NPs, or solid lipid NPs have been used to deliver various types of drugs such as growth factors (Wang, Cooke, Sachewsky, Morshead, & Shoichet, 2013; Zhao et al., 2011), erythropoietin (Chen et al., 2012), hemoglobin (Kawaguchi et al., 2007), antioxidant molecules (Reddy & Labhasetwar, 2009), or siRNA (Kim et al., 2010). Platinum (Takamiya et al., 2012) and ceria (Kim et al., 2012) NPs or carbon nanotubes (Lee et al., 2011) have also been investigated for their free radical scavenger properties.

4.3.1.1 BBB Crossing

Nanotechnologies in the setting of stroke have been mainly intended to improve the entry of the transported drug into the brain, usually by increasing the drug blood half-life time and promoting BBB translocation. The most investigated nanosystems in stroke preclinical studies are liposomes and polymeric NPs, especially PLGA NPs, for their biodegradability, low toxicity, and biocompatibility. Those systems are well characterized drug carriers, which usually have previously been shown to at least partially cross the BBB. Moreover, their scaling-up ability is already proven, especially in the case of liposomes. One investigation by Yun and colleagues (2013) compared the pharmacological efficiency of three different types of nanocarriers (PEGylated liposomes, PLA NPs, and PLGA NPs) encapsulating superoxide dismutase (SOD) as an antioxidant agent, with and without a targeting modality against NMDA receptor 1 (NR1). While the addition of a targeting moiety allowed to significantly decrease the infarct volume by more than 50%, no significant differences were observed in term of infarct volume reduction between the different nanocarrier types, when the animals were injected with the same quantity of encapsulated drug. Hence, the difference between these nanosystems will mainly arise from their drug loadings and encapsulation efficiencies, which will govern their relative systemic toxicity and manufacturability, together with their specific ADME profile.

4.3.1.2 Multitargeted Pleiotropic Molecules

As previously mentioned, given the complexity of the stroke pathological cascade, the use of multitargeted pleiotropic molecules represents an emerging treatment opportunity. In particular, the use of endogeneous molecules such as the vascular endothelial growth factor (VEGF), erythropoietin (EPO), or adenosine has been proposed. Indeed, those molecules have been previously demonstrated to be released after an ischemic event and to exert positive actions on the different cell types of the neurovascular unit. In particular, VEGF has been shown to confer neuroprotection, promote neurogenesis and angiogenesis. EPO has been shown to be cytoprotective in the brain, acting as one of the important factor for neuroprotection and neurodevelopment after an ischemic event. Finally, adenosine has been shown to decrease excitotoxicity, alleviate inflammatory processes, and promote vasodilatation and angiogenesis. However, given their endogenous trait, the intravenous administration of these molecules is limited by a very short blood half-life due to an immediate metabolization. Hence, VEGF has been delivered by Zhao et al. (2011), using PEGylated liposomes to increase systemic circulation time, and decorated with transferrin to enhance BBB penetration. Following intravenous administration in a rat model of permanent ischemia, the authors observed a long-term significant improvement of the motor score, likely resulting from functional neovascularization. Ishii et al. (2012) also developed a PEGylated liposomal formulation to deliver Asialo-EPO (AEPO), an EPO metabolite, able to bind to the EPO receptors without presenting its toxic hematopoietic effect. Although free AEPO suppressed the infarct volume by around 30%, AEPO-liposomes achieved to suppress more than 70%. The authors showed a clear accumulation of AEPO-liposomes in the infarcted hemisphere allowing a sustained exposure of the cells to a high concentration of the drug, which likely results from a prolonged circulation time in the blood. EPO has also been delivered to neonatal rats using PLGA NPs (Chen et al., 2012). The authors observed a significant decrease of the infarct volume and the use of PLGA NPs allowed decreasing by 16 × the dosage compared to the free drug, limiting the risk of toxicity.

4.3.2 Limits of NPs-Based Therapy for Stroke

These studies involve the encapsulation of the drug inside the nanocarrier and thus raise the question of drug loading and encapsulation efficiency to properly characterize the nanosystem. Reported encapsulation efficiency is usually high, ensuring a limited loss of raw material during the preparation process and thus allowing considering the industrial scaling-up with reasonable cost. However, drug loadings are usually low, likely inducing dose limitations due to side-toxicity of the carrier.

4.4 Novel NPs-Based Approaches for Stroke

4.4.1 Drug Loading Improvement

An alternate approach is the covalent linkage of the therapeutic molecule to the vector, followed by the formation of the nanoparticulate suspension. In this situation, drug loading is exclusively governed by the ratio between the molecular weights of the drug and the vector molecule. Gaudin et al. (2014) have investigated this strategy by coupling the adenosine to the squalene, forming an amphiphilic prodrug able to self-assemble as NPs. These NPs were shown to provide neuroprotection in a mouse model of cerebral ischemia and a rat model of spinal cord injury, by increasing the circulation time of adenosine and thus enhancing the exposure time to the drug. Interestingly, in this case, the molecule did not have to translocate across the BBB to exert its effect, as it was mainly acting on improving microcirculation.

4.4.2 Multiple Drugs Delivery

Another strategy to answer stroke pathology complexity is to deliver multiple drugs. Hence, the versatility of nanotechnologies, offering the possibility to deliver different molecules at different release rates, appears particularly appealing. A study by Wang et al. (2013) reports an original system including the delivery of two drugs, EPO and epidermal growth factor (EGF). In this formulation, EPO was encapsulated in PLGA NPs, which were then incorporated into bigger PLA microparticles (PLA MPs), whereas the EGF was also encapsulated in PLGA NPs but not incorporated into MPs. Both particulate formulations (EPO-PLGA NPs-PLA MPs and EGF-PLGA NPs) were then embedded together in a hyaluronan and methylcellulose (HAMC) hydrogel (Fig. 5A). It allowed the sequential release in vitro and in vivo (Fig. 5B) of first EGF, which promotes neurogenesis, followed by EPO, which prevents the apoptosis of the newly formed cells. The authors observed a decrease of inflammatory response and of number of apoptotic cells in the periinfarct region; however, regarding clinical translation, a major obstacle results from the need to implant the device at the cortical surface (epicortical delivery; Fig. 5A) and from the short diffusion length of the proteins into the brain tissue (typically 1 mm), which limits the volume of impregnated brain tissue.

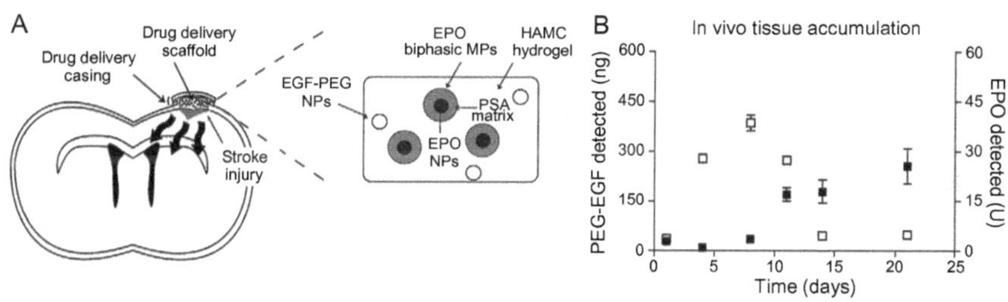

Fig. 5 Composite drug delivery system allowing the sequential release of EGF and EPO in vivo. (A) EPO was encapsulated in PLGA NPs, which were then incorporated in PLA microparticles (PLA MPs). PEGylated EGF (EGF-PEG) was encapsulated in PLGA NPs but without incorporation in microspheres. The two particulate formulations (EPO-PLGA NPs-PLA MPs and EGF-PEG-PLGA NPs) were then embedded together in a hyaluronan and methylcellulose (HAMC) hydrogel. For drug delivery, the gel was implanted at the cortical surface. (B) The in vivo drug release profile of EGF-PEG and EPO in the stroke-injured brain tissue showed that EGF-PEG was released first, with the peak concentration detected after 1 week, followed by EPO accumulation over 25 days. This should allow EGF to promote neurogenesis, followed by EPO, which will prevent the apoptosis of the newly formed cells. *Reprinted from Wang, Y., Cooke, M. J., Sachewsky, N., Morshead, C. M., & Shoichet, M. S. (2013). Bioengineered sequential growth factor delivery stimulates brain tissue regeneration after stroke.* Journal of Controlled Release, 172*(1), 1–11, with permission from Elsevier.*

Brain penetration across the BBB is a major limitation for the treatment of stroke, and a prolonged circulation does not necessarily ensure BBB crossing. Nanotechnology can address this issue via the decoration of NP surface with targeting ligands. Lu et al. (2014) and Agulla et al. (2013) clearly demonstrated the targeting ability of surface-functionalized solid lipid NPs and liposomes, respectively (Fig. 6). Compared to nontargeted nanocarriers, the authors

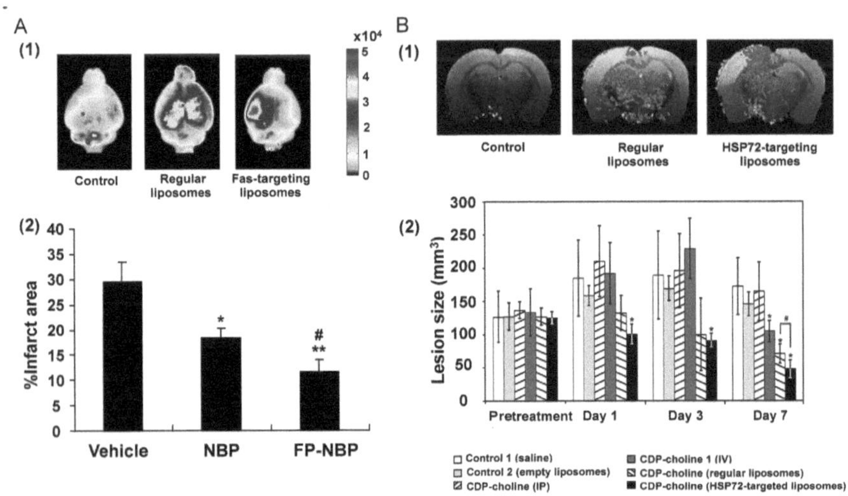

Fig. 6 Targeting of the ischemic area using a (A) FAS ligand or (B) a HSP72 ligand. (A—1) The conjugation of Fas ligand (protein belonging to the tumor necrosis factor family which receptor is a member of the death receptor family) to the surface of PEGylated liposomes clearly induced the targeting of the ischemic area 24 h after intravenous administration, compared to nontargeted liposomes. (A—2) When encapsulating butylphthalide, these targeted liposomes significantly reduced the infarct size by ca. 8% compared to the nontargeted ones. $*P < 0.05$, $**P < 0.01$ versus vehicle mice; $^{\#}P < 0.05$ versus regular NBP treated mice. FP-NBP, NBP encapsulated Fas ligand/PEG-nanoconjugate. (B—1) Similar results were obtained with liposomes targeted against HSP72, which is a member of the heat-shock proteins. Regular (nontargeted) liposomes appear to be randomly distributed along both hemispheres when targeted liposomes concentrate in the ischemic hemisphere. (B—2) This targeting allowed a significant reduction of the infarct size compared to nontargeted liposomes, 7 days postinjury. Noteworthy, in this study, the authors designed their targeting strategy by conducting a complete proteomic, immunoblot, and immunohistological study, ensuring the relevance of the target with the pathological setting. *Panel (A) reprinted from Lu, Y. M., Huang, J. Y., Wang, H., Lou, X. F., Liao, M. H., Hong, L. J., et al. (2014). Targeted therapy of brain ischaemia using Fas ligand antibody conjugated PEG-lipid nanoparticles.* Biomaterials, *35(1), 530–537, with permission from Elsevier and Panel (B) reprinted with permission from Agulla, J., Brea, D., Campos, F., Sobrino, T., Argibay, B., Al-Soufi, W., et al. (2013). In vivo theranostics at the peri-infarct region in cerebral ischemia.* Theranostics, *4(1), 90–105.*

observed in both studies a significant reduction of the infarct size thanks to the addition of the targeting ligand. However, the actual resulting benefit on the infarct size was only ca. 10%. This number might appear small in regard to the multiple steps of preparation and characterization required when adding a targeting moiety, and the cost that they imply. However, Agulla and collaborators (2013) explored another feature of codelivery using nanotechnology, which is the development of a theranostic nanoformulation, including both an imaging agent and a therapeutic molecule. In this case, even if the targeting moiety may not be profitable from the therapeutic point of view, it will definitely be essential concerning the diagnostic aspect. Hence, thanks to their targeted immunoliposomes loaded with citicoline and fluoresence and MR imaging probes, they were able to delineate the periinfarct region in vivo and to achieve superior therapeutic effect.

As previously shown when delivering multiple drugs, nanotechnology offers the possibility to manipulate kinetics of treatment, by exploiting release characteristics of nanocarriers. This property has also been investigated with single drug treatments, offering dosage regimen inaccessible with free drug. When establishing therapeutic windows, it has been shown that similarly to free drugs, the sooner the treatment, the better was the outcome. In particular, two reports investigating the use of liposomal FK506 (Ishii et al., 2013; an immunosuppressor) and PBCA NPs loaded with SOD (Yun et al., 2013) observed that delayed administrations at 3 h or 24 h after the reperfusion failed to significantly decrease the infarct volume, compared to an administration at the reperfusion. However, studies performing long-term evaluations (up to 7 days, 14 days, or 21 days postischemia) on functional outcome or on infarct evolution thanks to imaging technics assessed the sustained benefit of nanomedicines. For example, Reddy and Labhasetwar (2009) used PLGA NPs encapsulating SOD, and followed the neurological score and the infarct volume of the animals up to 28 days postinjury. In this investigation, functional improvement mainly appeared as soon as 6 h after treatment administration, but keep increasing up to 28 days after injury. This is likely due to the delayed release of SOD from PLGA NPs, which has been shown to last for more than 100 days in vitro.

4.5 Conclusions on the Application of Nanomedicine to Stroke Therapy

Nanomedicines represent highly promising opportunities for the treatment of stroke, as they open the way of enhanced drug delivery to the brain and controlled release of one or several drugs into the brain parenchyma. However, the biggest limitation of nanotechnologies for the treatment of stroke remains the penetration of the BBB. Even though improved half-life circulation and targeting strategies have been developed to enhance BBB crossing, most of the nanosystems enter the brain parenchyma in incremental amount, usually of a few percents. Hence, multiple injections or high doses of nanotherapeutic have to be administered, leading to toxicity issues. Local delivery strategies via intracortical injections have been explored, in particular in the case of nanocarriers delivering siRNA (Al-Jamal et al., 2011; Kim et al., 2010). Convection-enhanced delivery (CED) has also been suggested as a suitable technique to deliver NPs locally, especially in the case of brain tumors (Sawyer et al., 2011). In the ischemic brain, edematous ischemic tissue has been shown to alter the distribution of agents after CED because of a decreased extracellular space (ECS; Haar et al., 2010). Thus, CED of nanoparticulate systems might require using smaller sized nanocarriers than for CED in the normal or tumor-bearing brain. In those settings, it has been shown that NPs should be less than 100 nm to be able to navigate through the pore size (Zhou et al., 2013), and this size might have to be further reduced to allow circulation in the ischemic ECS. On the other hand, the ischemic brain appears to present decreased postinfusion clearance of the infusate, likely due to impaired microvasculature (Haar et al., 2010). Even though local delivery is more invasive and decreases patient compliancy, it might be a way to take advantage of the versatility of nanotechnologies for the treatment of stroke.

5. GENERAL CONCLUSION

It emerges from this review that colloidal systems can deliver drugs into the brain thanks to their ability to interact or cross the BBB. In many cases, NPs can reach the surface of the brain microvascular endothelial cells, only under the condition that the surface of these colloids is modified in a proper way (ie, by PEG). The prolonged blood circulation of these surface-modified NPs enhances exposure of the BBB which favors interaction and penetration into brain endothelial cells. NPs may further be modified with specific targeting molecules aiming to enhance their binding with surface receptors of the brain endothelial cells, thus promoting their transport across the BBB. Moreover, a recent work reveals the interest of another therapeutic strategy based on the targeting and treating of the endothelial cells themselves in order to preserve all cells of the neurovascular unit such as the neurons. Therefore, when drug is loaded, colloidal carriers may be helpful for the treatment of brain diseases (with or without disruption of BBB), because they offer clinical advantages such as decreased drug dose, reduced drug side effects, increased drug viability, noninvasive routes of administration, and improved patient quality of life. However, there is an urgent need to clarify the mechanisms which manage the carrier-mediated transport of the drugs to the brain but also to elucidate the physiopathology of the CNS diseases such as neurodegenerative disorders and then evidence the relevant targets. Efficient therapeutic approaches for CNS diseases should associate novel drugs to adequate nanoassemblies that can carrier these active molecules to the precise targeted brain site.

REFERENCES

Acharya, S., & Sahoo, S. K. (2011). PLGA nanoparticles containing various anticancer agents and tumour delivery by EPR effect. *Advanced Drug Delivery Reviews, 63*(3), 170–183.

Agulla, J., Brea, D., Campos, F., Sobrino, T., Argibay, B., Al-Soufi, W., et al. (2013). In vivo theranostics at the peri-infarct region in cerebral ischemia. *Theranostics, 4*(1), 90–105.

Ahmed, T., & Gilani, A. H. (2014). Therapeutic potential of turmeric in Alzheimer's disease: Curcumin or curcuminoids? *Phytotherapy Research, 28*(4), 517–525.

Alexandrov, A. V., Mikulik, R., Ribo, M., Sharma, V. K., Lao, A. Y., Tsivgoulis, G., et al. (2008). A pilot randomized clinical safety study of sonothrombolysis augmentation with ultrasound-activated perflutren-lipid microspheres for acute ischemic stroke. *Stroke, 39*(5), 1464–1469.

Al-Jamal, K. T., Gherardini, L., Bardi, G., Nunes, A., Guo, C., Bussy, C., et al. (2011). Functional motor recovery from brain ischemic insult by carbon nanotube-mediated siRNA silencing. *Proceedings of the National Academy of Sciences of the United States of America, 108*(27), 10952–10957.

Antunes, M., & Biala, G. (2012). The novel object recognition memory: Neurobiology, test procedure, and its modifications. *Cognitive Processing, 13*(2), 93–110.

Åslund, A. K., Berg, S., Hak, S., Mørch, Ý., Torp, S. H., Sandvig, A., et al. (2015). Nanoparticle delivery to the brain—By focused ultrasound and self-assembled nanoparticle-stabilized microbubbles. *Journal of Controlled Release, 220*, 287–294.

Bakota, L., & Brandt, R. (2016). Tau biology and tau-directed therapies for Alzheimer's disease. *Drugs, 76*, 301–313.

Balducci, C., Mancini, S., Minniti, S., La Vitola, P., Zotti, M., Sancini, G., et al. (2014). Multifunctional liposomes reduce brain β-amyloid burden and ameliorate memory impairment in Alzheimer's disease mouse models. *The Journal of Neuroscience, 34*(42), 14022–14031.

Balzeau, J., Pinier, M., Berges, R., Saulnier, P., Benoit, J. P., & Eyer, J. (2013). The effect of functionalizing lipid nanocapsules with NFL-TBS. 40-63 peptide on their uptake by glioblastoma cells. *Biomaterials, 34*(13), 3381–3389.

Barth, R. F. (1998). Rat brain tumor models in experimental neuro-oncology: The 9L, C6, T9, F98, RG2 (D74), RT-2 and CNS-1 gliomas. *Journal of Neuro-Oncology, 36*(1), 91–102.

Basile, L., Passirani, C., Huynh, N. T., Béjaud, J., Benoit, J. P., Puglisi, G., et al. (2012). Serum-stable, long-circulating paclitaxel-loaded colloidal carriers decorated with a new amphiphilic PEG derivative. *International Journal of Pharmaceutics*, *426*(1), 231–238.

Bastiancich, C., Vanvarenberg, K., Ucakar, B., Pitorre, M., Bastiat, G., Lagarce, F., et al. (2016). Lauroyl-gemcitabine-loaded lipid nanocapsule hydrogel for the treatment of glioblastoma. *Journal of Controlled Release*, *225*, 283–293.

Bertrand, N., Wu, J., Xu, X., Kamaly, N., & Farokhzad, O. C. (2014). Cancer nanotechnology: The impact of passive and active targeting in the era of modern cancer biology. *Advanced Drug Delivery Reviews*, *66*, 2–25.

Bloom, G. S. (2014). Amyloid-β and tau: The trigger and bullet in Alzheimer disease pathogenesis. *JAMA Neurology*, *71*(4), 505–508.

Bolhassani, A., Javanzad, S., Saleh, T., Hashemi, M., Aghasadeghi, M. R., & Sadat, S. M. (2014). Polymeric nanoparticles: Potent vectors for vaccine delivery targeting cancer and infectious diseases. *Human Vaccines & Immunotherapeutics*, *10*(2), 321–332.

Brambilla, D., Verpillot, R., Le Droumaguet, B., Nicolas, J., Taverna, M., Kóňa, J., et al. (2012). PEGylated nanoparticles bind to and alter amyloid-beta peptide conformation: Toward engineering of functional nanomedicines for Alzheimer's disease. *ACS Nano*, *6*(7), 5897–5908.

Brigger, I., Dubernet, C., & Couvreur, P. (2012). Nanoparticles in cancer therapy and diagnosis. *Advanced Drug Delivery Reviews*, *64*, 24–36.

Cannon, J. R., & Greenamyre, J. T. (2011). The role of environmental exposures in neurodegeneration and neurodegenerative diseases. *Toxicological Sciences*, *124*, 225–250. pii: kfr239.

Cavanaugh, S. E., Pippin, J. J., & Barnard, N. D. (2014). Animal models of Alzheimer disease: Historical pitfalls and a path forward. *Altex*, *31*(3), 279–302.

Chauhan, N., Narang, J., & Jain, U. (2015). Highly sensitive and rapid detection of acetylcholine using an ITO plate modified with platinum-graphene nanoparticles. *Analyst*, *140*(6), 1988–1994.

Chen, H., Spagnoli, F., Burris, M., Rolland, W. B., Fajilan, A., Dou, H., et al. (2012). Nanoerythropoietin is 10-times more effective than regular erythropoietin in neuroprotection in a neonatal rat model of hypoxia and ischemia. *Stroke*, *43*(3), 884–887.

Chen, C., Xu, T., Lu, Y., Chen, J., & Wu, S. (2013). The efficacy of temozolomide for recurrent glioblastoma multiforme. *European Journal of Neurology*, *20*(2), 223–230.

Cheng, Y. D., Al-Khoury, L., & Zivin, J. A. (2004). Neuroprotection for ischemic stroke: Two decades of success and failure. *NeuroRx*, *1*(1), 36–45.

Cheng, K. K., Chan, P. S., Fan, S., Kwan, S. M., Yeung, K. L., Wáng, Y. X. J., et al. (2015). Curcumin-conjugated magnetic nanoparticles for detecting amyloid plaques in Alzheimer's disease mice using magnetic resonance imaging (MRI). *Biomaterials*, *44*, 155–172.

Chiarelli, P. A., Kievit, F. M., Zhang, M., & Ellenbogen, R. G. (2015). Bionanotechnology and the future of glioma. *Surgical Neurology International*, *6*(Suppl. 1), S45.

Ciechanover, A., & Kwon, Y. T. (2015). Degradation of misfolded proteins in neurodegenerative diseases: Therapeutic targets and strategies. *Experimental & Molecular Medicine*, *47*(3), e147.

Clavreul, A., Montagu, A., Lainé, A. L., Tétaud, C., Lautram, N., Franconi, F., et al. (2015). Targeting and treatment of glioblastomas with human mesenchymal stem cells carrying ferrociphenol lipid nanocapsules. *International Journal of Nanomedicine*, *10*, 1259.

Cohen, Z. R., Ramishetti, S., Peshes-Yaloz, N., Goldsmith, M., Wohl, A., Zibly, Z., et al. (2015). Localized RNAi therapeutics of chemoresistant grade IV glioma using hyaluronan-grafted lipid-based nanoparticles. *ACS Nano*, *9*(2), 1581–1591.

Corace, G., Angeloni, C., Malaguti, M., Hrelia, S., Stein, P. C., Brandl, M., et al. (2014). Multifunctional liposomes for nasal delivery of the anti-Alzheimer drug tacrine hydrochloride. *Journal of Liposome Research*, *24*(4), 323–335.
Couvreur, P., & Vauthier, C. (2006). Nanotechnology: Intelligent design to treat complex disease. *Pharmaceutical Research*, *23*(7), 1417–1450.
Czupalla, C. J., Liebner, S., & Devraj, K. (2014). In vitro models of the blood–brain barrier. *Cerebral Angiogenesis: Methods and Protocols*, *1135*, 415–437.
Dawidczyk, C. M., Russell, L. M., & Searson, P. C. (2014). Nanomedicines for cancer therapy: State-of-the-art and limitations to pre-clinical studies that hinder future developments. *Frontiers in Chemistry*, *2*, 69.
Deardorff, W. J., Feen, E., & Grossberg, G. T. (2015). The use of cholinesterase inhibitors across all stages of Alzheimer's disease. *Drugs & Aging*, *32*(7), 537–547.
Dehay, B., Bourdenx, M., Gorry, P., Przedborski, S., Vila, M., Hunot, S., et al. (2015). Targeting α-synuclein for treatment of Parkinson's disease: Mechanistic and therapeutic considerations. *The Lancet Neurology*, *14*(8), 855–866.
De Jesus, M. B., & Zuhorn, I. S. (2015). Solid lipid nanoparticles as nucleic acid delivery system: Properties and molecular mechanisms. *Journal of Controlled Release*, *201*, 1–13.
Dirnagl, U., Iadecola, C., & Moskowitz, M. A. (1999). Pathobiology of ischaemic stroke: An integrated view. *Trends in Neurosciences*, *22*(9), 391–397.
Dixit, S., Hingorani, M., Achawal, S., & Scott, I. (2011). The sequential use of carmustine wafers (Gliadel®) and post-operative radiotherapy with concomitant temozolomide followed by adjuvant temozolomide: A clinical review. *British Journal of Neurosurgery*, *25*(4), 459–469.
Dorsey, E. R., George, B. P., Leff, B., & Willis, A. W. (2013). The coming crisis obtaining care for the growing burden of neurodegenerative conditions. *Neurology*, *80*(21), 1989–1996.
Drachman, D. A. (2014). The amyloid hypothesis, time to move on: Amyloid is the downstream result, not cause, of Alzheimer's disease. *Alzheimer's & Dementia*, *10*(3), 372–380.
Elnaggar, Y. S., Etman, S. M., Abdelmonsif, D. A., & Abdallah, O. Y. (2015). Intranasal piperine-loaded chitosan nanoparticles as brain-targeted therapy in Alzheimer's disease: Optimization, biological efficacy, and potential toxicity. *Journal of Pharmaceutical Sciences*, *104*(10), 3544–3556.
Eslami, M., Nikkhah, S. J., Hashemianzadeh, S. M., & Sajadi, S. A. S. (2016). The compatibility of Tacrine molecule with poly (n-butylcyanoacrylate) and Chitosan as efficient carriers for drug delivery: A molecular dynamics study. *European Journal of Pharmaceutical Sciences*, *82*, 79–85.
Field, K. M., Jordan, J. T., Wen, P. Y., Rosenthal, M. A., & Reardon, D. A. (2015). Bevacizumab and glioblastoma: Scientific review, newly reported updates, and ongoing controversies. *Cancer*, *121*(7), 997–1007.
Francis, P. T., Palmer, A. M., Snape, M., & Wilcock, G. K. (1999). The cholinergic hypothesis of Alzheimer's disease: A review of progress. *Journal of Neurology, Neurosurgery & Psychiatry*, *66*(2), 137–147.
Frost, B., Götz, J., & Feany, M. B. (2015). Connecting the dots between tau dysfunction and neurodegeneration. *Trends in Cell Biology*, *25*(1), 46–53.
Fuller, G. N., & Scheithauer, B. W. (2007). The 2007 revised World Health Organization (WHO) classification of tumours of the central nervous system: Newly codified entities. *Brain Pathology*, *17*(3), 304–307.
Gapp, K., Woldemichael, B. T., Bohacek, J., & Mansuy, I. M. (2014). Epigenetic regulation in neurodevelopment and neurodegenerative diseases. *Neuroscience*, *264*, 99–111.

Garcia-Garcia, E., Andrieux, K., Gil, S., & Couvreur, P. (2005). Colloidal carriers and blood-brain barrier (BBB) translocation: A way to deliver drugs to the brain? *International Journal of Pharmaceutics, 298*, 274–292.

Gaudin, A., Yemisci, M., Eroglu, H., Lepetre-Mouelhi, S., Turkoglu, O. F., Donmez-Demir, B., et al. (2014). Squalenoyl adenosine nanoparticles provide neuroprotection after stroke and spinal cord injury. *Nature Nanotechnology, 9*(12), 1054–1062.

Ginsberg, M. D. (2008). Neuroprotection for ischemic stroke: Past, present and future. *Neuropharmacology, 55*(3), 363–389.

Go, A. S., Mozaffarian, D., Roger, V. L., Benjamin, E. J., Berry, J. D., Blaha, M. J., et al. (2014). Executive summary: Heart disease and stroke statistics—2014 update: A report from the American Heart Association. *Circulation, 129*(3), 399–410.

Goldgaber, D., Lerman, M. I., McBride, W. O., Saffiotti, U., & Gajdusek, D. C. (1986). Isolation, characterization, and chromosomal localization of human brain cDNA clones coding for the precursor of the amyloid of brain in Alzheimer's disease, Down's syndrome and aging. *Journal of Neural Transmission. Supplementum, 24*, 23–28.

Goldstein, L. B. (2007). Acute ischemic stroke treatment in 2007. *Circulation, 116*(13), 1504–1514.

Gorelick, P. B., & Ruland, S. (2010). Cerebral vascular disease. *Disease-a-Month, 56*(2), 39–100.

Gozes, I. (2010). Tau pathology and future therapeutics. *Current Alzheimer Research, 7*(8), 685–696.

Grundke-Iqbal, I., Iqbal, K., Tung, Y. C., Quinlan, M., Wisniewski, H. M., & Binder, L. I. (1986). Abnormal phosphorylation of the microtubule-associated protein tau (tau) in Alzheimer cytoskeletal pathology. *Proceedings of the National Academy of Sciences of the United States of America, 83*(13), 4913–4917.

Haar, P. J., Broaddus, W. C., Chen, Z. J., Fatouros, P. P., Gillies, G. T., & Corwin, F. D. (2010). Quantification of convection-enhanced delivery to the ischemic brain. *Physiological Measurement, 31*(9), 1075–1089.

Haass, C., & Selkoe, D. J. (2007). Soluble protein oligomers in neurodegeneration: Lessons from the Alzheimer's amyloid β-peptide. *Nature Reviews. Molecular Cell Biology, 8*(2), 101–112.

Hacke, W., Kaste, M., Bluhmki, E., Brozman, M., Davalos, A., Guidetti, D., et al. (2008). Thrombolysis with alteplase 3 to 4.5 hours after acute ischemic stroke. *The New England Journal of Medicine, 359*(13), 1317–1329.

Hardy, J., & Allsop, D. (1991). Amyloid deposition as the central event in the aetiology of Alzheimer's disease. *Trends in Pharmacological Sciences, 12*, 383–388.

Henriksen, K., O'Bryant, S. E., Hampel, H., Trojanowski, J. Q., Montine, T. J., Jeromin, A., et al. (2014). The future of blood-based biomarkers for Alzheimer's disease. *Alzheimer's & Dementia, 10*(1), 115–131.

Henry, W., Querfurth, F. L., & LaFerla, F. M. (2010). Alzheimer's disease. *The New England Journal of Medicine, 362*, 329–344.

Hong, I. S., Jang, G. B., Lee, H. Y., & Nam, J. S. (2015). Targeting cancer stem cells by using the nanoparticles. *International Journal of Nanomedicine, 10*(Special issue), 251.

Huszthy, P. C., Daphu, I., Niclou, S. P., Stieber, D., Nigro, J. M., Sakariassen, P. Ø., et al. (2012). In vivo models of primary brain tumors: Pitfalls and perspectives. *Neuro-Oncology, 14*, 979–993. pii: nos135.

Huttner, A. (2012). Overview of primary brain tumors: Pathologic classification, epidemiology, molecular biology, and prognostic markers. *Hematology/Oncology Clinics of North America, 26*(4), 715–732.

Huynh, N. T., Morille, M., Bejaud, J., Legras, P., Vessieres, A., Jaouen, G., et al. (2011). Treatment of 9L gliosarcoma in rats by ferrociphenol-loaded lipid nanocapsules based on a passive targeting strategy via the EPR effect. *Pharmaceutical Research, 28*(12), 3189–3198.

Huynh, N. T., Passirani, C., Allard-Vannier, E., Lemaire, L., Roux, J., Garcion, E., et al. (2012). Administration-dependent efficacy of ferrociphenol lipid nanocapsules for the treatment of intracranial 9L rat gliosarcoma. *International Journal of Pharmaceutics*, *423*(1), 55–62.
Ishii, T., Asai, T., Oyama, D., Agato, Y., Yasuda, N., Fukuta, T., et al. (2013). Treatment of cerebral ischemia-reperfusion injury with PEGylated liposomes encapsulating FK506. *The FASEB Journal*, *27*(4), 1362–1370.
Ishii, T., Asai, T., Oyama, D., Fukuta, T., Yasuda, N., Shimizu, K., et al. (2012). Amelioration of cerebral ischemia-reperfusion injury based on liposomal drug delivery system with asialo-erythropoietin. *Journal of Controlled Release*, *160*(1), 81–87.
Jacobs, V. L., Valdes, P. A., Hickey, W. F., & De Leo, J. A. (2011). Current review of in vivo GBM rodent models: Emphasis on the CNS-1 tumour model. *ASN Neuro*, *3*(3). AN20110014.
Jacques, G., & Cormac, O. (2012). Central nervous system tumors. *Handbook of Clinical Neurology*, *112*, 931–958.
Jiang, X., Xin, H., Gu, J., Du, F., Feng, C., Xie, Y., et al. (2014). Enhanced antitumor efficacy by d-glucosamine-functionalized and paclitaxel-loaded poly (ethylene glycol)-co-poly (trimethylene carbonate) polymer nanoparticles. *Journal of Pharmaceutical Sciences*, *103*(5), 1487–1496.
Joshi, S. A., Chavhan, S. S., & Sawant, K. K. (2010). Rivastigmine-loaded PLGA and PBCA nanoparticles: Preparation, optimization, characterization, in vitro and pharmacodynamic studies. *European Journal of Pharmaceutics and Biopharmaceutics*, *76*(2), 189–199.
Kaluzova, M., Bouras, A., Machaidze, R., & Hadjipanayis, C. G. (2015). Targeted therapy of glioblastoma stem-like cells and tumor non-stem cells using cetuximab-conjugated iron-oxide nanoparticles. *Oncotarget*, *6*(11), 8788.
Kane, J. R., Miska, J., Young, J. S., Kanojia, D., Kim, J. W., & Lesniak, M. S. (2015). Sui generis: Gene therapy and delivery systems for the treatment of glioblastoma. *Neuro-Oncology*, *17*(Suppl. 2), ii24–ii36.
Kang, T., Jiang, M., Jiang, D., Feng, X., Yao, J., Song, Q., et al. (2015). Enhancing glioblastoma-specific penetration by functionalization of nanoparticles with an iron-mimic peptide targeting transferrin/transferrin receptor complex. *Molecular Pharmaceutics*, *12*(8), 2947–2961.
Kawaguchi, A. T., Fukumoto, D., Haida, M., Ogata, Y., Yamano, M., & Tsukada, H. (2007). Liposome-encapsulated hemoglobin reduces the size of cerebral infarction in the rat: Evaluation with photochemically induced thrombosis of the middle cerebral artery. *Stroke*, *38*(5), 1626–1632.
Kernie, S. G., & Parent, J. M. (2010). Forebrain neurogenesis after focal ischemic and traumatic brain injury. *Neurobiology of Disease*, *37*(2), 267–274.
Kieran, M. W., Walker, D., Frappaz, D., & Prados, M. (2010). Brain tumors: From childhood through adolescence into adulthood. *Journal of Clinical Oncology*, *28*(32), 4783–4789.
Kim, S. S., Harford, J. B., Pirollo, K. F., & Chang, E. H. (2015). Effective treatment of glioblastoma requires crossing the blood–brain barrier and targeting tumors including cancer stem cells: The promise of nanomedicine. *Biochemical and Biophysical Research Communications*, *468*(3), 485–489.
Kim, C. K., Kim, T., Choi, I. Y., Soh, M., Kim, D., Kim, Y. J., et al. (2012). Ceria nanoparticles that can protect against ischemic stroke. *Angewandte Chemie (International Ed in English)*, *51*(44), 11039–11043.
Kim, I. D., Lim, C. M., Kim, J. B., Nam, H. Y., Nam, K., Kim, S. W., et al. (2010). Neuroprotection by biodegradable PAMAM ester (e-PAM-R)-mediated HMGB1 siRNA delivery in primary cortical cultures and in the postischemic brain. *Journal of Controlled Release*, *142*(3), 422–430.

Kim, S. S., Rait, A., Kim, E., Pirollo, K. F., Nishida, M., Farkas, N., et al. (2014). A nanoparticle carrying the p53 gene targets tumors including cancer stem cells, sensitizes glioblastoma to chemotherapy and improves survival. *ACS Nano*, *8*(6), 5494–5514.

Kinghorn, A. D., Altmann, K. H., Höfle, G., Müller, R., & Prantz, K. (2009). In A. D. Kinghorn, H. Falk, & J. Kobayashi (Eds.), *The epothilones: An outstanding family of anti-tumor agents: From soil to the clinic*: *Vol. 90* (pp. 1–234). Vienna: Springer Science & Business Media.

Knowles, T. P., Vendruscolo, M., & Dobson, C. M. (2014). The amyloid state and its association with protein misfolding diseases. *Nature Reviews Molecular Cell Biology*, *15*(6), 384–396.

Koga, T., & Saito, N. (2012). Efficacy and limitations of stereotactic radiosurgery in the treatment of glioblastoma. *Neurologia Medico-Chirurgica*, *52*(8), 548–552.

Kreuter, J., Hekmatara, T., Dreis, S., Vogel, T., Gelperina, S., & Langer, K. (2007). Covalent attachment of apolipoprotein AI and apolipoprotein B-100 to albumin nanoparticles enables drug transport into the brain. *Journal of Controlled Release*, *118*(1), 54–58.

Krupinski, J., Kaluza, J., Kumar, P., Kumar, S., & Wang, J. M. (1994). Role of angiogenesis in patients with cerebral ischemic stroke. *Stroke*, *25*(9), 1794–1798.

Kumar, A., & Singh, A. (2015). A review on Alzheimer's disease pathophysiology and its management: An update. *Pharmacological Reports*, *67*(2), 195–203.

Lacoeuille, F., Garcion, E., Benoit, J. P., & Lamprecht, A. (2007). Lipid nanocapsules for intracellular drug delivery of anticancer drugs. *Journal of Nanoscience and Nanotechnology*, 7(12), 4612–4617.

Lamprecht, A., & Benoit, J. P. (2006). Etoposide nanocarriers suppress glioma cell growth by intracellular drug delivery and simultaneous P-glycoprotein inhibition. *Journal of Controlled Release*, *112*(2), 208–213.

Lannfelt, L., Relkin, N. R., & Siemers, E. R. (2014). Amyloid-ß-directed immunotherapy for Alzheimer's disease. *Journal of Internal Medicine*, *275*(3), 284–295.

Laserra, S., Basit, A., Sozio, P., Marinelli, L., Fornasari, E., Cacciatore, I., et al. (2015). Solid lipid nanoparticles loaded with lipoyl–memantine codrug: Preparation and characterization. *International Journal of Pharmaceutics*, *485*(1), 183–191.

Le Droumaguet, B., Nicolas, J., Brambilla, D., Mura, S., Maksimenko, A., De Kimpe, L., et al. (2012). Versatile and efficient targeting using a single nanoparticulate platform: Application to cancer and Alzheimer's disease. *ACS Nano*, *6*(7), 5866–5879.

Lee, H. J., Park, J., Yoon, O. J., Kim, H. W., Lee, D. Y., Kim do, H., et al. (2011). Amine-modified single-walled carbon nanotubes protect neurons from injury in a rat stroke model. *Nature Nanotechnology*, *6*(2), 121–125.

Levites, Y., O'Nuallain, B., Puligedda, R. D., Ondrejcak, T., Adekar, S. P., Chen, C., et al. (2015). A human monoclonal IgG that binds Aβ assemblies and diverse amyloids exhibits anti-amyloid activities in vitro and in vivo. *The Journal of Neuroscience*, *35*(16), 6265–6276.

Li, M., Deng, H., Peng, H., & Wang, Q. (2014). Functional nanoparticles in targeting glioma diagnosis and therapies. *Journal of Nanoscience and Nanotechnology*, *14*(1), 415–432.

Lima, F. R., Kahn, S. A., Soletti, R. C., Biasoli, D., Alves, T., da Fonseca, A. C. C., et al. (2012). Glioblastoma: Therapeutic challenges, what lies ahead. *Biochimica et Biophysica Acta (BBA)-Reviews on Cancer*, *1826*(2), 338–349.

Liu, Y., An, S., Li, J., Kuang, Y., He, X., Guo, Y., et al. (2016). Brain-targeted co-delivery of therapeutic gene and peptide by multifunctional nanoparticles in Alzheimer's disease mice. *Biomaterials*, *80*, 33–45.

Liu, Z., Gao, X., Kang, T., Jiang, M., Miao, D., Gu, G., et al. (2013). B6 peptide-modified PEG-PLA nanoparticles for enhanced brain delivery of neuroprotective peptide. *Bioconjugate Chemistry*, *24*(6), 997–1007.

Lo, E. H., Dalkara, T., & Moskowitz, M. A. (2003). Mechanisms, challenges and opportunities in stroke. *Nature Reviews. Neuroscience*, *4*(5), 399–415.

Lollo, G., Vincent, M., Ullio-Gamboa, G., Lemaire, L., Franconi, F., Couez, D., et al. (2015). Development of multifunctional lipid nanocapsules for the co-delivery of paclitaxel and CpG-ODN in the treatment of glioblastoma. *International Journal of Pharmaceutics*, *495*(2), 972–980.

Lovblad, K. O., Baird, A. E., Schlaug, G., Benfield, A., Siewert, B., Voetsch, B., et al. (1997). Ischemic lesion volumes in acute stroke by diffusion-weighted magnetic resonance imaging correlate with clinical outcome. *Annals of Neurology*, *42*(2), 164–170.

Lu, Y. M., Huang, J. Y., Wang, H., Lou, X. F., Liao, M. H., Hong, L. J., et al. (2014). Targeted therapy of brain ischaemia using Fas ligand antibody conjugated PEG-lipid nanoparticles. *Biomaterials*, *35*(1), 530–537.

Mahon, B. R., Nesbit, G. M., Barnwell, S. L., Clark, W., Marotta, T. R., Weill, A., et al. (2003). North American clinical experience with the EKOS MicroLysUS infusion catheter for the treatment of embolic stroke. *AJNR. American Journal of Neuroradiology*, *24*(3), 534–538.

Malam, Y., Lim, E. J., & Seifalian, A. M. (2011). Current trends in the application of nanoparticles in drug delivery. *Current Medicinal Chemistry*, *18*(7), 1067–1078.

Mancini, S., Minniti, S., Gregori, M., Sancini, G., Cagnotto, A., Couraud, P. O., et al. (2016). The hunt for brain Aβ oligomers by peripherally circulating multi-functional nanoparticles: Potential therapeutic approach for Alzheimer's disease. *Nanomedicine: Nanotechnology, Biology and Medicine*, *12*(1), 43–52.

Maresova, P., Mohelska, H., Dolejs, J., & Kuca, K. (2015). Socio-economic aspects of Alzheimer's disease. *Current Alzheimer Research*, *12*(9), 903–911.

Mathew, A., Fukuda, T., Nagaoka, Y., Hasumura, T., Morimoto, H., Yoshida, Y., et al. (2012). Curcumin loaded-PLGA nanoparticles conjugated with Tet-1 peptide for potential use in Alzheimer's disease. *PloS One*, 7(3), e32616.

Matsunaga, S., Kishi, T., & Iwata, N. (2015). Memantine monotherapy for Alzheimer's disease: A systematic review and meta-analysis. *PloS One*, *10*(4), e0123289.

Mattsson, N., Zetterberg, H., Hansson, O., Andreasen, N., Parnetti, L., Jonsson, M., et al. (2009). CSF biomarkers and incipient Alzheimer disease in patients with mild cognitive impairment. *JAMA*, *302*(4), 385–393.

Messaoudi, K., Clavreul, A., Danhier, F., Saulnier, P., Benoit, J. P., & Lagarce, F. (2015). Combined silencing expression of MGMT with EGFR or galectin-1 enhances the sensitivity of glioblastoma to temozolomide. *European Journal of Nanomedicine*, 7(2), 97–107.

Mikulca, J. A., Nguyen, V., Gajdosik, D. A., Teklu, S. G., Giunta, E. A., Lessa, E. A., et al. (2014). Potential novel targets for Alzheimer pharmacotherapy: II. Update on secretase inhibitors and related approaches. *Journal of Clinical Pharmacy and Therapeutics*, *39*(1), 25–37.

Mittal, G., Carswell, H., Brett, R., Currie, S., & Kumar, M. R. (2011). Development and evaluation of polymer nanoparticles for oral delivery of estradiol to rat brain in a model of Alzheimer's pathology. *Journal of Controlled Release*, *150*(2), 220–228.

Moghimi, S. M., Hunter, A. C., & Murray, J. C. (2001). Long-circulating and target-specific nanoparticles: Theory to practice. *Pharmacological Reviews*, *53*(2), 283–318.

Mohamed, N. V., Herrou, T., Plouffe, V., Piperno, N., & Leclerc, N. (2013). Spreading of tau pathology in Alzheimer's disease by cell-to-cell transmission. *European Journal of Neuroscience*, *37*(12), 1939–1948.

Mrugala, M. M. (2013). Advances and challenges in the treatment of glioblastoma: A clinician's perspective. *Discovery Medicine*, *15*(83), 221–230.

Mura, S., Nicolas, J., & Couvreur, P. (2013). Stimuli-responsive nanocarriers for drug delivery. *Nature Materials*, *12*(11), 991–1003.

Muresan, V., & Muresan, Z. L. (2015). Amyloid-β precursor protein: Multiple fragments, numerous transport routes and mechanisms. *Experimental Cell Research*, *334*(1), 45–53.

Murphy, T. H., & Corbett, D. (2009). Plasticity during stroke recovery: From synapse to behaviour. *Nature Reviews Neuroscience*, *10*(12), 861–872.

Nag, S. (Ed.), (2011). Morphology and properties of brain endothelial cells. In The blood-brain and other neural barriers: reviews and protocols (pp. 3–47). Vienna: Springer.

Nair, S. B., Dileep, A., & Rajanikant, G. K. (2012). Nanotechnology based diagnostic and therapeutic strategies for neuroscience with special emphasis on ischemic stroke. *Current Medicinal Chemistry*, *19*(5), 744–756.

Nayak, L., Lee, E. Q., & Wen, P. Y. (2012). Epidemiology of brain metastases. *Current Oncology Reports*, *14*(1), 48–54.

Neely, A., Perry, C., Varisli, B., Singh, A. K., Arbneshi, T., Senapati, D., et al. (2009). Ultrasensitive and highly selective detection of Alzheimer's disease biomarker using two-photon Rayleigh scattering properties of gold nanoparticle. *ACS Nano*, *3*(9), 2834–2840.

O'Regan, J., Lanctôt, K. L., Mazereeuw, G., & Herrmann, N. (2015). Cholinesterase inhibitor discontinuation in patients with Alzheimer's disease: A meta-analysis of randomized controlled trials. *The Journal of Clinical Psychiatry*, *76*(11), 1–478.

Oberoi, R. K., Parrish, K. E., Sio, T. T., Mittapalli, R. K., Elmquist, W. F., & Sarkaria, J. N. (2016). Strategies to improve delivery of anticancer drugs across the blood–brain barrier to treat glioblastoma. *Neuro-Oncology*, *18*(1), 27–36.

Ostrom, Q. T., Bauchet, L., Davis, F. G., Deltour, I., Fisher, J. L., Langer, C. E., et al. (2014). The epidemiology of glioma in adults: A "state of the science" review. *Neuro-Oncology*, *16*, 896–913. pii: nou087.

Pagar, K. P., Sardar, S. M., & Vavia, P. R. (2014). Novel L-lactide-depsipeptide polymeric carrier for enhanced brain uptake of rivastigmine in treatment of Alzheimer's disease. *Journal of Biomedical Nanotechnology*, *10*(3), 415–426.

Park, J. K., Hodges, T., Arko, L., Shen, M., Iacono, D. D., McNabb, A., et al. (2010). Scale to predict survival after surgery for recurrent glioblastoma multiforme. *Journal of Clinical Oncology*, *28*(24), 3838–3843.

Paulsen, J. S., Nance, M., Kim, J. I., Carlozzi, N. E., Panegyres, P. K., Erwin, C., et al. (2013). A review of quality of life after predictive testing for and earlier identification of neurodegenerative diseases. *Progress in Neurobiology*, *110*, 2–28.

Pereira, A. C., Martin, P. J., & Warburton, E. A. (2001). Thrombolysis in acute ischaemic stroke. *Postgraduate Medical Journal*, 77(905), 166–171.

Porru, M., Zappavigna, S., Salzano, G., Luce, A., Stoppacciaro, A., Balestrieri, M. L., et al. (2014). Medical treatment of orthotopic glioblastoma with transferrin-conjugated nanoparticles encapsulating zoledronic acid. *Oncotarget*, *5*(21), 10446–10459.

Rabanel, J., Aoun, V., Elkin, I., Mokhtar, M., & Hildgen, P. (2012). Drug-loaded nanocarriers: Passive targeting and crossing of biological barriers. *Current Medicinal Chemistry*, *19*(19), 3070–3102.

Raval, N., Mistry, T., Acharya, N., & Acharya, S. (2015). Development of glutathione-conjugated asiatic acid-loaded bovine serum albumin nanoparticles for brain-targeted drug delivery. *Journal of Pharmacy and Pharmacology*, *67*(11), 1503–1511.

Reddy, M. K., & Labhasetwar, V. (2009). Nanoparticle-mediated delivery of superoxide dismutase to the brain: An effective strategy to reduce ischemia-reperfusion injury. *The FASEB Journal*, *23*(5), 1384–1395.

Reitz, C., & Mayeux, R. (2014). Alzheimer disease: Epidemiology, diagnostic criteria, risk factors and biomarkers. *Biochemical Pharmacology*, *88*(4), 640–651.

Rizzi, L., Rosset, I., & Roriz-Cruz, M. (2014). Global epidemiology of dementia: Alzheimer's and vascular types. *BioMed Research International*, *2014*, Article ID 908915, 8 pages.

Rowinska-Zyrek, M., Salerno, M., & Kozlowski, H. (2015). Neurodegenerative diseases—Understanding their molecular bases and progress in the development of potential treatments. *Coordination Chemistry Reviews*, *284*, 298–312.

Ryu, S., Buatti, J. M., Morris, A., Kalkanis, S. N., Ryken, T. C., & Olson, J. J. (2014). The role of radiotherapy in the management of progressive glioblastoma. *Journal of Neuro-Oncology*, *118*(3), 489–499.

Sawyer, A. J., Saucier-Sawyer, J. K., Booth, C. J., Liu, J., Patel, T., Piepmeier, J. M., et al. (2011). Convection-enhanced delivery of camptothecin-loaded polymer nanoparticles for treatment of intracranial tumors. *Drug Delivery and Translational Research*, *1*(1), 34–42.

Schneider, L. S., Mangialasche, F., Andreasen, N., Feldman, H., Giacobini, E., Jones, R., et al. (2014). Clinical trials and late-stage drug development for Alzheimer's disease: An appraisal from 1984 to 2014. *Journal of Internal Medicine*, *275*(3), 251–283.

Stegurová, L., Dráberová, E., Bartos, A., Dráber, P., Řípová, D., & Dráber, P. (2014). Gold nanoparticle-based immuno-PCR for detection of tau protein in cerebrospinal fluid. *Journal of Immunological Methods*, *406*, 137–142.

Stupp, R., Roila, F., & ESMO Guidelines Working Group. (2009). Malignant glioma: ESMO clinical recommendations for diagnosis, treatment and follow-up. *Annals of Oncology*, *20*(Suppl. 4), iv126–iv128.

Suk, J. S., Xu, Q., Kim, N., Hanes, J., & Ensign, L. M. (2015). PEGylation as a strategy for improving nanoparticle-based drug and gene delivery. *Advanced Drug Delivery Reviews*, *99*, 28–51.

Sundar, S. J., Hsieh, J. K., Manjila, S., Lathia, J. D., & Sloan, A. (2014). The role of cancer stem cells in glioblastoma. *Neurosurgical Focus*, *37*(6), E6.

Takamiya, M., Miyamoto, Y., Yamashita, T., Deguchi, K., Ohta, Y., & Abe, K. (2012). Strong neuroprotection with a novel platinum nanoparticle against ischemic stroke- and tissue plasminogen activator-related brain damages in mice. *Neuroscience*, *221*, 47–55.

Thakkar, J. P., Dolecek, T. A., Horbinski, C., Ostrom, Q. T., Lightner, D. D., Barnholtz-Sloan, J. S., et al. (2014). Epidemiologic and molecular prognostic review of glioblastoma. *Cancer Epidemiology, Biomarkers & Prevention*, *23*(10), 1985–1996.

Tiwari, S. K., Agarwal, S., Seth, B., Yadav, A., Nair, S., Bhatnagar, P., et al. (2013). Curcumin-loaded nanoparticles potently induce adult neurogenesis and reverse cognitive deficits in Alzheimer's disease model via canonical Wnt/β-catenin pathway. *ACS Nano*, *8*(1), 76–103.

Toledo, J. B., Shaw, L. M., & Trojanowski, J. Q. (2013). Plasma amyloid beta measurements—A desired but elusive Alzheimer's disease biomarker. *Alzheimer's Research & Therapy*, *5*(2), 8.

Ulbrich, K., Knobloch, T., & Kreuter, J. (2011). Targeting the insulin receptor: Nanoparticles for drug delivery across the blood–brain barrier (BBB). *Journal of Drug Targeting*, *19*(2), 125–132.

Van der Kant, R., & Goldstein, L. S. (2015). Cellular functions of the amyloid precursor protein from development to dementia. *Developmental Cell*, *32*(4), 502–515.

Van Tellingen, O., Yetkin-Arik, B., De Gooijer, M. C., Wesseling, P., Wurdinger, T., & de Vries, H. E. (2015). Overcoming the blood–brain tumor barrier for effective glioblastoma treatment. *Drug Resistance Updates*, *19*, 1–12.

Venur, V. A., Peereboom, D. M., & Ahluwalia, M. S. (2015). Current medical treatment of glioblastoma. In J. Raizer & A. Parsa (Eds.), *Current understanding and treatment of gliomas* (pp. 103–115). Switzerland: Springer International Publishing.

Verma, J., Lal, S., & Van Noorden, C. J. (2014). Nanoparticles for hyperthermic therapy: Synthesis strategies and applications in glioblastoma. *International Journal of Nanomedicine*, *9*, 2863–2877.

Wagner, S., Zensi, A., Wien, S. L., Tschickardt, S. E., Maier, W., Vogel, T., et al. (2012). Uptake mechanism of ApoE-modified nanoparticles on brain capillary endothelial cells as a blood-brain barrier model. *PloS One*, 7(3), e32568.

Wang, Y., Cooke, M. J., Sachewsky, N., Morshead, C. M., & Shoichet, M. S. (2013). Bioengineered sequential growth factor delivery stimulates brain tissue regeneration after stroke. *Journal of Controlled Release*, *172*(1), 1–11.

Wang, X., Huang, X., Yang, Z., Gallego-Perez, D., Ma, J., Zhao, X., et al. (2014). Targeted delivery of tumor suppressor microRNA-1 by transferrin-conjugated lipopolyplex nanoparticles to patient-derived glioblastoma stem cells. *Current Pharmaceutical Biotechnology*, *15*(9), 839–846.

Wang, J. D., Khafagy, E. S., Khanafer, K. M., Takayama, S., & El-Sayed, M. E. (2016). Organization of endothelial cells, pericytes, and astrocytes into a 3D microfluidic in vitro model of the blood-brain barrier. *Molecular Pharmaceutics*, *13*, 895–906.

Wang, A. Z., Langer, R., & Farokhzad, O. C. (2012). Nanoparticle delivery of cancer drugs. *Annual Review of Medicine*, *63*, 185–198.

Wardlaw, J. M., Lindley, R. I., & Lewis, S. (2002). Thrombolysis for acute ischemic stroke: Still a treatment for the few by the few. *The Western Journal of Medicine*, *176*(3), 198–199.

Wasiak, T., Ionov, M., Nieznanski, K., Nieznanska, H., Klementieva, O., Granell, M., et al. (2012). Phosphorus dendrimers affect Alzheimer's (Aβ1–28) peptide and MAP-Tau protein aggregation. *Molecular Pharmaceutics*, *9*(3), 458–469.

Wingerchuk, D. M., & Carter, J. L. (2014). Multiple sclerosis: Current and emerging disease-modifying therapies and treatment strategies. *Mayo Clinic Proceedings*, *89*(2), 225–240.

Wischik, C. M., Staff, R. T., Wischik, D. J., Bentham, P., Murray, A. D., Storey, J., et al. (2015). Tau aggregation inhibitor therapy: An exploratory phase 2 study in mild or moderate Alzheimer's disease. *Journal of Alzheimer's Disease*, *44*(2), 705–720.

Wu, T. C., & Grotta, J. C. (2013). Hypothermia for acute ischaemic stroke. *Lancet Neurology*, *12*(3), 275–284.

Xu, Y. Y., Gao, P., Sun, Y., & Duan, Y. R. (2015). Development of targeted therapies in treatment of glioblastoma. *Cancer Biology & Medicine*, *12*(3), 223.

Yang, W., Chen, Y. H., Liu, H., & Qu, H. D. (2015). Neuroprotective effects of piperine on the 1-methyl-4-phenyl-1, 2, 3, 6-tetrahydropyridine-induced Parkinson's disease mouse model. *International Journal of Molecular Medicine*, *36*(5), 1369–1376.

Yang, L. J., Zhou, C. F., & Lin, Z. X. (2014). Temozolomide and radiotherapy for newly diagnosed glioblastoma multiforme: A systematic review. *Cancer Investigation*, *32*(2), 31–36.

Yepes, M., Roussel, B. D., Ali, C., & Vivien, D. (2009). Tissue-type plasminogen activator in the ischemic brain: More than a thrombolytic. *Trends in Neurosciences*, *32*(1), 48–55.

Yu, Y., Sun, X., Tang, D., Li, C., Zhang, L., Nie, D., et al. (2015). Gelsolin bound β-amyloid peptides (1–40/1–42): Electrochemical evaluation of levels of soluble peptide associated with Alzheimer's disease. *Biosensors and Bioelectronics*, *68*, 115–121.

Yun, X., Maximov, V. D., Yu, J., Zhu, H., Vertegel, A. A., & Kindy, M. S. (2013). Nanoparticles for targeted delivery of antioxidant enzymes to the brain after cerebral ischemia and reperfusion injury. *Journal of Cerebral Blood Flow and Metabolism*, *33*(4), 583–592.

Zaleska, M. M., Mercado, M. L., Chavez, J., Feuerstein, G. Z., Pangalos, M. N., & Wood, A. (2009). The development of stroke therapeutics: Promising mechanisms and translational challenges. *Neuropharmacology*, *56*(2), 329–341.

Zhang, J. H., Badaut, J., Tang, J., Obenaus, A., Hartman, R., & Pearce, W. J. (2012). The vascular neural network—A new paradigm in stroke pathophysiology. *Nature Reviews. Neurology*, *8*(12), 711–716.

Zhang, D., Fa, H. B., Zhou, J. T., Li, S., Diao, X. W., & Yin, W. (2015). The detection of β-amyloid plaques in an Alzheimer's disease rat model with DDNP-SPIO. *Clinical Radiology*, *70*(1), 74–80.
Zhang, B., Sun, X., Mei, H., Wang, Y., Liao, Z., Chen, J., et al. (2013). LDLR-mediated peptide-22-conjugated nanoparticles for dual-targeting therapy of brain glioma. *Biomaterials*, *34*(36), 9171–9182.
Zhang, C., Wan, X., Zheng, X., Shao, X., Liu, Q., Zhang, Q., et al. (2014). Dual-functional nanoparticles targeting amyloid plaques in the brains of Alzheimer's disease mice. *Biomaterials*, *35*(1), 456–465.
Zhao, H., Bao, X. J., Wang, R. Z., Li, G. L., Gao, J., Ma, S. H., et al. (2011). Postacute ischemia vascular endothelial growth factor transfer by transferrin-targeted liposomes attenuates ischemic brain injury after experimental stroke in rats. *Human Gene Therapy*, *22*(2), 207–215.
Zhou, J., Patel, T. R., Sirianni, R. W., Strohbehn, G., Zheng, M. Q., Duong, N., et al. (2013). Highly penetrative, drug-loaded nanocarriers improve treatment of glioblastoma. *Proceedings of the National Academy of Sciences of the United States of America*, *110*(29), 11751–11756.
Zülch, K. J. (2012). *Brain tumors: Their biology and pathology*. New York: Springer.

CHAPTER FOUR

Carbohydrate Nanoparticles for Brain Delivery

A. Lalatsa[1], E. Barbu
School of Pharmacy and Biomedical Sciences, University of Portsmouth, Portsmouth, United Kingdom
[1]Corresponding author: e-mail address: katerina.lalatsa@port.ac.uk

Contents

Abstract

Many brain tumors and neurological diseases can greatly benefit from the use of emerging nanotechnologies based on targeted nanomedicines that are able to noninvasively transport highly potent and specific pharmaceuticals across the blood–brain barrier. Carbohydrates have received considerable interest as materials for drug carriers due to their natural origin and inherent biodegradability and biocompatibility, as well as due to their hydrophilic character and ease of chemical modification combined with low cost and the possibility for large-scale manufacturing. This chapter provides an overview of the latest research involving the use of carbohydrate-based nanoparticles for drug delivery to the central nervous system. After reviewing the challenges posed by delivering drugs into the brain, the current state-of-the-art approaches for delivery of actives across the blood–brain barrier, including invasive and noninvasive strategies, are presented. A particular focus has been placed on chitosan polymers as they are among the most promising carbohydrate nanocarriers for the preparation and testing of chitosan-based nanomedicines that led, in preclinical proof-of-concept studies, to enhanced brain drug levels and increased pharmacodynamics responses after intravenous, nasal, and oral administration. While chitosan nanoparticles are to date among the most studied and most promising carriers, approaches based on other polysaccharides such as dextran, pullulan, and cellulose warrant further research in the attempt to advance the existing technologies for overcoming the blood–brain barrier.

International Review of Neurobiology, Volume 130
ISSN 0074-7742
http://dx.doi.org/10.1016/bs.irn.2016.05.004

1. DRUG DELIVERY TO THE BRAIN

Neurological diseases such as brain cancers (Lalatsa, Leite, & Pilkington, 2015), neurodegenerative conditions, infections, pain, and psychiatric disorders are a leading cause of disability, morbidity, and mortality (Lalatsa, Schatzlein, & Uchegbu, 2012). More than 1.5 billion people worldwide are affected by central nervous system (CNS) disorders, with the number predicted to rise to 1.9 billion by 2020 (Pardridge, 2010; Whiteford, Ferrari, Degenhardt, Feigin, & Vos, 2015). Drug delivery to the brain remains challenging due to the blood–brain barrier (BBB), which remains a bottleneck in the development of CNS therapies (Pardridge, 2005) and leads to long drug development times and a high failure rate in clinical trials (Kola & Landis, 2004).

1.1 The BBB

The BBB was originally identified by Paul Ehrlich's experimental observation that most peripheral organs were stained after intravenous injection of a water-soluble dye, with the exception of the brain and the spinal cord (Ehrlich, 1885). Follow-up experiments by Edwin Goldman confirmed the presence of the BBB by injection of trypan blue directly into the cerebrospinal fluid, which resulted in staining deeply all brain cell types but left peripheral organs unstained (Goldmann, 1913). The hypothesis that the cerebral capillaries provide the anatomical basis for a physiological barrier between the brain and the remaining body led to the important concept of a barrier system protecting the brain from circulating toxins or infectious agents, which was later termed as the "BBB" by Lina Stern in 1921 (Bradbury, 1979).

The BBB is formed at the level of the endothelial cells of the cerebral capillaries and is comprised of a series of two plasma membranes (namely, the luminal and the abluminal membranes of the brain capillary endothelium) that are separated by about 0.3 μm of endothelial cytosol (Pardridge, 1991). The capillary walls act as a continuous lipid bilayer and prevent the entry of microscopic, large polar, or lipid-insoluble molecules into the brain, except for the exchange of gasses. Anatomically, the endothelial cells of the BBB are distinguished from those in the periphery by increased mitochondrial content, a lack of fenestrations, minimal pinocytotic activity, and the presence of tight junctions (Lalatsa, Schatzlein, & Uchegbu, 2012). These tight junctions between the capillary cells are produced by the interaction of several transmembrane proteins that project into and seal the paracellular pathway resulting in the BBB's impermeability and a high transepithelial electrical resistance in the region of 1500–2000 Ω cm^2. Pericytes and endothelial cells are sheathed by the basal lamina, a membrane 30–40-nm thick which is continuous with the plasma membranes of astrocytic end-foot processes, which sheath the cerebral capillaries (Hawkins & Davis, 2005). The plasma bilayer is high in cholesterol, which allows for a high packing density of membrane components; therefore, the resistance to passive diffusion of large lipophilic molecules is greater compared to that of noncerebral capillaries (Schirmacher et al., 2000). As well as preventing the entry of solutes into the brain from the blood, the BBB does not allow free diffusional movement of solutes out of the CNS either, hindering free diffusion in a bidirectional way and restricting movement to transcellular transport (passive diffusion, carrier transport, or endocytotic processes). Despite human brain capillaries having an estimated total length of almost 650 km and a total surface are of about 12 m^2 (Misra, Ganesh, Shahiwala, & Shah, 2003), the BBB is very efficient in making the brain practically inaccessible to polar molecules. As a result, the endothelial cells are required to maintain a high level of expression of transport proteins for essential polar metabolites such as glucose and amino acids in order to facilitate their entry into the brain (Lalatsa, Schatzlein, & Uchegbu, 2012).

The diffusion of drugs from the blood into the brain depends mainly upon the ability of the biologically active molecule to traverse lipid membranes. Important factors in drug transport across the BBB include the concentration gradient of drug/polymer, the flexibility and conformation of drug/polymer, the molecular weight ($<$500 Da), the molecular or surface charge, the lipophilicity of the drug (polar surface area below 60–70 Å; Kelder, Grootenhuis, Bayada, Delbressine, & Ploemen, 1999; Lennernas & Lundgren, 2004), possible sequestration by other cells, affinity for efflux proteins, affinity for receptors or carriers, pathological status, cerebral blood flow, systemic enzymatic stability, metabolism by other tissues, clearance rate of drug/polymer, and cellular enzymatic stability (Table 1). Several strategies to overcome the BBB, having different degrees of invasiveness, are under development.

1.2 Invasive Strategies for Drug Delivery to the Brain

1.2.1 Disruption of the BBB

Drug delivery across the BBB may be achieved by its temporary disruption using hypertonic media, surface-active agents, or other chemicals, as well as by using ultrasound.

Table 1 Factors Affecting the Drug Delivery to CNS (Lalatsa, Schatzlein, & Uchegbu, 2011)

Compound Properties	BBB Characteristics	Pharmacokinetic Parameters
Molecular weight	Membrane charge	Protein binding
Ionization	Carrier systems	Brain tissue binding
Lipophilicity	Direction of carrier systems	Volume of distribution
Hydrogen bonding	Receptors	Metabolic stability
Aggregation	Lipid composition	Clearance rate
Concentration gradient	Cerebral blood flow	Bioavailability
Amino acid composition (peptide/proteins)	Pathology	
Conformation		
Flexibility		
Folding		
Efflux proteins substrate		
Receptor/carrier affinity		
Intracellular enzymatic stability		

The opening of the BBB clinically by an intracarotid injection of hypertonic media (usually a 25% mannitol solution) results in the transient disruption of the BBB, which enables drug delivery for a limited period of time; however, this is not suitable for chronic use as the repetitive opening of the BBB with hypertonic media has been associated with a high incidence of seizures as well as with chronic neuropathologic changes in the brain. The hypertonic solution is thought to osmotically withdraw water from the brain vascular endothelial cells, causing cell shrinkage (Fortin, Gendron, Boudrias, & Garant, 2007), which in turn results in the disengagement of the extracellular domains of the proteins forming and regulating the tight junctions.

Surface-active agents also increase permeability across the BBB. Coadministration of polysorbate 80 with kyotorphin (an analgesic oligopeptide that does not cross the BBB) results in analgesia in mice, possibly due to a membrane-destabilizing effect caused by the detergent, resulting in BBB disruption and drug permeation (Sakane et al., 1989). Other agents that have been used to chemically disrupt the BBB are bradykinin (acting by activation of bradykinin B2 receptors through a calcium-mediated mechanism) (Borlongan & Emerich, 2003). Coadministration of RMP-7 (Cereport©, a synthetic bradykinin analog) with carboplatin resulted in successful reduction in tumor volume and stabilization of tumor growth in a rat model, although the approach was abandoned due to a lack of efficacy in phase II and III studies. Additionally, intracarotid administration of short-chain alkylglycerols has been shown to shortly (30 min) and reversibly modulate the BBB tight junctions allowing the delivery of small- and large-molecular-weight hydrophilic drugs in healthy and glioblastoma animal brains (Erdlenbruch et al., 2003, 2005; Hulper et al., 2013).

Ultrasound has also been shown to be capable of BBB disruption, though the usefulness of this method is limited by poor reliability and potential harmful effects. Local disruption of the BBB with ultrasound is associated with focal damage or transient alterations of neuronal function (Jalali, Huang, Dumont, & Hynynen, 2010). Recently, a novel method combining pulsed sonication with a preceding intravenous injection of an ultrasound contrast agent in the form of 2–6 μm microbubbles increased the distribution of Herceptin (trastuzumab) in brain tissue (Hynynen, 2008; Kinoshita, McDannold, Jolesz, & Hynynen, 2006); however, the actual mechanism of BBB disruption is not clear. Ultrasound waves are known to cause the microbubbles to expand and contract in the capillaries, resulting in stretching of the capillary walls that leads to tight junction opening. Microbubble expansion and contraction could also lead to local transient ischemia, triggering BBB opening; alternatively, the collapse of the microbubbles during sonication could cause mechanical opening of the BBB.

The temporary disruption of the BBB is associated with a low therapeutic efficiency and substantial side effects, while having the potential of being highly traumatic. The opening of the tight junctions of the BBB permits many undesirable substances, even neurotoxic by-products of normal metabolic processes, to cross the BBB and cause irreversible neuronal damage.

1.2.2 Convection-Enhanced Delivery

In convection-enhanced delivery (CED), a cannula is stereotactically introduced directly into the CNS interstitial spaces with its tip at the target site.

Controlled infusions induce bulk flows, which distribute infusate over large volumes within the extracellular space and are capable of achieving multi-centimeter depths of penetration (Lonser, Sarntinoranont, Morrison, & Oldfield, 2015). The infusion is continued for several days and the catheters are removed at the bedside. Drug transport depends on bulk flow, as well as on diffusion and clearance losses. The minimization of the cannula tip diameter, which reduces tissue expansion, is a way to minimize back leakage during infusion and has been demonstrated by mathematical modeling and animal data (Morrison, Chen, Chadwick, Lonser, & Oldfield, 1999). Unlike diffusion-driven distribution, which results in limited tissue penetration from the diffusive surface (1–2 mm) and a steep reduction in concentration over that distance (250–1000-fold decrease), bulk flow of the infusate during CED permits homogeneous distribution of drug through a perfused tissue region following a "square-shaped" distribution pattern in gray and white matter consistent with a uniform high concentration of infusate over a perfused CNS region (Lonser et al., 2015). CED has been shown to deliver high-molecular-weight proteins 2 cm from the injection site in the brain parenchyma after as little as 2 h of continuous infusion (Bobo et al., 1994). Although targeted drug delivery can be achieved via CED, success of delivery depends critically on precise placement of catheters as some areas of the brain and particularly infiltrated tissues surrounding a cavity are difficult to fully saturate with the infusate (Vandergrift, Patel, Nicholas, & Varma, 2006).

1.2.3 Direct Injection (Intraventricular and Intracerebral) and the Use of Implants

The bolus injection of chemotherapy and the placement of biodegradable, chemotherapeutic (carmustine, paclitaxel) impregnated wafers into a tumor resection cavity rely both on diffusion to drive the drug into the infiltrated brain. Relevant drug concentration is reached though only within the first 2 mm from the polymer implant, and drug concentration was shown to fall exponentially with distance (Fung et al., 1998). Drug delivery directly to the brain interstitium using polymeric devices implanted intracranially leads to high levels of active present at the target over a sustained period of time. This has been illustrated for stereotactically implanted biodegradable microspheres (20–200 μm) made from poly(lactide-*co*-glycolide) (PLGA) (Benoit, Faisant, Venier-Julienne, & Menei, 2000). Unfortunately, invasive techniques have been associated with increased risk of infection and high neurosurgical costs.

1.3 Noninvasive Strategies for Drug Delivery to the Brain

1.3.1 Chemical Modification of Drugs

The most common modification involves the enhancement of lipophilicity, typically by grafting lipidic moieties or by conjugation to a lipid-soluble carrier (Lalatsa et al., 2011). Amidation or esterification of amino, carboxyl, and hydroxyl groups increases the drug lipid solubility and enhances the brain uptake. In some cases the lipidized drug acts as a prodrug (Lalatsa, Lee, et al., 2012) that is converted into the parent molecule via enzymatic cleavage, ideally at the target site (eg, in the brain) (Serrano Lopez & Lalatsa, 2013). Apart from small-molecular-weight drugs, this strategy has been successfully employed with peptides (Lalatsa, Schatzlein, & Uchegbu, 2014).

1.3.2 Inhibition of Adenosine Triphosphate-Binding Cassette Efflux Transporters

Coadministration of an efflux transporter inhibitor, or drug analogs that are not substrates for efflux transporters, has been used as strategies to overcome drug efflux across the BBB. Coadministration of the *p*-glycoprotein inhibitor valspodar (SDZ PSC-833) with paclitaxel resulted in reductions in tumor volume in mice after oral administration (Fellner et al., 2002). However, the long-term safety effects are unknown as toxic substrates may also gain entry by *p*-glycoprotein inhibition (Fellner et al., 2002). Polymer conjugation is another way of generating molecules with low affinity for the efflux transporters and is gaining popularity; PEGylation of the opioid peptide DPDPE with a 2-kDa PEG (poly(ethylene glycol)) moiety on the terminal tyrosine decreased clearance, reduced plasma protein binding, first-pass metabolism, and *p*-glycoprotein binding, while it enhanced the analgesic effect after intravenous administration (Witt, Gillespie, Huber, Egleton, & Davis, 2001). The main disadvantage of PEGylation is that the drugs are made more hydrophilic and larger in size, which can in turn increase the delivery challenge. Furthermore, there is a potential risk of activity loss associated with an improper choice of PEG (length, branching), linker, and/or attachment site (Lalatsa, Schatzlein, & Uchegbu, 2012).

1.3.3 Carrier-Mediated Transport

Various nutrient and hormone transport proteins or carrier systems at the BBB level have been exploited for brain delivery. An excellent example is the case of levodopa, which utilizes the large neutral amino acid carrier system for the treatment of Parkinson's disease; the presence of the carboxyl and α-amino groups of the lipid-insoluble precursor of dopamine enables its interaction with the carrier protein and therefore its permeation across the BBB.

Some carriers are however very selective in their stereochemical substrate requirements; an example of this kind is the GLUT-1 carrier, as only molecules that closely mimic D-glucose are transported. Glycosylated analogs of methionine^5-enkephalin are transported via the hexose transporter GLUT-1 (Polt et al., 1994), and endocytotic mechanisms have also been shown to be involved in the transport of glycosylated peptides. There are a number of examples of analgesia being enhanced following glycosylation of some peptide opioid agonists (eg, glycosylated deltorphin and glycosylated leucine^5-enkephalin amide, as well as the L-serynyl-β-D-glucoside analogs of methionine^5-enkephalin); this is thought to be due to an increase in metabolic stability, reduced clearance, and improved BBB transport (Lalatsa et al., 2011).

Though not properly exploited yet for nanoparticulate delivery across the BBB, the hexose, and the large neutral amino acid carriers have a high drug capacity and presently are considered as very good candidates for the delivery of substrates to the brain. However, the success of such a strategy depends on a well-informed selection of both the drug/peptide and the transporter (Lalatsa, Schatzlein, & Uchegbu, 2012).

Glutathione (GSH) has been used as a vector for both simple drugs and nanoparticulate carrier systems (Gaillard, 2010); GSH-tagged PEGylated liposomes have been shown to deliver antiviral drugs to the brain via the GSH transporter (Gaillard, 2008), and GSH-coated PLGA nanoparticles of paclitaxel have been studied for the treatment of brain cancers (Geldenhuys, Mbimba, Bui, Harrison, & Sutariya, 2011).

1.3.4 Receptor-Mediated Transport

Endocytosis remains the main route of cellular entry for large-molecular-weight compounds including peptides, proteins, and antibodies, as well as for nanoparticulate carriers that can be loaded with a wide range of drugs. Binding of the ligand to its specific membrane receptor on the cell surface induces a modification of the receptor protein and triggers the formation of invaginations that can be clathrin coated (coated pits that can be visualized using electron microscopy). These caveoli trigger the formation of endocytotic vesicles that may fuse with an endosome (prelysosomal compartment with an acidic pH), followed by the dissociation of the ligand from the receptor, which allows the free receptor to be recycled back to the cell surface. Receptors that have been studied in the transport of large biomacromolecules or nanoparticles include those that are specific for transferrin, melanotransferrin, insulin, epidermal growth factor, leptin, immunoglobulin G, cytokines (such as interleukins as IL1a, IL1b, IL6), nicotinic acetylcholine (nAchR), diphtheria toxin, and low-density lipoprotein (LDL), all previously reviewed (Lalatsa, Schatzlein, & Uchegbu, 2012). The LDL receptor is a high-capacity receptor that is receiving significant preclinical interest for the delivery of nanomedicines that can be readily functionalized with specific ligands (eg, LDL 1 and 2, lactoferrin, apolipoprotein E, or designed peptide ligands such as Angiopeps (Angiopep-2) (TFFYGGSRGKRNNFKTEEY), positive charge (+2); designed by Angiochem Inc.) (Lalatsa, Schatzlein, & Uchegbu, 2012; Lalatsa et al., 2014).

1.3.5 Nanocarrier-Mediated Delivery

Nanoparticulate delivery involving liposomes, solid lipid nanoparticles, nanocapsules, nanospheres, polymeric nanoparticles, dendrimers, or nanoemulsions has received considerable interest; drug transport across the BBB can be achieved in these cases either via passive or active endocytotic mechanisms involving also carrier- or receptor-mediated processes (Chen & Liu, 2012).

Ideally, nanoparticles need to be nontoxic, biodegradable, and biocompatible, with a particle diameter preferably between 50 and 100 nm; they need to be stable in the blood and to avoid opsonization in order to ensure a long circulation half-life. Ability to easily functionalize nanoparticles with ligands for carrier- or receptor-mediated endocytosis is desirable as it can allow the uptake across the BBB as well as targeting specific receptors within the brain parenchyma or brain tumor. The possibility to tailor the release profile of the cargo, scalability, and cost-effective manufacturing are also desirable requirements.

Several mechanisms have been proposed for the ability of nanoparticles to cross the BBB. These involve: (i) an increased retention of the nanoparticles in the brain blood capillaries via adsorption to the capillary walls, which creates a concentration gradient that enhances transport across the endothelial cell layer and therefore achieves delivery into the brain (Lalatsa, Garrett, et al., 2012; Lalatsa, Schatzlein, et al., 2015); (ii) a general toxic effect on the brain microvasculature that leads to the permeabilization of the brain blood vessel endothelial cells; (iii) a general surfactant effect characterized by the solubilization of the endothelial cell membrane lipids that leads to membrane fluidization and an enhanced drug permeability through the BBB; (iv) opening of the tight junctions between the endothelial cells enables the drug to permeate through, either in free form or nanoparticle bound; (v) endocytosis by the endothelial cells, followed by drug release within the cells and subsequent delivery to the brain; (vi) transcytosis of the drug-bound nanoparticles through the endothelial cells; and (vii) a possible combination of the above mechanisms (Kreuter, 2013).

1.4 Carbohydrate Nanoparticles for Brain Delivery

1.4.1 Carbohydrates: General Considerations

Carbohydrates are the third most abundant group of organic compounds in the human body after proteins and lipids. Most carbohydrates are found naturally as polysaccharides and comprise of more than 10 monosaccharide units interlinked via a glycosidic bond. Two of the most abundant natural homopolysaccharides, cellulose and chitin, function as structural elements in plant cell walls and in the exoskeleton of arthropods (eg, insects, lobsters, crabs). Heteropolysaccharides such as hyaluronates, keratan sulfate, and chondroitin sulfate ensure extracellular support for bacteria and mammals, providing protection, support, and shape to cells, tissues, and organs (Glasser, 2008). Among all natural polymers, polysaccharides are the main focus in drug delivery due to their biocompatibility, biodegradability, hydrophilicity, stability, safety, and adhesive properties (Liu, Jiao, Wang, Zhou, & Zhang, 2008). Polysaccharides can have multiple roles in nanoparticle formulations, from rheology modifiers and emulsion stabilizers to surface modifiers. The variety of the functional groups present in polysaccharides allows for facile chemical modifications, which can result in a diversity of structures able to finely tune the properties of nanoparticles prepared from chemically modified carbohydrates (Fig. 1). Chemical modifications such as carboxymethylation and grafting (with a variety of pendent chains) are among the most common modifications employed for optimizing the required properties.

Carbohydrates can also act as polyelectrolytes, having either positive (eg, chitosan) or negative charges (alginate, heparin, hyaluronic acid, pectin, etc.; Liu et al., 2008). Additionally, carbohydrates are substrates for cell surface receptors, which can facilitate the design of targeted nanoparticulate delivery systems (eg, mannose–lectin-targeted nanoparticles for immunity; Carrillo-Conde et al., 2011). The presence of hydrophilic groups (hydroxyl, carboxylic, amino) favors interactions with biological

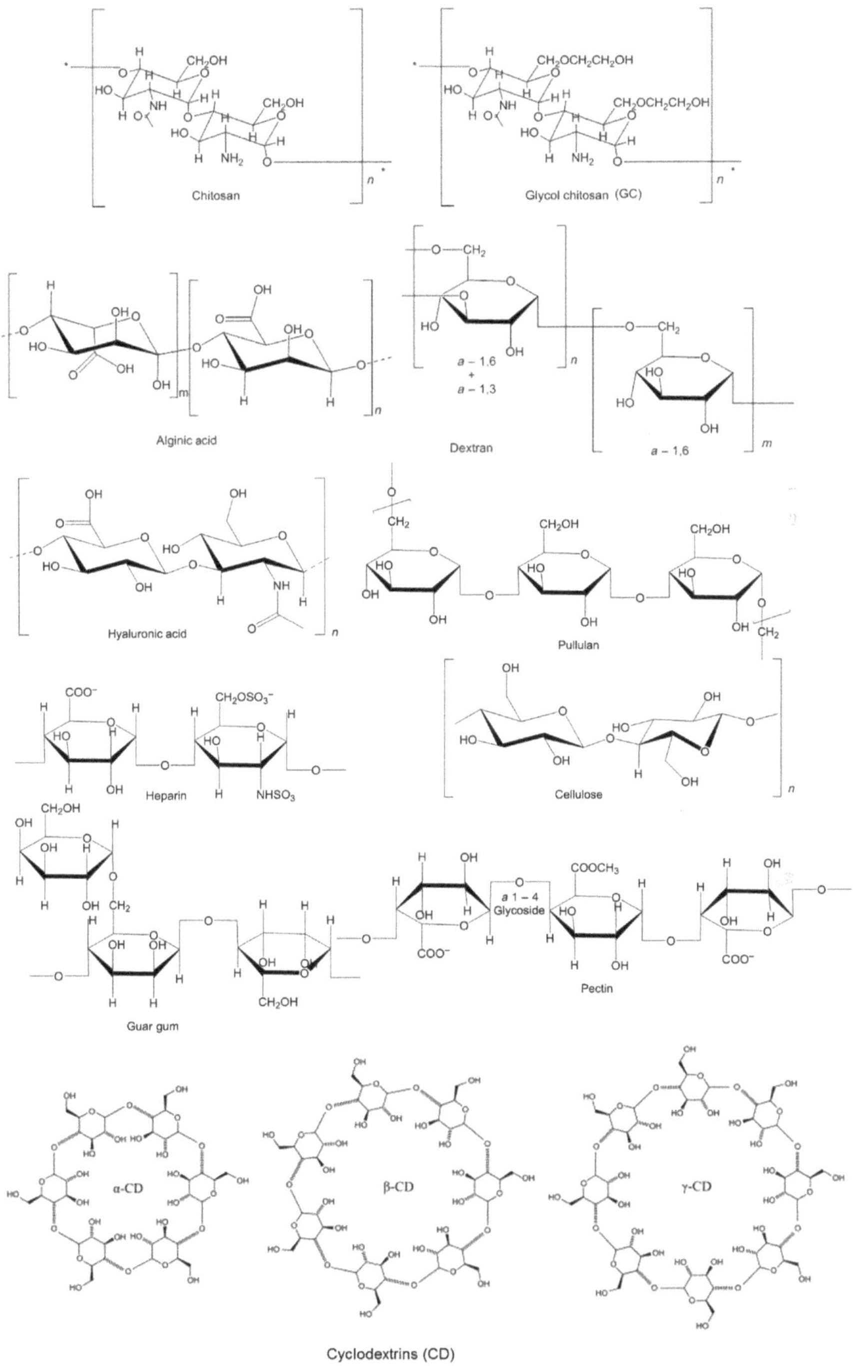

Fig. 1 Structures of polysaccharides and cyclodextrins commonly used in drug delivery.

targets and bioadhesion to biological tissue via noncovalent bonds (Lee, Park, & Robinson, 2000).

1.4.2 Methods of Preparation of Carbohydrate Nanoparticles

Gelation of emulsion droplets, covalent or ionic cross-linking, self-assembly, nanoprecipitation, or a combination of the above are typical methods used for preparation of carbohydrate nanoparticles. Many polysaccharides are easily dissolved in water, and their ability to thicken depending on concentration makes them ideal for the preparation of nanoparticles using the gelation method. Although particular methods depend on the type of polysaccharide, generally an insoluble gelling agent is dispersed into an aqueous polysaccharide solution, before emulsification with an oil phase that contains an appropriate concentration of surfactant that can ensure the formation of spherical particulates of appropriate size (30–300 nm). This method has been used successfully for the preparation of chitosan (Yoksan, Jirawutthiwongchai, & Arpo, 2010), alginate (Hosseini et al., 2013), and dextran (Reis, Ribeiro, Houng, Veiga, & Neufeld, 2007) nanoparticles.

The formation of chemical links between polysaccharidic chains in order to form a three-dimensional network of connected macromolecules prevents dissolution in water. The cross-linking density, in combination with the molecular structure of the polysaccharide, can alter the swelling characteristics of the network and therefore control the release of the loaded (either coated or encapsulated) drug. Cross-linking can be achieved by either chemical bonds (via addition or condensation reactions) or physical interaction between the polysaccharidic chains. Chemical cross-linking agents can be either homo- or heterobifunctional and lead to the formation of inter- and intramolecular bonds; usually, glutaraldehyde, dialdehydes, or other bifunctional reagents such as genipin or epichlorohydrin are commonly used (Liu et al., 2008). The reaction usually takes place under heterogeneous conditions, frequently employing adequate surfactants at a set ratio of reactants.

When ionic cross-linking or ionic gelation is used to prepare nanoparticles, the procedure commonly requires only an aqueous environment, under mild conditions (room temperature, neutral pH). Ionic cross-linking involves electrostatic interactions between charged polysaccharides and selected counter ions. Anionic cross-linkers such as sodium sulfate and sodium tripolyphosphate are typically used with chitosan (Csaba, Koping-Hoggard, & Alonso, 2009; Li, Guo, Wen, & Zhang, 2013), while cationic cross-linkers such as calcium, potassium, or barium salts (carbonates, chlorides, etc.) are used with alginates (Jay & Saltzman, 2009).

Self-assembly has also been utilized for the preparation of carbohydrate nanoparticles by either formation of polyelectrolyte complexes between oppositely charged polymers or use of polysaccharides grafted with lipid moieties imparting amphiphilic character; in both cases, core–shell nanoparticles can be formed. The stability of polyelectrolyte complexes is dependent on a number of factors, including: the degree of ionization of the polymers, their charge density, concentration, molar ratio, the nature and position of the ionic groups on the polymeric chains, the polymer chain flexibility, the molecular weight, temperature and the ionic strength, and pH of the medium (Mihai et al., 2011). For amphiphilic polymers, the self-assembly process is driven by the entropy gain associated with the dehydration of the hydrophobic moieties and the resultant hydrophobic associations (Lalatsa, Schatzlein, Mazza, Le, & Uchegbu, 2012). The morphology of the resulting nanoparticles is defined by the critical packing parameter and thus largely depends on the molecular weight, the molecular volume and size of the hydrophobic moiety grafted, as well as on the hydrophilic head group area, while the stability of nanoparticles depends on their critical aggregate concentration (Lalatsa, Schatzlein, Mazza, et al., 2012). Thus, the stability of polymer self-assembled nanoparticles upon dilution with biological fluids is a critical parameter for their success as drug delivery systems. Stable nanoparticles from chitosan (Lalatsa, Garrett, et al., 2012; Lalatsa, Lee, et al., 2012; Uchegbu et al., 2001), alginate (Li, Jia, Cheng, Pan, & Jiang, 2011; Simsek-Ege, Bond, & Stringer, 2003), or cellulose (Zhu, Lin, & Yao, 2010) have been reported.

Nanoprecipitation of a polymer from an aqueous solution following interaction with a water-miscible nonsolvent has also been exploited for the preparation of carbohydrate nanoparticles. The fast diffusion of the aqueous polymer solution in the nonsolvent results in the instantaneous formation of nanoparticles, without the need for any energy input (such as from milling, sonication, or high-pressure homogenization). This methods have been used for dextrins (Kaewprapan, Inprakhon, Marie, & Durand, 2012; Toman et al., 2015), cellulose (Geissler, Biesalski, Heinze, & Zhang, 2014), and cyclodextrins (CDs) (Cirpanli et al., 2011). The particle size and shape can be varied; however, complete removal of the cosolvent is critical for their safe use in vivo (Schubert, Delaney, & Schubert, 2011).

1.4.3 Preclinical Studies of Carbohydrate Nanoparticles

Targeted nanoparticle-based technologies are the only technologies to date to have shown promise in the delivery of actives across the BBB. Natural polymers with high biocompatibility and biodegradability such as carbohydrates and peptides/proteins are ideal biomaterials for the development of nanomedicines aimed at treating neurological disorders. Carbohydrate nanoparticles have been proven to exhibit low protein binding and limited uptake by the mononuclear phagocyte system and thus are ideal carriers for intravenous delivery (Lalatsa, Garrett, et al., 2012; Lalatsa, Schatzlein, et al., 2015; Osterberg et al., 1995). Nanoparticles with sizes between 15 and 100 nm possess a long circulation time compared to smaller particles (10–20 nm; that are rapidly filtered via the kidneys) or larger ones (>150 nm; that are uptaken by the reticuloendothelial system). When the particle size of carbohydrate nanoparticles is above 50 nm, they will also escape renal clearance (Jiang, Kim, Rutka, & Chan, 2008; Nance et al., 2012; Win & Feng, 2005). Even nanoparticles of optimal size will eventually be uptaken by the liver, but prior to that they would have had a long circulation half-life (2–40 h) (Owens & Peppas, 2006). Additionally, the recognition of certain saccharides by cell surface receptors can enable the targeting of carbohydrate nanoparticles to specific cell types; a typical example is that of hyaluronic acid, which specifically binds cluster determinant 44 (CD44) receptors—expressed at low levels in haematopoietic, epithelial, and neuronal cells, but at significantly higher levels in tumor cells such brain and lung tumors, melanomas, or lymphomas (Ganau, 2014).

Among the carbohydrates intensively researched for drug delivery to the brain are chitosan (Lalatsa, Lee, et al., 2012), alginates (Haque, Md, Sahni, Ali, & Baboota, 2014), dextrans (Kaewprapan et al., 2012; Toman et al., 2015), pullulans (Singh, Kaur, & Kennedy, 2015), hyaluronic acid (Mittapalli et al., 2013), heparin (Wang, Kouyoumdjian, Zhu, & Huang, 2015), cellulose (Dong, Cho, Lee, & Roman, 2014), and CDs (Tao et al., 2013).

The majority of proposed carbohydrate-based nanomedicines with in vivo proof of concept (Table 2) were developed from chitosan polymers. Chitosan comprises of linear 1-4-linked 2-acetamido-2-deoxy-β-D-glucopyranose (GlcNAc) and 2-amino-β-D-glucopyranose (GlcN) units derived from chitin (main component of crab shells, shrimp, and krill) by alkaline N-deacetylation (Kas, 1997). If the degree of deacetylation is above 50%, polymers become soluble in acidic solutions (Ledung, Milas, Rinaudo, & Desbrieres, 1994) and are known as chitosans. While chitosan has demonstrated biocompatibility, its clinical translation is still limited by its poor water solubility at physiological pH (Knight, Shapka, & Amsden, 2007). On the other hand, glycol chitosan (GC, 4–210 kDa) (Lalatsa,

Table 2 Summary of Noninvasive Carbohydrate-Based Nanomedicines Investigated in Preclinical Studies for Drug Delivery to the Brain

Polymer/Formulation Method	Drug/Dose	Route	Size (Zeta Potential)/ Morphology	Findings	References
Chitosan/glycol chitosan (cationic)					
Glycol chitosan—5β-cholanic acid$_{(150)}$–Cys 5.5$_{(4.8)}$ (Mw: 306 kDa) nanoparticles, self-assembly	Cys 5.5 (5 mg/mL of conjugate)	IV	264 ± 30 nm, quasispherical	U87 (5×10^5 cells/5 μL) brain and flank tumors in mice: after 21 days both brain and flank tumors showed strong NIRF signals—permeation linked to BBB partial opening due to the tumor and particle accumulation to the tumor site, explained by the angiogenic vasculature-specific targeting of nanoparticles	Na et al. (2011)
Quaternary ammonium palmitoyl glycol chitosan (GCPQ) (Mw: 10 kDa) self-assembly	Lomustine (L) (1.04 mg/mL)/13 mg/kg	IV	336 ± 0.44 nm, spherical	In vivo IV administration in CD-1 mice resulted in a twofold enhancement in brain levels and reduced liver and bone marrow levels. The L-GCPQ formulation significantly increased mice (U-87 MG tumors) survival times	Fisusi et al. (2016)

Continued

Table 2 Summary of Noninvasive Carbohydrate-Based Nanomedicines Investigated in Preclinical Studies for Drug Delivery to the Brain—cont'd

Polymer/Formulation Method	Drug/Dose	Route	Size (Zeta Potential)/ Morphology	Findings	References
				(L-GCPQ, ethanolic lomustine, and untreated mean survival times of 33.2, 22.5, and 21.3 days, respectively)	
SLN (PA:Chol:TPGS: Curcumin 20:2:6:1, w/w): *N*-trimethyl chitosan (Mw: 15 kDa, DQ: 48%) (50:1, w/w), complexation, self-assembly	Curcumin/50 mg/kg	PO	412 ± 80 nm (35.7 ± 1.03 mV), quasispherical	Enhanced plasma AUC_{0-8} (6.23 ± 0.75 vs 0.24 ± 0.05 h μg/mL) and enhanced brain levels in both brain homogenate and supernatant fractions	Ramalingam and Ko (2015)
Chitosan (Mw: 110 kDa): TPP (0.20%: 0.07%, ie, 2.7:1, w/w) nanoparticles Ionic gelation	Dopamine 4.05 ± 0.35 mg/mL/ 6–12 mg/kg	IP	110 ± 3 nm (27 ± 1 mV), quasispherical	In vivo microdialysis in male Wistar rats: after 4 h of a 12 mg/kg single injection, basal extracellular concentrations of dopamine (440 ± 62 pg/mL) were detected (439 % increase of baseline striatal levels). Increase was dose dependent	De Giglio, Trapani, Cafagna, Sabbatini, and Cometa (2011) and Trapani et al. (2011)

Chitosan (medium Mw, DA: 85%): TPP (3.3:1, w/w)	Bromocriptine (0.5 mg/mL)/2 mg/kg	IN	161 ± 5 nm (40.32 ± 2.78 mV), quasispherical	IN delivery reduced catalepsy and akinesia in swiss albino mice pretreated with haloperidol orally (2 mg/kg) and a 0.16% dose of ^{99}TC-labeled bromocriptine nanoparticles were recovered in the brain	Md et al. (2013)
Chitosan (Mn: 54 kDa, Mw: 80 kDa)–PEG_{5000} (Mn: 5055, Mw: 5088)–BIO/SA-TfRMab: TPP (2.5:1, w/w) Ionic gelation	Z-DEVD-FMK 200 ng/mL (50 ng/mg nanospheres) or 800 ng/mL 188 ng/mg nanospheres	IV	650 ± 2 nm (20 ± 4 mV) or 637 ± 2 nm (18 ± 4 mV), quasispherical	In vivo administration of Nile red-loaded (CS-PEG_{5000}-BIO/SA-TfRMab) mice resulted in a twofold increase in brain extravascular Nile red concentration (1 h after administration). Z-DEVD-FMK-loaded nanospheres inhibit caspase-3 activity and provided neuroprotection as shown by decrease in infarct volume compared to peptide-free nanospheres in a dose-dependent manner (50 or 188 ng/mg nanospheres)	Karatas et al. (2009)

Continued

Table 2 Summary of Noninvasive Carbohydrate-Based Nanomedicines Investigated in Preclinical Studies for Drug Delivery to the Brain—cont'd

Polymer/Formulation Method	Drug/Dose	Route	Size (Zeta Potential)/ Morphology	Findings	References
Chitosan (Mw: 10 kDa)–pluronic 127 (12.6 kDa)–PEG (Mw: 2.1 kDa)–RVG29-Cys photopolymerization	β-Galactosidase	IV	148 ± 31 nm (25°C), 63 ± 32 nm (37°C) (12.1 ± 0.8 mV), quasispherical	After 48 h of IV injection of nanoparticles, all regions of the brain showed strong β-galactosidase activity compared to enzyme alone, which showed no activity, as well as 25% of injection dose per gram of brain tissue	Kim, Choi, Kim, and Tae (2013)
Chitosan (Mw: 750 kDa): TPP (1.5:2, w/w) Ionic gelation	Rivastigmine/0.068 mg/kg	IN	186 ± 8 nm (38.4 ± 2.85 mV), quasispherical	In vivo administration in Wistar rats resulted in a 2.22-fold increase in AUC and 1.9-fold increase in brain C_{max} after 1 h of administration	Fazil et al. (2012)
Chitosan (medium Mw, DA: 96%): TPP (1.5:2, w/w) Ionic gelation	Thymoquinone 9.5 mg/mL/^{99m}Tc-TQ (100 mCi/100 mL) 5 mg/kg	IN	172 ± 7 nm (30.3 ± 2.15 mV), spherical	IN administration (0.5 mg; 25 μL per nostril) in male Wistar rats resulted in double AUC for TQ nanoparticles compared to TQ solution	Alam et al. (2012)

Chitosan (Mv: 123 kDa, 0.12%): TPP (3:1, w/w) coated with polysorbate 80 (1.49%) Ionic gelation	Rivastigmine/1.5 mg/kg	IV	154 ± 2 nm (26.6 ± 0.9 mV), quasispherical	Treatment significantly reversed the increase in TL and TSTQ induced by SA in mice as compared to SC alone in EPM and MWM test, respectively, with no effects on locomotor activity. MTD: 2.5 mg/kg	Nagpal, Singh, and Mishra (2013c)
Chitosan (Mv: 123 kDa): TPP (3:1, w/w) coated with polysorbate 80 Ionic gelation	Minocycline 100 mg/kg	IV	162 ± 13 nm (33.23 ± 0.37 mV), quasispherical	Increased levels (sixfold) at C_{max} compared to drug alone and a longer circulation half-life and resulted in memory enhancement effects [reduction of TL in EPM test (TL: 6.54 ± 1.62 s compared to 19.09 ± 2.09 s of drug alone) and higher TSTQ in MWM test (TSTQ: 141.69 ± 2.63 vs 117.92 ± 2.86 s of drug alone)]. Antidepressant effects also reported in the	Nagpal, Singh, and Mishra (2013a, 2013b)

Continued

Table 2 Summary of Noninvasive Carbohydrate-Based Nanomedicines Investigated in Preclinical Studies for Drug Delivery to the Brain—cont'd

Polymer/Formulation Method	Drug/Dose	Route	Size (Zeta Potential)/ Morphology	Findings	References
				tail suspension test and despair swim test with no changes in locomotor activity. MTD: 350 mg/kg	
Quaternary ammonium palmitoyl glycol chitosan (GCPQ) (Mw: 6–14 kDa) self-assembly	Leucine enkephalin, TPLENK, IV: 4 mg/mL, PO: 15 mg/mL/IV: 14 mg/kg, PO: 70 mg/kg	IV, PO	71 ± 7 nm (16 ± 1.27 mV), spherical; 136 ± 1 nm (19.6 ± 0.97 mV), nanofibers	In vivo IV administration resulted in a 50% increase in leucine encephalin levels (0.2% of dose) and an enhanced antinociceptive effect in CD-1 mice (tail-flick model) over 4 h. Oral administration in CD-1 mice resulted in a 67% increase in brain levels and antinociception lasting for 8 h after administration. Leucine-enkephalin-GCPQ nanoparticles oral bioavailability is 21%	Lalatsa, Lee, et al. (2012)

N-trimethyl chitosan (TMC, DQ: 25%, 500 kDa, DA: 85%): TPP (5.5:1, w/w) Ionic gelation	Leucine enkephalin (0.25 mg/mL)/1 μg/mouse	IN	443 ± 23 nm (15 ± 2 mV), quasispherical	Enhanced antinociception in hot-plate test (thermal pain) and in acetic acid-induced writhing reaction (inflammatory pain) after IN administration of 5 μL (0.1 mg/mL) in each nostril of Balb/c mice	Kumar et al. (2013)
Thiolated chitosan (60 or 450 kDa): sodium alginate (4:1, w/w) (LMTCNPs/MWTCNPs) self-assembly	Titanidine (TZ) HCl 1.32–1.51 mg/mL/0.00104 mg/g	IN	263 ± 16 nm (17.8 ± 2.1 mV) 276 ± 14 nm (18.3 ± 1.4 mV), quasispherical	In vivo intranasal delivery of LMTCNPs/MWTCNPs at a dose of 0.00104 mg/g TZ in Swiss albino mice (10 μL in each nostril) resulted a threefold increase in C_{max} of ^{99m}TC-TZ% dose/g of brain as well as a threefold enhancement in the antinociceptive effect compared to drug alone. The mice reaction time less than 12 s and more than 18 s before treatment were excluded from the study. An arbitrary cutoff was taken at 45 s	Patel, Naik, and Misra (2012)
Chitosan (low Mw, DA: 95%): TPP (8.3:1, w/w) with and without	Methotrexate/25 mg/kg	IV	119 ± 16 nm (21.92 ± 0.83 mV) and overcoated 107 ± 7 nm	Enhanced methotrexate plasma and brain levels at all time points within 48 h	Azadi, Hamidi, and Rouini (2013)

Continued

Table 2 Summary of Noninvasive Carbohydrate-Based Nanomedicines Investigated in Preclinical Studies for Drug Delivery to the Brain—cont'd

Polymer/Formulation Method	Drug/Dose	Route	Size (Zeta Potential)/ Morphology	Findings	References
1% (w/w) polysorbate 80 overcoating Ionic gelation			(16.13 ± 1.56 mV), spherical	postadministration with a 13- and 7.4-fold increase in brain levels after 4 h for uncoated and overcoated nanoparticles in male Sprague–Dawley rats	
Glycol chitosan (Mw: 250 kDa, DA: 82.7%)–Ferulic acid (DS: 12.8)	Ferulic acid	IV	236 nm, spherical	Using a spinal cord contusion injury model, significant recovery in locomotor function was observed in rats that received IV nanoparticles at 2 h postinjury, as compared to no improvement following methylprednisolone administration	Wu et al. (2014)
Chitosan (Mw: 10 kDa)–PEG (Mw: 2 kDa)–TAT: CPP (1–1.4, w/w) Ionic gelation	siRNA/0.5 mg/kg	IN	5–10 nm, spherical	Successful delivery of siRNA to hippocampus, thalamus, hypothalamus, and Purkinje cells in cerebellum 4 h postnasal delivery	Malhotra, Tomaro-Duchesneau, and Prakash (2013) and Malhotra, Tomaro-Duchesneau, Saha, and Prakash (2013)

Hyaluronic acid (HA) (anionic)					
HA (4–5 kDa)–paclitaxel	Paclitaxel/6 mg/kg	IV		NuNu female mice inoculated with 1.75×10^5 human 231Br cells/100 mL into the left cardiac ventricle in serum-free media were treated 3 days after tumor cell injection with once a week IV injection of formulations. The median survival time was 37, 42, and 49 days for control, paclitaxel, and HA–paclitaxel-treated animals, respectively	Mittapalli et al. (2013)
Alginate					
Alginate (75–100 kDa)–chitosan glutamate (DA: 75–85%) Ionic gelation ($CaCl_2$) and cross-linking (mannuronic to guluronic acid ratio of 1.5)	Venflaxine (0.6075 mg/kg) (1.35 = 0.6075 mg)	IN	174 ± 3 nm (37.40 ± 1.74 mV), quasispherical	Wistar depressed rats tested in the force swimming and locomotor activity test illustrated an increased latency in swimming, climbing, and immobility time and limited improvement in locomotor activity compared to control, oral VFX, IN VFX, and IN	Haque et al. (2014)

Continued

Table 2 Summary of Noninvasive Carbohydrate-Based Nanomedicines Investigated in Preclinical Studies for Drug Delivery to the Brain—cont'd

Polymer/Formulation Method	Drug/Dose	Route	Size (Zeta Potential)/ Morphology	Findings	References
				VFC nanoparticles. IN administration resulted in doubling brain AUC_{0-8} and C_{max} compared to VFX alone	
Cyclodextrin					
HP-β-CD: PLGA (Mw: 31 kDa, 50:50) complexation	Puerarin/5 mg/kg	PO	165 nm (37.40 ± 1.74 mV), quasispherical	In a middle cerebral artery occlusion/reperfusion Wistar rat model, the infarction volume was significantly decreased on day 3 and 7 after administration compared to control and puerarin alone	Tao et al. (2013)

Key: *BIO*, biotin; *Chol*, cholesterol; C_{max}, maximum plasma/serum concentration; *DA*, degree of deacetylation; *DQ*, degree of quaternization; *DS*, degree of substitution; *EPM*, elevated plus maize test; *HP-β-CD*, hydroxypropyl-beta-cyclodextrin; *IN*, intranasal; *IP*, intraperitoneal; *IV*, intravenous; *L*, lomustine; *NIRF*, near-infrared fluorescence; *MTD*, maximum tolerated dose; *Mw*, weight-average molecular weight; *MWM*, Morris water maze test; *Mv*, viscosity-average molecular weight; *PA*, palmitic acid; *PLGA*, poly(lactide-co-glycolide); *PO*, oral; *RHT*, rivastigmine; *RVG29-Cys*, rabies virus glycoprotein cysteine peptide (YTWMPENRPPGTPCDOFTNSRGKRASNGGGGGGC); *SA*, streptavidin; *SLN*, solid lipid nanoparticles; *TAT*, transactivator of transcription (TAT) of human immunodeficiency virus cell-penetrating peptide (RKKRRQRRR); *TL*, transfer latency; *TPGS*, D-α-tocopherol polyethylene glycol 1000 succinate; *TPLENK*, tyrosyl palmitoyl leucine enkephalin; *TPP*, tripolyphosphate; *TQ*, thymoquinone; *TSTQ*, mean time spent in the target quadrant; *TZ*, titanidine; *Z-DEVD-FMK*, caspase inhibitor peptide (*N*-benzyloxycarbonyl-Asp(OMe)-Glu(OMe)-Val-Asp(OMe)-fluoromethyl ketone); *VFX*, venflaxine.

Garrett, et al., 2012) or hydroxyethyl chitosan, compared to other water-soluble chitosans produced by carboxymethylation, dicarboxymethylation, or PEGylation, is soluble at physiological pH while maintaining the free amino groups, which can be used for further conjugation with targeting moieties or biological interactions (Knight et al., 2007).

The free amino groups in chitosan represent a critical feature for drug delivery, as they confer a cationic charge at physiological pH, allowing chitosan to interact with the negatively charged glycocalyx (heparin sulfate proteoglycan, chondroitin sulfate proteoglycan, hyaluronic acid) of epithelial cells including those forming the BBB (Henry & Duling, 1999) and to avoid opsonization. Adhesion to endothelial cells of the BBB and cerebral capillaries has been reported for glycol chitosan nanoparticles that are able to target the brain after systemic injection, even with an intact BBB (Lalatsa, Garrett, et al., 2012; Lalatsa, Schatzlein, et al., 2015; Na et al., 2011) (Table 2). However, careful particle size control is needed as particles above 250 nm are unlikely to spread into the brain beyond the BBB endothelium. Indeed, recent research results illustrated the ability of densely coated PEGylated nanoparticles with near neutral surface charge and a particle size $\leq$115 nm to accumulate within the brain after systemic administration, and to efficiently spread by diffusion within the brain parenchyma (Nance et al., 2012).

A near neutral charge is also preferred for chitosan nanoparticles, as it is well documented that chitosans can modulate tight junctions. Chitosan hydrochloride and chitosan glutamate are known to reversibly reduce the transepithelial electrical resistance across Caco-2 cell monolayers at a pH of 6.2 (van der Merwe, Verhoef, Verheijden, Kotze, & Junginger, 2004). However, this effect is largely dependent on the surface charge of chitosan (which can be enhanced by quaternization as in the case of *N*-trimethyl chitosan chloride), and on the concentration and molecular weight of the polymers (Jonker-Venter et al., 2006; Opanasopit et al., 2007). Highly quaternized low-molecular-weight chitosans have been shown to impact more significantly on TEER. However, recent studies with quaternized amphiphilic low-molecular-weight (<20 kDa) glycol chitosan nanoparticles have reported no effects on TEER across Caco-2 monolayers (Siew et al., 2012).

The reported chitosan-based nanomedicines for brain delivery were formulated using a wide variety of molecular weights (ranging between 6 and 306 kDa), and in the majority of cases were prepared by ionic gelation (Table 2). Ionic gelation yields particles that are stable for intranasal administration, where dilution with biological fluids is minimal, but can become unstable in the case of intravenous or oral administration. As particle size is linked to the molecular weight of the chitosan, systems designed for intravenous delivery will benefit for low-molecular-weight chitosans, ideally prepared by self-assembly. Quaternary ammonium palmitoyl glycol chitosan (GCPQ, *N*-palmitoyl-*N*-monomethyl-*N*,*N*-dimethyl-*N*,*N*,*N*-trimethyl-6-*O*-glycolchitosan) nanoparticles prepared from glycol chitosan polymers of low-molecular weight (6–20 kDa) have shown promising results for intravenous delivery across the BBB (Fisusi et al., 2016; Lalatsa, Lee, et al., 2012) as well as for oral-to-brain delivery of both hydrophilic and hydrophobic drugs (Lalatsa, Garrett, et al., 2012; Lalatsa, Lee, et al., 2012). Nanomerics Ltd. (St Albans, UK) is currently exploiting the resulting technology (MET), which enables to tailor the nanoscale polymeric aggregates able to package or specifically interact with peptides. The technology has confirmed preclinically the ability to deliver leucine^5-enkephalin (an endogenous opioid peptide with a plasma half-life of 3 min) across the BBB, with antinociceptive activity demonstrated in a rodent acute pain animal model, after both intravenous and oral administration (Lalatsa, Garrett, et al., 2012; Lalatsa, Lee, et al., 2012; Lalatsa, Schatzlein, et al., 2015) (Table 2). The antinociceptive response in these studies lasted for up to 8 h postoral administration. The technology works by protecting the peptide from degradation in the gastrointestinal tract, and the nanoparticles being uptaken by enterocytes; the absorbed particles stabilize the peptide and the peptide prodrug against plasma degradation. The particles adhering to the endothelial cells of the BBB enable the peptide and peptide prodrug to cross the BBB. Entrapment of lomustine within MET nanoparticles resulted in only modest brain targeting (brain/bone AUC_{0-4} ratios compared to intravenous administration of ethanolic lomustine of 0.90 and 0.53, respectively). The MET formulation significantly increased mice (U87MG orthotopic tumors) survival times (33.2 vs 22.5 days with ethanolic lomustine and 21.3 days with lomustine alone) without major bone marrow toxicity (Fisusi et al., 2016) (Table 2).

Active transport has also been attempted with chitosan nanoparticles that exploit receptor-mediated endocytosis by either coating with polysorbate 80 (Nagpal et al., 2013c) or conjugation to OX26 monoclonal antibody (Karatas et al., 2009) or rabies virus glycoprotein cysteine peptide (Kim et al., 2013), as illustrated by pharmacodynamics studies reported in the literature (Table 2). Chitosan nanospheres conjugated with poly(ethylene glycol) bearing OX26 monoclonal antibody—that has affinity for the transferrin receptor—were able to deliver an anticaspase peptide (Z-DEVD-FMK) into the brain, reducing vulnerability to neuronal cell death and increasing neuronal cell survival following cerebral ischemia (Karatas et al., 2009). This study illustrated that after intravenous administration fluorescently labeled nanoparticles were located in significant amounts in the brain, outside of the intravascular compartment; it was the first study of decorated chitosan nanoparticles for delivery across the BBB. Additionally, the nanoparticles were selectively directed to the brain and were not measurably taken up by the liver and spleen, which has been also shown for the GCPQ nanoparticles (Fisusi et al., 2016; Lalatsa, Garrett, et al., 2012; Serrano et al., 2015).

Chitosan has been widely used to deliver to the brain various therapeutics including drugs, peptides, and nerve growth factors, after intranasal administration (Lalatsa et al., 2014). Chitosan's mucoadhesive properties are the result of interactions between the free amine groups and the sialic acid residues present on the mucosal layer, and reduce particle clearance from the mucosal cavity. Three different pathways across the olfactory epithelium may be envisaged: (i) a transcellular pathway, especially across the sustentacular cells, most likely via receptor-mediated endocytosis, fluid-phase endocytosis, or passive diffusion; (ii) a paracellular pathway through the tight junctions between the sustentacular cells and olfactory neurons, which is relevant for chitosans able to modulate the tight junctions; and (iii) the olfactory nerve pathway, where the drug is taken up into the neurons by endocytotic or pinocytotic mechanisms and then is transported by intracellular axonal transport to the olfactory bulb (Illum, 2000). The transneuronal pathway is very slow and agents reach the CNS as late as 24 h after administration; hence, transport via neuronal routes cannot explain the rapid appearance of drug in the cerebrovascular fluid that is seen for a range of molecular weight drugs or biomacromolecules (Illum, 2000).

Recently, PEGylated chitosan nanoparticles functionalized with a cell-penetrating peptide were shown to be able to deliver successfully siRNA intranasally to the hippocampus, thalamus, hypothalamus, and Purkinje cells in the cerebellum, 4 h after administration (Table 2) (Malhotra, Tomaro-Duchesneau, & Prakash, 2013; Malhotra, Tomaro-Duchesneau, Saha, Kahouli, & Prakash, 2013; Malhotra, Tomaro-Duchesneau, Saha, & Prakash, 2013). Polymer complexes were formed between the positively charged chitosan and negatively charged siRNA, with a size of 5–10 nm. The optimal dose was determined at 0.5 mg/kg and particles had no visible, local toxic effects (Malhotra, Tomaro-Duchesneau, & Prakash, 2013; Malhotra, Tomaro-Duchesneau, Saha, Kahouli, et al., 2013; Malhotra, Tomaro-Duchesneau, Saha, & Prakash, 2013). Intranasal delivery of peptides using chitosan has also been reported. Leucine^5-enkephalin-loaded *N*-trimethyl chitosan (500 kDa) nanoparticles prepared using the ionic gelation method were evaluated as vehicles for delivery to the brain via the nasal route. Using *N*-trimethyl chitosan nanoparticulate gels, a significant improvement in the observed antinociceptive effect of leucine^5-enkephalin was found, as evidenced by the hot plate and acetic acid-induced writhing bioassay (Kumar et al., 2013). Chitosan nanoparticles resulted in at least doubling of the AUC after intranasal administration of a variety of molecules and an enhancement of pharmacodynamics activity (Alam et al., 2012; Fazil et al., 2012; Patel et al., 2012).

Alginates, derived from marine brown algae cell walls, are anionic biodegradable and nontoxic polysaccharides consisting of a chain of (1-4)-linked β-D-mannuronic acid and α-L-guluronic acid in different arrangements (Sarei, Dounighi, Zolfagharian, Khaki, & Bidhendi, 2013). Alginate nanoparticles administered nasally or orally possess mucoadhesive properties and are able to reduce drug degradation and enhance its permeability. Due to their anionic charge, alginates can be easily combined with chitosan using electrostatic complexation. Alginate–chitosan nanoparticles loaded with venflaxine resulted in doubling the brain AUC_{0-8} after intranasal administration and improved pharmacodynamic effects in depressed Wistar rats, as illustrated by the force swimming and locomotor activity tests (Table 2) (Haque et al., 2014).

Hyaluronic acid (with molecular weight between 1 and 10,000 kDa) is a water-soluble biocompatible polysaccharide that produces aqueous gels that can behave as a lubricant with various viscosities, depending on the chain length, degree of entanglement, cross-linking, pH, and chemical modification (while the absolute concentration seems to be much less important; Jin, Ubonvan, & Kim, 2010). HA-based materials have been used for the sustained delivery of peptide/protein drugs in subcutaneous formulations, for ocular and nasal delivery, and for targeting specific intracellular delivery of genes and drugs. HA has been exploited in the development of targeted drug polymer conjugates for the treatment of brain tumors, particularly as it can be readily functionalized and conjugated with chemotherapeutics such as doxorubicin or paclitaxel (Jin et al., 2012; Yang et al., 2013). As the BBB is compromised in patients with brain tumors, HA–paclitaxel conjugates can paracellularly diffuse through the blood tumor barrier openings to the tumor interstitial spaces, where the conjugate can actively target metastatic cancer cells by CD44 receptor-mediated endocytosis (Mittapalli et al., 2013). Additionally, HA–paclitaxel conjugates were shown to reduce *p*-glycoprotein efflux of paclitaxel, increasing further its brain and tumor distribution (Mittapalli et al., 2013). This strategy resulted in a significant increase in survival in animals developing brain tumors after intracardiac injection of 231Br cells (Table 2).

CDs are cyclic oligosaccharides derived from starch. The relatively hydrophobic cavity formed by the ring shape of the molecule allows noncovalent encapsulation of suitably sized drug moieties. Hydroxypropyl-beta-cyclodextrin (HP-β-CD) is a modified seven glucose unit CD with high aqueous solubility, a proven high solubilizing capacity and safe profile, especially for parenteral use. Complexes with poly(D,L-lactic-*co*-glycolic acid) nanoparticles loaded with puerarin, resulted in a significant decrease in infarction volume in a middle cerebral artery Wistar rat model (Table 2) (Tao et al., 2013). Quaternary ammonium–cyclodextrin nanoparticles (QA-β-CD) (65–88 nm) were proposed also as drug delivery carriers for doxorubicin across a bovine BCEC monolayer (Gil, Li, Xiao, & Lowe, 2009). Adsorptive-mediated endocytosis was speculated due to the near neutral cationic charge. Interestingly, complexation of doxorubicin with the QA-β-CD nanoparticles preserved its cytotoxicity against human glioblastoma cell line (U87MG), but attenuated cytotoxicity against the endothelial cells (Gil et al., 2009).

1.4.4 Promising Novel Carbohydrate Nanocarriers for Brain Delivery

There is a wide range of novel carbohydrate nanoparticulate systems under investigation as promising delivery carriers across the BBB, with many based on pullulans and dextrans; however, most lack the in vivo proof of concept.

Pullulan is a natural water-soluble polysaccharide consisting of maltotriose units (a trisaccharide consisting of three glucose molecules linked via 1,4-glycosidic bonds) and has been utilized for the development of nanoparticles for the detection and treatment of Alzheimer's disease. Nanogels composed of cholesterol-bearing pullulan (CHP) prevented Aβ oligomerization of toxic Aβ oligomeric species (Nazem & Mansoori, 2008). A significant reduction in $A\beta_{42}$ toxicity in the primary cortical cell culture and microglial cell culture was reported after exposure to CHP nanogels (Boridy, Takahashi, Akiyoshi, & Maysinger, 2009). Both neutral and cationic CHP nanoparticles interacted with $A\beta_{(1-42)}$ monomers and oligomers, particularly in primary cortical cultures, with neutral nanoparticles proven to be safer. These results suggest that CHP nanogels could provide a valid complementary strategy to antibody immunotherapy in neurological disorders characterized by the formation of soluble toxic aggregates, as in the case of Alzheimer's disease.

Dextrans, a family of polysaccharide macromolecular carriers composed of a-D-glucans with anhydro-D-glucopyranose units that are not substrates for selective transport proteins, are one of the most promising carriers for a wide variety of therapeutic agents, due to their excellent physicochemical properties and physiological acceptance. Thus, dextrans have been used as coatings or stabilizers for delivery across the BBB. Dextran-coated nanostructured lipid carriers were shown to have an enhanced encapsulation efficiency for mitoxantrone hydrochloride and their use resulted in higher amounts of drug in brain, liver, and spleen; interestingly, the results indicated lower drug levels in the kidney after intravenous administration (Zhang et al., 2012). Modified dextrans able to self-assemble have also been developed for drug delivery to the brain. Alkylglyceryl-modified dextran-*graft*-poly(lactic acid) polymers were used to prepare stable nanoparticles of 100–200 nm diameter using both solvent displacement and electrospraying methods (Toman et al., 2015). Transwell-model experiments employing mouse (bEnd3) and human (hCMEC/D3) brain endothelial cell monolayers revealed enhanced permeation of Rhodamine B entrapped within the nanoparticles. Electric cell-substrate impedance sensing results suggested a transient decrease of the barrier function in an in vitro BBB model following incubation with these nanoparticles (Toman et al., 2015). These materials look promising but need further exploring to assess their efficacy and safety following in vivo administration.

Another promising delivery system was developed from cellulose nanocrystals (CNCs), which are elongated glucose-based nanoparticles with mean aspect ratios of 10–100 nm (depending on the cellulose source and the experimental conditions used in their preparation; Beck-Candanedo, Roman, & Gray, 2005). Their long-axial geometry and high abundance of surface hydroxyl groups (3 per repeating unit) make CNCs very promising as drug delivery vectors. The hydroxyl groups enable easy coupling of imaging, targeting, and therapeutic agents to the CNC surface and render the particle surface hydrophilic, which results in a delayed clearance from the blood (Dong et al., 2014). Interestingly, CNCs are nontoxic to human brain endothelial cells and exhibit nearly no specific cellular uptake (Roman, Dong, Hirani, & Lee, 2009). Folic acid-conjugated CNCs have been shown to possess a 46-fold higher uptake by folate receptor-positive rat C6 brain tumor cells than nontargeted CNCs, likely via caveolae-mediated endocytosis (Dong et al., 2014). CNCs have very low toxicity and a high surface area amenable to surface modification or functionalization; however, further studies are needed to improve their aqueous solubility and make intravenous administration possible, in order to unlock their full in vivo potential.

1.5 Concluding Remarks

Considerable research efforts have been focused recently toward the development of versatile nanomedicines for overcoming biological barriers and in particular for achieving delivery across the BBB. Carbohydrates originating from various natural sources are ideal candidate materials for this challenging task due to their intrinsic recognition by the biological environment and the subsequent conveyance into metabolic degradation. Their biocompatibility and hydrophilicity, combined with the ease of chemical modification offered by the variety of functional groups present in their structures, substantiate the efforts of tailoring these promising materials in terms of their processability into nanoparticles with expected size and shape, stability, drug release profile and, in turn, with specific biological responses. Amphiphilic carbohydrates in particular are receiving significant interest.

Both active and passive strategies employing carbohydrate-based nanoparticles have been reported in the literature for drug delivery across the BBB. Based on results obtained in preclinical studies, chitosan-based nanomedicines are currently most advanced and have been reported to achieve a significant increase in brain concentration of various drugs, peptides, and proteins, as well as genetic material. Several companies are currently focusing on preparing carbohydrate nanocarriers under GMP conditions and on assessing their full toxicological profile, in order to translate these technologies into clinical trials and successfully test them in humans (phase I and II).

An increasing number of other polysaccharides are investigated for their potential as nanoparticles capable to deliver drugs into the brain, and research efforts into the proof of concept of these novel approaches are currently intensifying. To date, alternative strategies employing dextran, pullulan, or CNCs seem most promising for overcoming the BBB and represent emerging technologies that warrant further preclinical research.

REFERENCES

Alam, S., Khan, Z. I., Mustafa, G., Kumar, M., Islam, F., Bhatnagar, A., & Ahmad, F. J. (2012). Development and evaluation of thymoquinone-encapsulated chitosan nanoparticles for nose-to-brain targeting: A pharmacoscintigraphic study. *International Journal of Nanomedicine*, 7, 5705–5718. http://dx.doi.org/10.2147/IJN.S35329.

Azadi, A., Hamidi, M., & Rouini, M. R. (2013). Methotrexate-loaded chitosan nanogels as 'Trojan Horses' for drug delivery to brain: Preparation and in vitro/in vivo characterization. *International Journal of Biological Macromolecules*, *62*, 523–530. http://dx.doi.org/10.1016/j.ijbiomac.2013.10.004.

Beck-Candanedo, S., Roman, M., & Gray, D. G. (2005). Effect of reaction conditions on the properties and behavior of wood cellulose nanocrystal suspensions. *Biomacromolecules*, *6*(2), 1048–1054. http://dx.doi.org/10.1021/bm049300p.

Benoit, J. P., Faisant, N., Venier-Julienne, M. C., & Menei, P. (2000). Development of microspheres for neurological disorders: From basics to clinical applications. *Journal of Controlled Release*, *65*(1–2), 285–296.

Bobo, R. H., Laske, D. W., Akbasak, A., Morrison, P. F., Dedrick, R. L., & Oldfield, E. H. (1994). Convection-enhanced delivery of macromolecules in the brain. *Proceedings of the National Academy of Sciences of the United States of America*, *91*(6), 2076–2080.

Boridy, S., Takahashi, H., Akiyoshi, K., & Maysinger, D. (2009). The binding of pullulan modified cholesteryl nanogels to Abeta oligomers and their suppression of cytotoxicity. *Biomaterials*, *30*(29), 5583–5591. http://dx.doi.org/10.1016/j.biomaterials.2009.06.010.

Borlongan, C. V., & Emerich, D. F. (2003). Facilitation of drug entry into the CNS via transient permeation of blood brain barrier: Laboratory and preliminary clinical evidence from bradykinin receptor agonist, Cereport. *Brain Research Bulletin*, *60*(3), 297–306.

Bradbury, M. W. B. (1979). *The concept of the blood-brain barrier*. Chichester: Wiley.

Carrillo-Conde, B., Song, E. H., Chavez-Santoscoy, A., Phanse, Y., Ramer-Tait, A. E., Pohl, N. L., … Narasimhan, B. (2011). Mannose-functionalized "pathogen-like" polyanhydride nanoparticles target C-type lectin receptors on dendritic cells. *Molecular Pharmaceutics*, *8*(5), 1877–1886. http://dx.doi.org/10.1021/mp200213r.

Chen, Y., & Liu, L. (2012). Modern methods for delivery of drugs across the blood-brain barrier. *Advanced Drug Delivery Reviews*, *64*(7), 640–665. http://dx.doi.org/10.1016/j.addr.2011.11.010.

Cirpanli, Y., Allard, E., Passirani, C., Bilensoy, E., Lemaire, L., Calis, S., & Benoit, J. P. (2011). Antitumoral activity of camptothecin-loaded nanoparticles in 9 L rat glioma model. *International Journal of Pharmaceutics*, *403*(1–2), 201–206. http://dx.doi.org/10.1016/j.ijpharm.2010.10.015.

Csaba, N., Koping-Hoggard, M., & Alonso, M. J. (2009). Ionically crosslinked chitosan/tripolyphosphate nanoparticles for oligonucleotide and plasmid DNA delivery. *International Journal of Pharmaceutics*, *382*(1–2), 205–214. http://dx.doi.org/10.1016/j.ijpharm.2009.07.028.

De Giglio, E., Trapani, A., Cafagna, D., Sabbatini, L., & Cometa, S. (2011). Dopamine-loaded chitosan nanoparticles: Formulation and analytical characterization. *Analytical and Bioanalytical Chemistry*, *400*(7), 1997–2002. http://dx.doi.org/10.1007/s00216-011-4962-y.

Dong, S., Cho, H. J., Lee, Y. W., & Roman, M. (2014). Synthesis and cellular uptake of folic acid-conjugated cellulose nanocrystals for cancer targeting. *Biomacromolecules*, *15*(5), 1560–1567. http://dx.doi.org/10.1021/bm401593n.

Ehrlich, P. (1885). *Das Sauerstoff-Bedürfniss des Organismus: Eine farbenanalytische Studie*. Berlin: Hirschwald.

Erdlenbruch, B., Alipour, M., Fricker, G., Miller, D. S., Kugler, W., Eibl, H., & Lakomek, M. (2003). Alkylglycerol opening of the blood-brain barrier to small and large fluorescence markers in normal and C6 glioma-bearing rats and isolated rat brain capillaries. *British Journal of Pharmacology*, *140*(7), 1201–1210. http://dx.doi.org/10.1038/sj.bjp.0705554.

Erdlenbruch, B., Kugler, W., Schinkhof, C., Neurath, H., Eibl, H., & Lakomek, M. (2005). Blood-brain barrier opening with alkylglycerols: Biodistribution of 1-O-pentylglycerol after intravenous and intracarotid administration in rats. *Journal of Drug Targeting*, *13*(3), 143–150. http://dx.doi.org/10.1080/10611860400029085.

Fazil, M., Md, S., Hague, S., Kumar, M., Baboota, S., Sahni, J. K., & Ali, J. (2012). Development and evaluation of rivastigmine loaded chitosan nanoparticles for brain targeting. *European Journal of Pharmaceutical Sciences*, *47*(1), 6–15. http://dx.doi.org/10.1016/j.ejps.2012.04.013.

Fellner, S., Bauer, B., Miller, D. S., Schaffrik, M., Fankhanel, M., Spruss, T., … Fricker, G. (2002). Transport of paclitaxel (Taxol) across the blood-brain barrier in vitro and in vivo. *The Journal of Clinical Investigation*, *110*(9), 1309–1318. http://dx.doi.org/10.1172/JCI15451.

Fisusi, F. A., Siew, A., Chooi, K. W., Okubanjo, O., Garrett, N., Lalatsa, K., … Uchegbu, I. F. (2016). Lomustine nanoparticles enable both bone marrow sparing and high brain drug levels—A strategy for brain cancer treatments. *Pharmaceutical Research*, *33*, 1289–1303. http://dx.doi.org/10.1007/s11095-016-1872-x.

Fortin, D., Gendron, C., Boudrias, M., & Garant, M. P. (2007). Enhanced chemotherapy delivery by intraarterial infusion and blood-brain barrier disruption in the treatment of cerebral metastasis. *Cancer*, *109*(4), 751–760. http://dx.doi.org/10.1002/cncr.22450.

Fung, L. K., Ewend, M. G., Sills, A., Sipos, E. P., Thompson, R., Watts, M., … Saltzman, W. M. (1998). Pharmacokinetics of interstitial delivery of carmustine, 4-hydroperoxycyclophosphamide, and paclitaxel from a biodegradable polymer implant in the monkey brain. *Cancer Research*, *58*(4), 672–684.

Gaillard, P. J. (2008). Targeted Intracellular delivery of antiviral agents, WO/2008/118013, PCT/NL2008/050166.

Gaillard, P. J. (2010). Glutathione-based drug delivery system, WO/2010/095940, PCT/NL2010/050082.

Ganau, M. (2014). Tackling gliomas with nanoformulated antineoplastic drugs: Suitability of hyaluronic acid nanoparticles. *Clinical & Translational Oncology*, *16*(2), 220–223. http://dx.doi.org/10.1007/s12094-013-1114-1.

Geissler, A., Biesalski, M., Heinze, T., & Zhang, K. (2014). Formation of nanostructured cellulose stearoyl esters via nanoprecipitation. *Journal of Materials Chemistry A*, *2*(4), 1107–1116. http://dx.doi.org/10.1039/c3ta13937a.

Geldenhuys, W., Mbimba, T., Bui, T., Harrison, K., & Sutariya, V. (2011). Brain-targeted delivery of paclitaxel using glutathione-coated nanoparticles for brain cancers. *Journal of Drug Targeting*, *19*(9), 837–845. http://dx.doi.org/10.3109/1061186X.2011.589435.

Gil, E. S., Li, J. S., Xiao, H. N., & Lowe, T. L. (2009). Quaternary ammonium beta-cyclodextrin nanoparticles for enhancing doxorubicin permeability across the in vitro blood-brain barrier (vol. 10, pg 506, 2009). *Biomacromolecules*, *10*(4), 1020. http://dx.doi.org/10.1021/bm900279c.

Glasser, W. G. (2008). 6.3 Cellulose and associated heteropolysaccharides. In B. O. Fraser-Reid, K. Tatsuka, & J. Thiem (Eds.), *Glycoscience: Vol. 1* (pp. 1473–1512). Berlin, Heidelberg: Springer-Verlag.

Goldmann, E. (1913). Vitalfarbung am zentralnervensystem. *Abhandl Konigl preuss Akad Wiss*, *1*, 1–60.

Haque, S., Md, S., Sahni, J. K., Ali, J., & Baboota, S. (2014). Development and evaluation of brain targeted intranasal alginate nanoparticles for treatment of depression. *Journal of Psychiatric Research*, *48*(1), 1–12. http://dx.doi.org/10.1016/j.jpsychires.2013.10.011.

Hawkins, B. T., & Davis, T. P. (2005). The blood-brain barrier/neurovascular unit in health and disease. *Pharmacological Reviews*, *57*(2), 173–185. http://dx.doi.org/10.1124/pr.57.2.4.
Henry, C. B., & Duling, B. R. (1999). Permeation of the luminal capillary glycocalyx is determined by hyaluronan. *The American Journal of Physiology*, *277*(2 Pt. 2), H508–H514.
Hosseini, S. M., Hosseini, H., Mohammadifar, M. A., Mortazavian, A. M., Mohammadi, A., Khosravi-Darani, K., … Khaksar, R. (2013). Incorporation of essential oil in alginate microparticles by multiple emulsion/ionic gelation process. *International Journal of Biological Macromolecules*, *62*, 582–588. http://dx.doi.org/10.1016/j.ijbiomac.2013.09.054.
Hulper, P., Veszelka, S., Walter, F. R., Wolburg, H., Fallier-Becker, P., Piontek, J., … Deli, M. A. (2013). Acute effects of short-chain alkylglycerols on blood-brain barrier properties of cultured brain endothelial cells. *British Journal of Pharmacology*, *169*(7), 1561–1573. http://dx.doi.org/10.1111/bph.12218.
Hynynen, K. (2008). Ultrasound for drug and gene delivery to the brain. *Advanced Drug Delivery Reviews*, *60*(10), 1209–1217. http://dx.doi.org/10.1016/j.addr.2008.03.010.
Illum, L. (2000). Transport of drugs from the nasal cavity to the central nervous system. *European Journal of Pharmaceutical Sciences*, *11*(1), 1–18.
Jalali, S., Huang, Y., Dumont, D. J., & Hynynen, K. (2010). Focused ultrasound-mediated bbb disruption is associated with an increase in activation of AKT: Experimental study in rats.. *BMC Neurology*, *10*, 114. http://dx.doi.org/10.1186/1471-2377-10-114.
Jay, S. M., & Saltzman, W. M. (2009). Controlled delivery of VEGF via modulation of alginate microparticle ionic crosslinking. *Journal of Controlled Release*, *134*(1), 26–34. http://dx.doi.org/10.1016/j.jconrel.2008.10.019.
Jiang, W., Kim, B. Y., Rutka, J. T., & Chan, W. C. (2008). Nanoparticle-mediated cellular response is size-dependent. *Nature Nanotechnology*, *3*(3), 145–150. http://dx.doi.org/10.1038/nnano.2008.30.
Jin, Y. J., Termsarasab, U., Ko, S. H., Shim, J. S., Chong, S., Chung, S. J., … Kim, D. D. (2012). Hyaluronic acid derivative-based self-assembled nanoparticles for the treatment of melanoma. *Pharmaceutical Research*, *29*(12), 3443–3454. http://dx.doi.org/10.1007/s11095-012-0839-9.
Jin, Y.-J., Ubonvan, T., & Kim, D. D. (2010). Hyaluronic acid in drug delivery systems. *Journal of Pharmaceutical Investigation*, *40*(Special Issue), 33–43.
Jonker-Venter, C., Snyman, D., Janse van Rensburg, C., Jordaan, E., Schultz, C., Steenekamp, J. H., … Kotze, A. F. (2006). Low molecular weight quaternised chitosan (11): In vitro assessment of absorption enhancing properties. *Pharmazie*, *61*(4), 301–305.
Kaewprapan, K., Inprakhon, P., Marie, E., & Durand, A. (2012). Enzymatically degradable nanoparticles of dextran esters as potential drug delivery systems. *Carbohydrate Polymers*, *88*(3), 875–881. http://dx.doi.org/10.1016/j.carbpol.2012.01.030.
Karatas, H., Aktas, Y., Gursoy-Ozdemir, Y., Bodur, E., Yemisci, M., Caban, S., … Dalkara, T. (2009). A nanomedicine transports a peptide caspase-3 inhibitor across the blood-brain barrier and provides neuroprotection. *The Journal of Neuroscience*, *29*(44), 13761–13769. http://dx.doi.org/10.1523/JNEUROSCI.4246-09.2009.
Kas, H. S. (1997). Chitosan: Properties, preparations and application to microparticulate systems. *Journal of Microencapsulation*, *14*(6), 689–711.
Kelder, J., Grootenhuis, P. D., Bayada, D. M., Delbressine, L. P., & Ploemen, J. P. (1999). Polar molecular surface as a dominating determinant for oral absorption and brain penetration of drugs. *Pharmaceutical Research*, *16*(10), 1514–1519.
Kim, J. Y., Choi, W. I., Kim, Y. H., & Tae, G. (2013). Brain-targeted delivery of protein using chitosan- and RVG peptide-conjugated, pluronic-based nano-carrier. *Biomaterials*, *34*(4), 1170–1178. http://dx.doi.org/10.1016/j.biomaterials.2012.09.047.
Kinoshita, M., McDannold, N., Jolesz, F. A., & Hynynen, K. (2006). Targeted delivery of antibodies through the blood-brain barrier by MRI-guided focused ultrasound.

Biochemical and Biophysical Research Communications, *340*(4), 1085–1090. http://dx.doi.org/10.1016/j.bbrc.2005.12.112.
Knight, D. K., Shapka, S. N., & Amsden, B. G. (2007). Structure, depolymerization, and cytocompatibility evaluation of glycol chitosan. *Journal of Biomedical Materials Research. Part A*, *83*(3), 787–798. http://dx.doi.org/10.1002/jbm.a.31430.
Kola, I., & Landis, J. (2004). Can the pharmaceutical industry reduce attrition rates? *Nature Reviews. Drug Discovery*, *3*(8), 711–715. http://dx.doi.org/10.1038/nrd1470.
Kreuter, J. (2013). Mechanism of polymeric nanoparticle-based drug transport across the blood-brain barrier (BBB). *Journal of Microencapsulation*, *30*(1), 49–54. http://dx.doi.org/10.3109/02652048.2012.692491.
Kumar, M., Pandey, R. S., Patra, K. C., Jain, S. K., Soni, M. L., Dangi, J. S., & Madan, J. (2013). Evaluation of neuropeptide loaded trimethyl chitosan nanoparticles for nose to brain delivery. *International Journal of Biological Macromolecules*, *61*, 189–195. http://dx.doi.org/10.1016/j.ijbiomac.2013.06.041.
Lalatsa, A., Garrett, N. L., Ferrarelli, T., Moger, J., Schatzlein, A. G., & Uchegbu, I. F. (2012). Delivery of peptides to the blood and brain after oral uptake of quaternary ammonium palmitoyl glycol chitosan nanoparticles. *Molecular Pharmaceutics*, *9*(6), 1764–1774. http://dx.doi.org/10.1021/mp300068j.
Lalatsa, A., Lee, V., Malkinson, J. P., Zloh, M., Schatzlein, A. G., & Uchegbu, I. F. (2012). A prodrug nanoparticle approach for the oral delivery of a hydrophilic peptide, leucine (5)-enkephalin, to the brain. *Molecular Pharmaceutics*, *9*(6), 1665–1680. http://dx.doi.org/10.1021/mp300009u.
Lalatsa, A., Leite, D. M., & Pilkington, G. J. (2015). Nanomedicines and the future of glioma. *Oncology News*, *10*(2), 51–57.
Lalatsa, A., Schatzlein, A. G., Garrett, N. L., Moger, J., Briggs, M., Godfrey, L., … Uchegbu, I. F. (2015). Chitosan amphiphile coating of peptide nanofibres reduces liver uptake and delivers the peptide to the brain on intravenous administration. *Journal of Controlled Release*, *197*, 87–96. http://dx.doi.org/10.1016/j.jconrel.2014.10.028.
Lalatsa, A., Schatzlein, A. G., Mazza, M., Le, T. B., & Uchegbu, I. F. (2012a). Amphiphilic poly(L-amino acids)—New materials for drug delivery. *Journal of Controlled Release*, *161*(2), 523–536. http://dx.doi.org/10.1016/j.jconrel.2012.04.046.
Lalatsa, A., Schatzlein, A. G., & Uchegbu, I. F. (2011). 5.08 Drug delivery across the blood-brain barrier. In M. Murray Moo-Young, M. Butler, C. Webb, A. Moreira, B. Grodzinski, & Z. Cui (Eds.), *Comprehensive biotechnology: Vol. 5* (XPATH ERROR: missing expression component.,pp. 657–668). Amsterdam: Elsevier.
Lalatsa, A., Schatzlein, A., & Uchegbu, I. (2012b). Nanostructures overcoming the blood-brain barrier: Physiological considerations and mechanistic issues. In M. J. Alonso & N. S. Csaba (Eds.), *Nanostructured biomaterials for overcoming biological barriers* (pp. 329–363). London: RSC Publishing.
Lalatsa, A., Schatzlein, A. G., & Uchegbu, I. F. (2014). Strategies to deliver peptide drugs to the brain. *Molecular Pharmaceutics*, *11*(4), 1081–1093. http://dx.doi.org/10.1021/mp400680d.
Ledung, P., Milas, M., Rinaudo, M., & Desbrieres, J. (1994). Water-soluble derivatives obtained by controlled chemical modifications of chitosan. *Carbohydrate Polymers*, *24*(3), 209–214.
Lee, J. W., Park, J. H., & Robinson, J. R. (2000). Bioadhesive-based dosage forms: The next generation. *Journal of Pharmaceutical Sciences*, *89*(7), 850–866. http://dx.doi.org/10.1002/1520-6017(200007)89:7<850::AID-JPS2>3.0.CO;2-G.
Lennernas, H., & Lundgren, E. (2004). Intestinal and blood-brain drug transport: Beyond involvement of a single transport function. *Drug Discovery Today. Technologies*, *1*(4), 417–422. http://dx.doi.org/10.1016/j.ddtec.2004.11.010.
Li, G., Guo, L., Wen, Q., & Zhang, T. (2013). Thermo- and pH-sensitive ionic-crosslinked hollow spheres from chitosan-based graft copolymer for 5-fluorouracil release.

International Journal of Biological Macromolecules, *55*, 69–74. http://dx.doi.org/10.1016/j.ijbiomac.2012.12.048.

Li, Y. F., Jia, H. P., Cheng, Q. L., Pan, F. S., & Jiang, Z. Y. (2011). Sodium alginate-gelatin polyelectrolyte complex membranes with both high water vapor permeance and high permselectivity. *Journal of Membrane Science*, *375*(1–2), 304–312. http://dx.doi.org/10.1016/j.memsci.2011.03.058.

Liu, Z., Jiao, Y., Wang, Y., Zhou, C., & Zhang, Z. (2008). Polysaccharides-based nanoparticles as drug delivery systems. *Advanced Drug Delivery Reviews*, *60*(15), 1650–1662. http://dx.doi.org/10.1016/j.addr.2008.09.001.

Lonser, R. R., Sarntinoranont, M., Morrison, P. F., & Oldfield, E. H. (2015). Convection-enhanced delivery to the central nervous system. *Journal of Neurosurgery*, *122*(3), 697–706. http://dx.doi.org/10.3171/2014.10.JNS14229.

Malhotra, M., Tomaro-Duchesneau, C., & Prakash, S. (2013). Synthesis of TAT peptide-tagged PEG-ylated chitosan nanoparticles for siRNA delivery targeting neurodegenerative diseases. *Biomaterials*, *34*(4), 1270–1280. http://dx.doi.org/10.1016/j.biomaterials.2012.10.013.

Malhotra, M., Tomaro-Duchesneau, C., Saha, S., Kahouli, I., & Prakash, S. (2013). Development and characterization of chitosan-PEG-TAT nanoparticles for the intracellular delivery of siRNA. *International Journal of Nanomedicine*, *8*, 2041–2052. http://dx.doi.org/10.2147/IJN.S43683.

Malhotra, M., Tomaro-Duchesneau, C., Saha, S., & Prakash, S. (2013). Intranasal, siRNA delivery to the brain by TAT/MGF tagged PEG-ylated chitosan nanoparticles. *Journal of Pharmaceutics (Cairo)*, *2013*, 812387. http://dx.doi.org/10.1155/2013/812387.

Md, S., Khan, R. A., Mustafa, G., Chuttani, K., Baboota, S., Sahni, J. K., & Ali, J. (2013). Bromocriptine loaded chitosan nanoparticles intended for direct nose to brain delivery: Pharmacodynamic, pharmacokinetic and scintigraphy study in mice model. *European Journal of Pharmaceutical Sciences*, *48*(3), 393–405. http://dx.doi.org/10.1016/j.ejps.2012.12.007.

Mihai, M., Ghiorghita, C. A., Stoica, I., Nita, L., Popescu, I., & Fundueanu, G. (2011). New polyelectrolyte complex particles as colloidal dispersions based on weak synthetic and/or natural polyelectrolytes. *Express Polymer Letters*, *5*(6), 506–511.

Misra, A., Ganesh, S., Shahiwala, A., & Shah, S. P. (2003). Drug delivery to the central nervous system: A review. *Journal of Pharmacy & Pharmaceutical Sciences*, *6*(2), 252–273.

Mittapalli, R. K., Liu, X. L., Adkins, C. E., Nounou, M. I., Bohn, K. A., Terrell, T. B., … Lockman, P. R. (2013). Paclitaxel-hyaluronic nanoconjugates prolong overall survival in a preclinical brain metastases of breast cancer model. *Molecular Cancer Therapeutics*, *12*(11), 2389–2399. http://dx.doi.org/10.1158/1535-7163.MCT-13-0132.

Morrison, P. F., Chen, M. Y., Chadwick, R. S., Lonser, R. R., & Oldfield, E. H. (1999). Focal delivery during direct infusion to brain: Role of flow rate, catheter diameter, and tissue mechanics. *The American Journal of Physiology*, *277*(4 Pt. 2), R1218–R1229.

Na, J. H., Koo, H., Lee, S., Min, K. H., Park, K., Yoo, H., … Kim, K. (2011). Real-time and non-invasive optical imaging of tumor-targeting glycol chitosan nanoparticles in various tumor models. *Biomaterials*, *32*(22), 5252–5261. http://dx.doi.org/10.1016/j.biomaterials.2011.03.076.

Nagpal, K., Singh, S. K., & Mishra, D. (2013a). Evaluation of safety and efficacy of brain targeted chitosan nanoparticles of minocycline. *International Journal of Biological Macromolecules*, *59*, 20–28. http://dx.doi.org/10.1016/j.ijbiomac.2013.04.011.

Nagpal, K., Singh, S. K., & Mishra, D. N. (2013b). Formulation, optimization, in vivo pharmacokinetic, behavioral and biochemical estimations of minocycline loaded chitosan nanoparticles for enhanced brain uptake. *Chemical & Pharmaceutical Bulletin (Tokyo)*, *61*(3), 258–272.

Nagpal, K., Singh, S. K., & Mishra, D. N. (2013c). Optimization of brain targeted chitosan nanoparticles of Rivastigmine for improved efficacy and safety. *International Journal of Biological Macromolecules*, *59*, 72–83. http://dx.doi.org/10.1016/j.ijbiomac.2013.04.024.

Nance, E. A., Woodworth, G. F., Sailor, K. A., Shih, T. Y., Xu, Q., Swaminathan, G., … Hanes, J. (2012). A dense poly(ethylene glycol) coating improves penetration of large polymeric nanoparticles within brain tissue. *Science Translational Medicine*, *4*(149), 149ra119. http://dx.doi.org/10.1126/scitranslmed.3003594.

Nazem, A., & Mansoori, G. A. (2008). Nanotechnology solutions for Alzheimer's disease: Advances in research tools, diagnostic methods and therapeutic agents. *Journal of Alzheimer's Disease*, *13*(2), 199–223.

Opanasopit, P., Aumklad, P., Kowapradit, J., Ngawhiranpat, T., Apirakaramwong, A., Rojanarata, T., & Puttipipatkhachorn, S. (2007). Effect of salt forms and molecular weight of chitosans on in vitro permeability enhancement in intestinal epithelial cells (Caco-2). *Pharmaceutical Development and Technology*, *12*(5), 447–455. http://dx.doi.org/10.1080/10837450701555901.

Osterberg, E., Bergstrom, K., Holmberg, K., Schuman, T. P., Riggs, J. A., Burns, N. L., … Harris, J. M. (1995). Protein-rejecting ability of surface-bound dextran in end-on and side-on configurations: Comparison to PEG. *Journal of Biomedical Materials Research*, *29*(6), 741–747. http://dx.doi.org/10.1002/jbm.820290610.

Owens, D. E., & Peppas, N. A. (2006). Opsonization, biodistribution, and pharmacokinetics of polymeric nanoparticles. *International Journal of Pharmaceutics*, *307*(1), 93–102. http://dx.doi.org/10.1016/j.ijpharm.2005.10.010.

Pardridge, W. M. (1991). *Peptide drug delivery to the brain*. New York: Raven Press Ltd.

Pardridge, W. M. (2005). The blood-brain barrier: Bottleneck in brain drug development. *NeuroRx*, *2*(1), 3–14. http://dx.doi.org/10.1602/neurorx.2.1.3.

Pardridge, W. M. (2010). *Brain drug targeting: The future of brain drug development* (1st ed.). Cambridge: Cambridge University Press.

Patel, D., Naik, S., & Misra, A. (2012). Improved transnasal transport and brain uptake of tizanidine HCl-loaded thiolated chitosan nanoparticles for alleviation of pain. *Journal of Pharmaceutical Sciences*, *101*(2), 690–706. http://dx.doi.org/10.1002/jps.22780.

Polt, R., Porreca, F., Szabo, L. Z., Bilsky, E. J., Davis, P., Abbruscato, T. J., … Hruby, V. J. (1994). Glycopeptide enkephalin analogues produce analgesia in mice: Evidence for penetration of the blood-brain barrier. *Proceedings of the National Academy of Sciences of the United States of America*, *91*(15), 7114–7118.

Ramalingam, P., & Ko, Y. T. (2015). Enhanced oral delivery of curcumin from N-trimethyl chitosan surface-modified solid lipid nanoparticles: Pharmacokinetic and brain distribution evaluations. *Pharmaceutical Research*, *32*(2), 389–402. http://dx.doi.org/10.1007/s11095-014-1469-1.

Reis, C. P., Ribeiro, A. J., Houng, S., Veiga, F., & Neufeld, R. J. (2007). Nanoparticulate delivery system for insulin: Design, characterization and in vitro/in vivo bioactivity. *European Journal of Pharmaceutical Sciences*, *30*(5), 392–397. http://dx.doi.org/10.1016/j.ejps.2006.12.007.

Roman, M., Dong, S., Hirani, A., & Lee, Y. W. (2009). Cellulose nanocrystals for drug delivery. In K. J. Edgar, T. Heinze, & C. M. Buchanan (Eds.), *Polysaccharide materials: Performance by design* (pp. 81–91).Washington, DC: American Chemical Society. Chapter 4.

Sakane, T., Tanaka, C., Yamamoto, A., Hashida, M., Sezaki, H., Ueda, H., & Takagi, H. (1989). The effect of polysorbate 80 on brain uptake and analgesic effect of D-kyotorphin. *International Journal of Pharmaceutics*, *57*(1), 77–83.

Sarei, F., Dounighi, N. M., Zolfagharian, H., Khaki, P., & Bidhendi, S. M. (2013). Alginate nanoparticles as a promising adjuvant and vaccine delivery system. *Indian Journal of Pharmaceutical Sciences*, *75*(4), 442–449. http://dx.doi.org/10.4103/0250-474X.119829.

Schirmacher, A., Winters, S., Fischer, S., Goeke, J., Galla, H. J., Kullnick, U., ... Stogbauer, F. (2000). Electromagnetic fields (1.8 GHz) increase the permeability to sucrose of the blood-brain barrier in vitro. *Bioelectromagnetics*, *21*(5), 338–345.
Schubert, S., Delaney, J. T., & Schubert, U. S. (2011). Nanoprecipitation and nanoformulation of polymers: From history to powerful possibilities beyond poly(lactic acid). *Soft Matter*, 7(5), 1581–1588. http://dx.doi.org/10.1039/c0sm00862a.
Serrano Lopez, D. R., & Lalatsa, A. (2013). Peptide pills for brain diseases? Reality and future perspectives. *Therapeutic Delivery*, *4*(4), 479–501. http://dx.doi.org/10.4155/tde.13.5.
Serrano, D. R., Lalatsa, A., Dea-Ayuela, M. A., Bilbao-Ramos, P. E., Garrett, N. L., Moger, J., ... Uchegbu, I. F. (2015). Oral particle uptake and organ targeting drives the activity of amphotericin B nanoparticles. *Molecular Pharmaceutics*, *12*(2), 420–431. http://dx.doi.org/10.1021/mp500527x.
Siew, A., Le, H., Thiovolet, M., Gellert, P., Schatzlein, A., & Uchegbu, I. (2012). Enhanced oral absorption of hydrophobic and hydrophilic drugs using quaternary ammonium palmitoyl glycol chitosan nanoparticles. *Molecular Pharmaceutics*, *9*(1), 14–28. http://dx.doi.org/10.1021/mp200469a.
Simsek-Ege, F. A., Bond, G. M., & Stringer, J. (2003). Polyelectrolye complex formation between alginate and chitosan as a function of pH. *Journal of Applied Polymer Science*, *88*(2), 346–351. http://dx.doi.org/10.1002/App.11989.
Singh, R. S., Kaur, N., & Kennedy, J. F. (2015). Pullulan and pullulan derivatives as promising biomolecules for drug and gene targeting. *Carbohydrate Polymers*, *123*, 190–207. http://dx.doi.org/10.1016/j.carbpol.2015.01.032.
Tao, H. Q., Meng, Q., Li, M. H., Yu, H., Liu, M. F., Du, D., ... Peng, H. S. (2013). HP-beta-CD-PLGA nanoparticles improve the penetration and bioavailability of puerarin and enhance the therapeutic effects on brain ischemia-reperfusion injury in rats. *Naunyn-Schmiedeberg's Archives of Pharmacology*, *386*(1), 61–70. http://dx.doi.org/10.1007/s00210-012-0804-5.
Toman, P., Lien, C. F., Ahmad, Z., Dietrich, S., Smith, J. R., An, Q., ... Barbu, E. (2015). Nanoparticles of alkylglyceryl-dextran-graft-poly(lactic acid) for drug delivery to the brain: Preparation and in vitro investigation. *Acta Biomaterialia*, *23*, 250–262. http://dx.doi.org/10.1016/j.actbio.2015.05.009.
Trapani, A., De Giglio, E., Cafagna, D., Denora, N., Agrimi, G., Cassano, T., ... Trapani, G. (2011). Characterization and evaluation of chitosan nanoparticles for dopamine brain delivery. *International Journal of Pharmaceutics*, *419*(1–2), 296–307. http://dx.doi.org/10.1016/j.ijpharm.2011.07.036.
Uchegbu, I. F., Sadiq, L., Arastoo, M., Gray, A. I., Wang, W., Waigh, R. D., & Schatzlein, A. G. (2001). Quaternary ammonium palmitoyl glycol chitosan—A new polysoap for drug delivery. *International Journal of Pharmaceutics*, *224*(1–2), 185–199.
van der Merwe, S. M., Verhoef, J. C., Verheijden, J. H., Kotze, A. F., & Junginger, H. E. (2004). Trimethylated chitosan as polymeric absorption enhancer for improved peroral delivery of peptide drugs. *European Journal of Pharmaceutics and Biopharmaceutics*, *58*(2), 225–235. http://dx.doi.org/10.1016/j.ejpb.2004.03.023.
Vandergrift, W. A., Patel, S. J., Nicholas, J. S., & Varma, A. K. (2006). Convection-enhanced delivery of immunotoxins and radioisotopes for treatment of malignant gliomas. *Neurosurgical Focus*, *20*(4), E13. http://dx.doi.org/10.3171/foc.2006.20.4.8.
Wang, P., Kouyoumdjian, H., Zhu, D. C., & Huang, X. (2015). Heparin nanoparticles for beta amyloid binding and mitigation of beta amyloid associated cytotoxicity. *Carbohydrate Research*, *405*, 110–114. http://dx.doi.org/10.1016/j.carres.2014.07.020.
Whiteford, H. A., Ferrari, A. J., Degenhardt, L., Feigin, V., & Vos, T. (2015). The global burden of mental, neurological and substance use disorders: An analysis from the Global Burden of Disease Study 2010. *PloS One*, *10*(2), e0116820. http://dx.doi.org/10.1371/journal.pone.0116820.

Win, K. Y., & Feng, S. S. (2005). Effects of particle size and surface coating on cellular uptake of polymeric nanoparticles for oral delivery of anticancer drugs. *Biomaterials*, *26*(15), 2713–2722. http://dx.doi.org/10.1016/j.biomaterials.2004.07.050.

Witt, K. A., Gillespie, T. J., Huber, J. D., Egleton, R. D., & Davis, T. P. (2001). Peptide drug modifications to enhance bioavailability and blood-brain barrier permeability. *Peptides*, *22*(12), 2329–2343.

Wu, W., Lee, S. Y., Wu, X. B., Tyler, J. Y., Wang, H., Ouyang, Z., … Cheng, J. X. (2014). Neuroprotective ferulic acid (FA)-glycol chitosan (GC) nanoparticles for functional restoration of traumatically injured spinal cord. *Biomaterials*, *35*(7), 2355–2364. http://dx.doi.org/10.1016/j.biomaterials.2013.11.074.

Yang, X. Y., Li, Y. X., Li, M., Zhang, L., Feng, L. X., & Zhang, N. (2013). Hyaluronic acid-coated nanostructured lipid carriers for targeting paclitaxel to cancer. *Cancer Letters*, *334*(2), 338–345. http://dx.doi.org/10.1016/j.canlet.2012.07.002.

Yoksan, R., Jirawutthiwongchai, J., & Arpo, K. (2010). Encapsulation of ascorbyl palmitate in chitosan nanoparticles by oil-in-water emulsion and ionic gelation processes. *Colloids and Surfaces. B, Biointerfaces*, *76*(1), 292–297. http://dx.doi.org/10.1016/j.colsurfb.2009.11.007.

Zhang, P., Ling, G., Pan, X., Sun, J., Zhang, T., Pu, X., … He, Z. (2012). Novel nanostructured lipid-dextran sulfate hybrid carriers overcome tumor multidrug resistance of mitoxantrone hydrochloride. *Nanomedicine*, *8*(2), 185–193. http://dx.doi.org/10.1016/j.nano.2011.06.007.

Zhu, L. Y., Lin, D. Q., & Yao, S. J. (2010). Biodegradation of polyelectrolyte complex films composed of chitosan and sodium cellulose sulfate as the controllable release carrier. *Carbohydrate Polymers*, *82*(2), 323–328. http://dx.doi.org/10.1016/j.carbpol.2010.04.062.

CHAPTER FIVE

Gold Nanoparticles for Imaging and Drug Transport to the CNS

D. Male[1], R. Gromnicova, C. McQuaid
The Open University, Milton Keynes, United Kingdom
[1]Corresponding author: e-mail address: david.male@open.ac.uk

Contents

Abstract

Gold nanoparticles with a core size of 2 nm covalently coated with glycans to maintain solubility, targeting molecules for brain endothelium, and cargo molecules hold great potential for delivery of therapies into the CNS. They have low toxicity, pass through brain endothelium in vitro and in vivo, and move rapidly through the brain parenchyma. Within minutes of infusion the nanoparticles can be detected in neurons and glia. These

International Review of Neurobiology, Volume 130
ISSN 0074-7742
http://dx.doi.org/10.1016/bs.irn.2016.05.003

nanoparticles are relatively easy to synthesize in association with their surface ligands. They can be detected by electron microscopy, ICP-mass spectrometry, and spectroscopy. However, modification of the basic gold nanoparticle is required for in vivo imaging by MR or radioactive methods. Depending on their surface coat, the nanoparticles cross the brain endothelium by the plasma membrane/cytosolic route (passive transport) or by vesicular transcytosis (active transport). A primary aim of current research is to improve the biodistribution of the nanoparticles for CNS drug delivery. Smaller gold nanoparticles are removed rapidly via the kidney, while larger nanoparticles are taken up by mononuclear phagocytes in various tissues. Receptors selectively located on brain endothelium can act as targets for the nanoparticles, to increase their delivery to the brain.

1. INTRODUCTION

The blood–brain barrier (BBB), formed by microvascular endothelium, pericytes, and astrocytes, prevents the movement of most large hydrophilic molecules (>1 kDa) and many pharmaceuticals into the brain. The key elements of the barrier are the continuous tight junctions between endothelial cells, which prevent molecules from diffusing into the brain by the paracellular route, and ABC transporters that actively pump xenobiotics out of the brain (Sarkadi et al., 2006; Wolburg & Lippoldt, 2002). In addition, brain endothelial cells have enzymes and cytochromes which degrade many small drugs, and only low levels of pinocytotic activity and vesicular transcytosis (Marchesi, 1965), which could potentially transport larger molecules across the cells.

The BBB presents a particular problem for the treatment of many CNS diseases. More than 95% of the drugs that could potentially treat neurological diseases do not cross the endothelial component of the barrier. Consequently, many of the treatments that have been used successfully to treat animal models of disease have not been translated into treatments in humans (Baker et al., 2011; Baker & Hankey, 2003; Sloane et al., 2009). In many cases this is due to the difficulty in delivering the treatment to the brain, particularly where repeated or regular treatment is required.

The problem presented by the BBB has become increasingly acute in the last 10 years as biological therapeutic agents have been developed. Antibodies, antibody domains, cytokines, and oligonucleotides including miRNA and siRNA are hydrophilic and range in size from 6 to 160 kDa; they are all effectively excluded from the CNS, since they cannot penetrate the continuous tight junctions between brain endothelial cells.

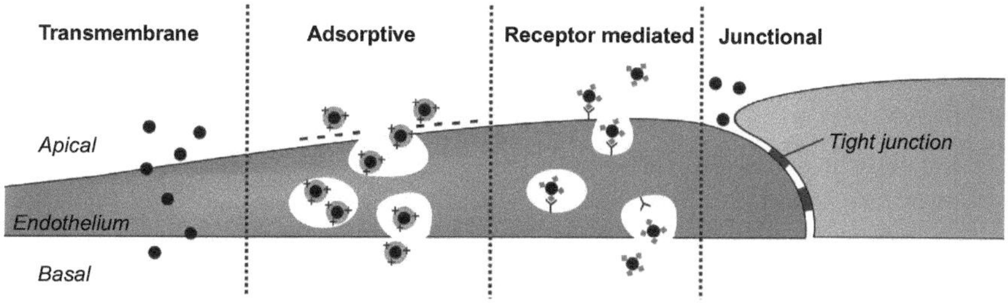

Fig. 1 Possible routes for movement of gold nanoparticles across the brain endothelium. Small nanoparticles may cross the apical and basal plasma membranes and move through the cytosol. Adsorptive endocytosis depends on nonspecific interactions between the nanoparticles and the cell surface, which may depend on the charge of the interacting components. Endocytosis of nanoparticles carrying specific ligands may be enhanced by binding to receptors present on brain endothelium (transferrin, insulin, LDL, etc.). The continuous tight junctions of brain endothelium effectively prevent movement of gold nanoparticles via the paracellular route.

This has led to an intensive search for carriers which could transport these biomolecules across brain endothelium (Fig. 1).

Gold nanoparticles have many potential advantages in this area. They can be easily synthesized; they are chemically stable, and they have now been tested in nanomedicine for both diagnosis and therapy (Kanwar et al., 2012). Generally, they show low immunogenicity and toxicity (reviewed in De Jong & Borm, 2008). These nanoparticles consist of a gold core with covalently or noncovalently attached surface ligands. The gold core is inert, but it does interact with biological material and can have biological effects. To address this, a variety of sizes and surface modifications have been investigated which affect the specific behavior of the nanoparticles (Chen et al., 2009; Etame et al., 2011; Sonavane, Tomoda, & Makino, 2008).

2. THE CHEMISTRY AND PROPERTIES OF GOLD NANOPARTICLES

Gold nanoparticles have a long history. Since medieval times, where panes of glass in churches were stained with gold nanoparticles, people have been intrigued by properties of nanogold. For example, gold nanoparticles are red in color rather than gold as their bulk relative. They have interesting optical properties where from the color of the solution we can easily determine their size due to the quantum effect. In this section, we will discuss the properties of the gold core and how they affect the overall nanoparticle stability, and ligands that can be attached to gold nanoparticles.

2.1 The Gold Core

Gold nanoparticles have several distinct properties which must be considered, before selecting a potential nanocarrier. These factors include (a) the shape of the gold core; (b) the size, which determines the surface coverage of ligands; and (c) the way the gold core has been synthesized which will influence the stability of these nanoparticles. All of these elements then determine the chemical and physical properties of the nanoparticles.

Gold nanoparticles exist in several shapes, such as nanospheres, nanorods, nanocages, and nanoshells. The majority of applied research has focused on the use of nanospheres or sometimes nanorods; these two types will be detailed here. For information on the chemical properties of gold nanocages and nanoshells, there are a number of excellent reviews (Hu et al., 2006; Jelveh & Chithrani, 2011; Xia et al., 2011).

Spherical gold nanoparticles have been prepared in various sizes—for biological applications, most of the nanoparticles used have diameters of 1.5–20 nm. The size determines how many atoms on the surface of the nanoparticle have the ability to react with a molecule and form a covalent bond. This depends on the steric ability of ligands to contact the surface of the gold core and can thus be calculated. For example, Hostetler et al. (1996) calculated that gold nanoparticles of diameter of 2.2 nm have about 162 surface atoms but not all of them can form bonds—in this example 92 atoms would be able to form bonds. Larger nanoparticles, such as 5.6 nm in diameter, have 1472 surface atoms but only 603 would be able to react with ligands (Hostetler et al., 1998). This example shows the relationship of surface-to-volume area where small nanoparticles have over 50% of their surface available for reaction. The functional capacity of ligand binding is also dependent on the stereochemistry and bulk of the ligands themselves. Furthermore, the size of the nanoparticles has a major effect on their tissue distribution, mechanisms of cellular uptake, and biological effects.

Spherical gold nanoparticles can be synthesized in two main ways. Generally, a gold salt is reduced causing formation of gold nanoparticles. In the first method, discovered by Turkevich, Stevenson, and Hillier (1951), the gold salt is reduced and stabilized by sodium citrate, yielding gold nanoparticles of 20 nm in a dilute solution. A broader size range can be obtained by varying the ratio of gold to sodium citrate (Frens, 1973). This second method, also called the "Brust–Schiffrin" method (Brust et al., 1994), revolutionized the field because it generated very stable nanoparticles. Here, the reducing agent is sodium borohydride and tetraoctylammonium bromide functions as a stabilizing agent, producing nanoparticles of about 5 nm. Many modifications of this method have resulted in varied size, from very small (1.5 nm) upward. Also a ligand coating of monolayer-protected thiols has been used to stabilize and coat the nanoparticles during the synthesis (Ackerson, Jadzinsky, & Kornberg, 2005; Templeton et al., 1998).

Rod-shaped gold nanoparticles can be synthesized in three ways. First, in a template method (Hulteen & Martin, 1997; Martin, 1994), gold is deposited into pores of alumina or polycarbon membranes. The membrane is then dissolved and the gold nanorods are released. The size of the membrane pore determines the size of the resulting nanorod. The second method, electrochemical synthesis (Perez-Juste et al., 2005), yields large quantities of gold nanorods. Here, two electrochemical plates are needed: gold as an anode and platinum as a cathode. The electrolytic solution contains surfactants and stabilizers, and the process also involves ultrasonication to help release the forming nanorods. The third method is based on synthesis of spherical gold nanoparticles that are allowed to grow while controlling for their aspect ratio in this seeded growth method (Perez-Juste et al., 2005).

2.2 Physical Properties of the Gold Core

Unlike other nanoparticles, gold nanoparticles have very useful and tuneable optical properties, called plasmonic properties, due to the gold having a surface plasmon resonance (SPR) (Fig. 2). After light strikes the nanoparticle, the forming oscillating field induces resonation of surface electrons (conduction-band electrons). This causes separation of the charge from the core electrons leading to a formation of an ionic core. A dipole oscillation forms along the direction of the original light radiation—this is called

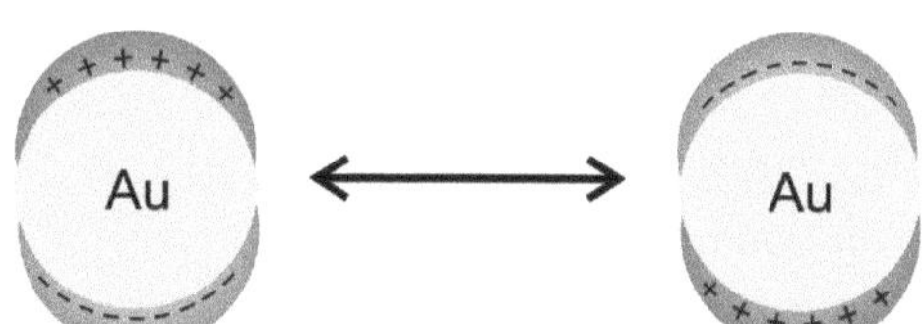

Fig. 2 Plasmonic nanoparticles. Plasmon resonance occurs due to oscillations of free electrons within the nanoparticle. Electromagnetic waves interact with the nanoparticle forcing the electrons to oscillate with the same frequency. This is called the resonant frequency and it depends on the size of the nanoparticle. This property causes gold nanoparticles of defined size/shape to absorb distinct wavelengths of light causing them to have distinctive colors and the ability to quench locally generated fluorescence emissions.

the SPR. It can be measured by a spectrophotometer in the UV–visible range as it induces a strong absorption band at around 520 nm. SPR depends on the size, composition, shape, structure, or environment of the nanoparticle (Huang et al., 2007). For spherical gold nanoparticles smaller than 10 nm, the absorbance peak is less apparent which is due to collisions of surface electrons that cause phase changes. Also, the peak disappears for nanoparticles of about 1.6 nm (Hostetler et al., 1998). The shape of the nanoparticles also affects the spectrum; gold nanorods have two oscillations along both of their axes, resulting in two absorbance peaks (Huang & El-Sayed, 2010; Huang et al., 2007). The position of the SPR peak can change by addition of various ligands onto nanoparticles. Moreover, the change in color of small nanoparticles after aggregation is also due to plasmonic coupling of nanoparticles where the color can change from red to gray-purple (Hu et al., 2006).

SPR also enhances several optical properties of spherical gold nanoparticles, and according to Gustav Mie's theory (Mie, 1908), this effect would depend on the size of the nanoparticle. Nanoparticles larger than 40 nm absorb light which they can either reemit as fluorescence or scatter it (Huang et al., 2007; Swierczewska, Lee, & Chen, 2011). There are different types of scattering, such as surface-enhanced Raman scattering or luminescent scattering. Ligands bound to the nanoparticles may affect this scattering too. These optical properties of gold nanoparticles are useful for gold detection, measuring gold concentration, and characterization of nanoparticle size, shape, or ligand coating (Huang et al., 2007). On the other hand, nanoparticles smaller than 40 nm still absorb light, but rather have an increased ability to quench fluorescence (Swierczewska et al., 2011). When a fluorophore is at a certain distance from a gold nanoparticle (Dulkeith et al., 2005), its excitation/emission property is changed—the electrons that would emit light are changed by the SPR (Kang et al., 2011).

Another useful behavior of gold nanoparticles is to generate heat after absorbing light. The nanoparticle can either dissipate the heat to its surroundings, melt, in the case of nanorods (Link et al., 1999), or cause a structural change and ablation of nanoparticles. The heating property has been widely used for photothermal therapy (Huang & El-Sayed, 2010). For example, gold–silica nanoshells have been shown to enter gliomas and reduce the tumor mass after infrared irradiation of mice (Day et al., 2011).

Gold nanoparticles also have subtle magnetic properties (Nealon et al., 2012) which can be detected by a superconducting quantum interference device; however, the explanation for this unusual phenomenon is still controversial (Nealon et al., 2012). More often magnetic nanoparticles of iron oxide have been coated with gold and used for imaging purposes (Ahmad et al., 2012).

2.3 Ligands

Ligands, molecules attached to the nanoparticle, are important for designing a successful nanocarrier. Ligands may serve as targeting molecules, therapeutic cargo, or imaging/tracking molecules. The way in which ligands are attached affects their stability and the rate of release of cargo molecules, which may be nucleic acids, proteins, or drugs, as discussed in this section.

Ligands also provide stability to gold nanoparticles in water. Bare colloidal spherical nanoparticles can stabilize themselves by forming a negative charge that keeps them separate; however, they can still aggregate over time and form larger nanoparticles. Unlike its bulk relative, nanosized gold is reactive, thus its use as a catalyst in engineering (Hvolbæk et al., 2007). In order to improve stability and solubility of gold nanoparticles, the gold core has to be coated with ligand molecules, which can be charged; neutral (Gromnicova et al., 2013), negative (Schleh et al., 2012), or positive (Schleh et al., 2012) are all possible to use. A difficulty may occur when ligands of opposite charges are used to coat nanoparticles as they cause the nanoparticles to aggregate, rendering them useless for biological studies. For example, adding negatively charged nucleic acid onto a positively charged coated nanoparticle will result in aggregation. It is possible to go around this problem by coating the nanoparticles layer by layer (Elbakry et al., 2009) or by using long neutrally charged molecules such as polyethylene glycol to keep the nanoparticles separate (Lee et al., 2008).

Ligands can be attached to gold nanoparticles in several ways: covalently, noncovalently, or by encapsulation. The covalent approach results in a ligand that is more stably attached in various conditions. Noncovalent or steric attachment is also sometimes used for delivering cargo due to faster release from the nanoparticle. The major covalent linkage to the gold core is a thiol bond and its variations. For example, sulfur-containing ligands such as xanthates, disulfides, dithiols, trithiols, or tetrathiols can be used. But also other covalent ligand attachments have been prepared (with phosphine or amine) (Daniel & Astruc, 2004). The strength of the covalent bond is crucial for delivery applications. The thiol bond is a strong linkage and thus employing several thiols to attach a single molecule to the gold core or using cyclic thiols such as a 1,2-dithiane end group (Letsinger et al., 2000) increases the strength of ligand binding and thus cargo stability.

Another covalent type of bond for attaching ligands is the amine bond. This bond is much weaker than the thiol bond and is rather comparable in its strength to noncovalent attachment of ligands (Hoft et al., 2007). Amine–gold bond can be sometimes problematic, for example, when attaching antibiotics to gold nanoparticles where an aggregation may occur. Many antibiotics contain several amines and thus when mixed with gold nanoparticles, they aggregate (Nirmala Grace & Pandian, 2007). On the other hand, when a molecule with a single amine group is attached, the resulting nanoparticle is stable against aggregation and can be characterized (Rai, Prabhune, & Perry, 2010).

Thiol-bound ligands to the gold core can exchange with other thiolated ligands via a very simple ligand-exchange reaction first described by Murphy (Hostetler et al., 1996; Templeton, Wuelfing, & Murray, 2000). The excess concentration of thiolated molecules can react with the gold core and covalently bind, releasing the original thiol-bound ligand (Fig. 3). This reaction is used for coating of gold nanoparticles with ligands that may be problematic

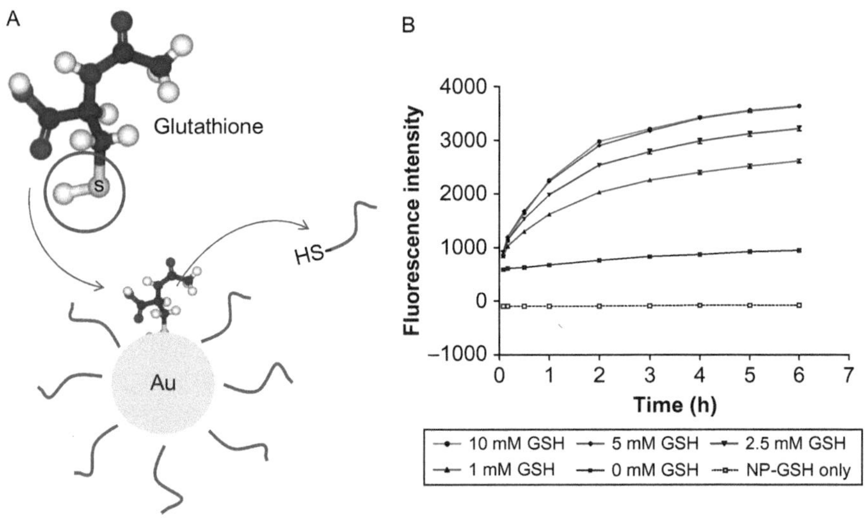

Fig. 3 Effect of reduced glutathione on release of fluorescent ligand from gold nanoparticles coated with glutathione-FITC. (A) Scheme of thiolated ligand release from gold nanoparticles in the presence of glutathione by exchange reaction. (B) Rate of release of glutathione-FITC from nanoparticles was measured by release from fluorescence quenching in the presence of 1–10 mM glutathione (GSH). Background fluorescence of nanoparticles with ligand glutathione only (NP-GSH). Representative graph from two independent experiments, the points are means ± SEM of four technical replicates; fluorescence intensity is on arbitrary scale. Release of the ligand is time dependent and related to redox potential (GSH concentration).

to add to bare nanoparticles during the synthesis. The reaction works by simply mixing the thiolated ligand and the stable coated nanoparticle with the ligand in molar excess. Over time the place-exchange reaction proceeds and results in a mixture of nanoparticles with various proportions of the new ligand. For example, a fluorescent dye alkanethiol-BODIPY was attached this way to 2 nm gold nanoparticles (Atukorale et al., 2015). Moreover, this ability of thiolated molecules to exchange from the nanoparticle can also result in release of the ligand. The relatively high cytosolic concentration of glutathione and other thiols can cause release of ligands and cargo from the gold core, inside the cell (see Section 5.3) (Hong et al., 2006; Verma et al., 2004).

Noncovalent attachment is mostly based on electrostatic interactions of cargo molecules with a charged ligand which is covalently attached to the gold core. Mixed alkanethiol-covered nanoparticles, termed as monolayer mixed-protected gold clusters or MMPC, are based on monolayer-protected clusters (Hostetler et al., 1996, 1998; Templeton et al., 1998) and can be used to carry a charge. For example, Rotello's group (McIntosh et al., 2001) prepared gold nanoparticles (2 nm gold core) coated with neutral alkanethiol and with positively charged trimethylammoniumundecanethiol. The positive amine groups were able to attract and hold DNA strands (McIntosh et al., 2001), and the resulting nanoparticles were used to transfect plasmids into human embryonic kidney cells (293T) (Sandhu et al., 2002). Similarly, polyethylenimine (PEI) attached to the gold nanoparticle can be effective in noncovalently holding plasmid DNA and transfecting cells (Thomas & Klibanov, 2003).

Another way that gold nanoparticles can hold ligands is to encapsulate them. For example, doxorubicin has been transported in gold nanocages into cancer cells (MDA-MB-231) and released by radiation (Wang et al., 2014). However, the encapsulation approach may be limited to target cancer cells only, due to the toxicity associated with heating of the nanoparticles after irradiation.

2.4 Therapeutic Cargo Molecules and Targeting Molecules

Gold nanoparticles can deliver several therapeutically useful molecules, such as nucleic acids, proteins, drugs, or enzymes. Nucleic acids that have been attached to gold nanoparticles include plasmids, DNA, and RNA, both single (ss) or double (ds) stranded. Nucleic acids cannot penetrate the cell normally due to their negative charge. Gold nanoparticles can potentially carry nucleic acids as they can easily attach to them, covalently or noncovalently. Interestingly, covalently attached ssDNA on gold nanoparticles was transported into cells even though no other surface-bound ligands were present and the resulting nanoparticle still had a negative charge (Rosi et al., 2006). These nanoparticles successfully transfected the cells and were also able to change the gene expression. Also, more stable locked nucleic acids have been delivered by covalent attachment to gold nanoparticles into A549 lung carcinoma cells (Seferos et al., 2007). DNA-functionalized nanoparticles can be also used as a delivery system, for example, platinum (IV) prodrug has been attached to amine-functionalized DNA, covalently bound to nanoparticles (Dhar et al., 2009). Another advantage of using nucleic acid-coated gold nanoparticles is that the gold core may be dissolved and thus removed, as shown in the work of Cutler et al. (2011), where a gold nanoparticle was coated with a cross-linked shell of modified oligonucleotides and the gold core was then dissolved, leaving a hollow nanoparticle (Cutler et al., 2011). Strictly speaking, in this case the gold nanoparticles are a route for the synthesis of another class of nanoparticle. However, the use of such a nanoparticle may be particularly relevant in chronic diseases where several doses a day are necessary and which solve the potential problem of clearance of gold out of the body.

Delivering RNA is also very relevant for therapeutic purposes, even though the preparation of these RNA nanoparticles may be more challenging due to nuclease degradation. Gold nanoparticles have been shown to deliver siRNAs, hairpin RNAs, or RNA aptamers. siRNA for gene silencing can be attached to the gold nanoparticles either covalently (Giljohann et al., 2009; Oishi et al., 2006) or noncovalently (Guo et al., 2010; Niikura et al., 2014; Zhao et al., 2012). The transfection efficiency for these nanoparticles was similar or higher in both types of attachment of siRNA to the nanoparticle. For example, firefly luciferase siRNA covalently attached to gold nanoparticles had a similar silencing effect to a standard siRNA silencing approach using Lipofectamine (Giljohann et al., 2009). These nanoparticles seemed to convey stability to the RNA because they were better at silencing at 4 days postaddition of nanoparticles. Also, lamin A/C siRNA attached noncovalently to positively charged PEI-coated gold nanoparticles, showed higher silencing efficiency than Lipofectamine transfection (Guo et al., 2010). The release of covalently bound siRNA from gold nanoparticles inside cells is facilitated by the place-exchange reaction with intracellular glutathione.

Another way to facilitate the release of nucleic acids in target cells is by attaching them to gold nanoshells. Upon laser irradiation the cargo is released and escapes from vesicles through pores. An example of this was shown when irradiating gold nanoshells coated with siRNA against green fluorescent protein (Braun et al., 2009). Small hairpin RNA (shRNA) has also been successfully delivered into target cells; shRNA against p53 was attached by hybridizing with covalently bound oligonucleotide to the nanoparticle and was able to reduce expression of p53 (Ryou et al., 2010). The same research group has also shown delivery of RNA aptamer by gold nanoparticles into the nucleus of A549 lung carcinoma cells (Ryou et al., 2011).

Gold nanoparticles can also facilitate transport of proteins into cells. The principle of protein delivery has been shown in delivering transferrin, which has been either attached covalently to the gold core (Yang et al., 2005) or attached covalently via a linker (Choi et al., 2010). These studies showed transport of transferrin into cells (Yang et al., 2005) or into organs (Choi et al., 2010). A cargo protein that has been widely investigated for delivery using gold nanoparticles is tumor necrosis factor alpha (TNFα). TNFα is a potent cytokine which can be used for cancer treatment. Gold nanoparticles carrying this cytokine were able to enter tumor cells in vivo (Visaria et al., 2006) and have since proceeded into clinical trials (Libutti et al., 2010). Insulin for treating diabetes has been also successfully delivered into the body via oral, transmucosal (Bhumkar et al., 2007; Joshi et al., 2006), or nasal (Bhumkar et al., 2007) delivery routes.

Drugs can be also effectively attached to gold nanoparticles and transported into cells. For example, paclitaxel (Gibson, Khanal, & Zubarev, 2007) or doxorubicin (Kim et al., 2010)-coated gold nanoparticles have been prepared; the latter were also able to enter tumor cylindroids (Kim et al., 2010). By designing the attachment of the drugs it is possible to modulate release within cells, for example, in the anticancer drug 5-fluorouracil has been attached to the nanoparticles via a linker containing *o*-nitrobenzyl group, which can be cleaved by UV irradiation (Agasti et al., 2009).

3. BIOLOGICAL PROPERTIES OF GOLD NANOPARTICLES

Gold nanoparticles are often used as carriers or imaging reagents because of their low level of interaction with biological molecules. Many studies have investigated toxicity of gold nanoparticles for cells or whole organisms, although there is relatively little on their actions on neurons and glia. Another important consideration is how gold nanoparticles interact with vascular endothelium in different tissues, as this has a major effect on their tissue distribution.

3.1 Tissue Distribution

In developing nanocarriers for the CNS, the tissue distribution of the nanoparticles is a critical consideration. Most studies (Khlebtsov & Dykman, 2011) have used intravascular injection or intraperitoneal injection. The diverse nanoparticles and protocols used in the different studies make generalization difficult, but it is clear that nanoparticle size and surface coatings have a major effect on their biodistribution (Table 1). For nanoparticles of 10 nm, levels of 0.2% of the injected dose were found in the brain after 24 h. For larger nanoparticles (50 and 250 nm), the levels in the brain were very low, and a high proportion was taken up by phagocytic cells in the liver and spleen.

In our study using 5 nm nanoparticles coated with glucose, 10 min after intracarotid injection, the highest level of uptake was seen in kidney with lower levels in the liver, and comparable levels in lung and cerebral cortex (Gromnicova et al., 2016). Although the amount within the brain corresponded to ~1.5% of the serum level, this is still considerably higher than the equilibrium CSF:serum ratio of proteins of similar size (CSF:serum ratio <0.1%); this fact may be accounted for by the transport of nanoparticles into the brain across the endothelium, avoiding the tight junctions of the paracellular diffusion route. Nanoparticle levels within the cortex were

Table 1 Biodistribution of Gold Nanoparticles

Tissue/Organ	10 nm	50 nm	250 nm
Blood	36	31	16
Liver	46	21	31
Spleen	2.2	1.3	1.2
Lungs	0.3	2.3	<0.1
Kidney	1.0	0.2	0.1
Brain	0.2	ND	ND

Percentage of injected dose in mice at 24 h following intravenous administration of 0.3–0.5 μg gold per animal. ND, not detectable. Based on data calculated in Khlebtsov and Dykman (2011).

significantly higher than in the cerebellum, and there was evidence for local variation in the levels within the cortex. This effect may be partly caused by streaming of the injected dose, which occurs with intracarotid infusion (Saris et al., 1988). Similar studies using PEG-amine-coated nanoparticles showed higher levels within the CNS, but a similar overall distribution (Fig. 4).

Similarly, albumin-coated 15 nm gold nanoparticles showed accumulation in the head within 30 min of infusion peaking by 19 h (Sousa et al., 2010). The nanoparticles, detected by fluorescence imaging, had crossed the BBB; however, the route of entry could not be determined. Within the brain higher levels were present in the cortex than in the cerebellum, similar to our study of the glucose-coated 5 nm gold nanoparticles.

There is a consensus from several in vivo studies that 20 nm is the upper size limit for entry of untargeted nanoparticles into the brain (Khlebtsov & Dykman, 2011), and the limit is probably lower for paracellular movement. While these levels of CNS penetration indicate great promise for gold nanocarriers, there is room for improvement: (a) reducing uptake for

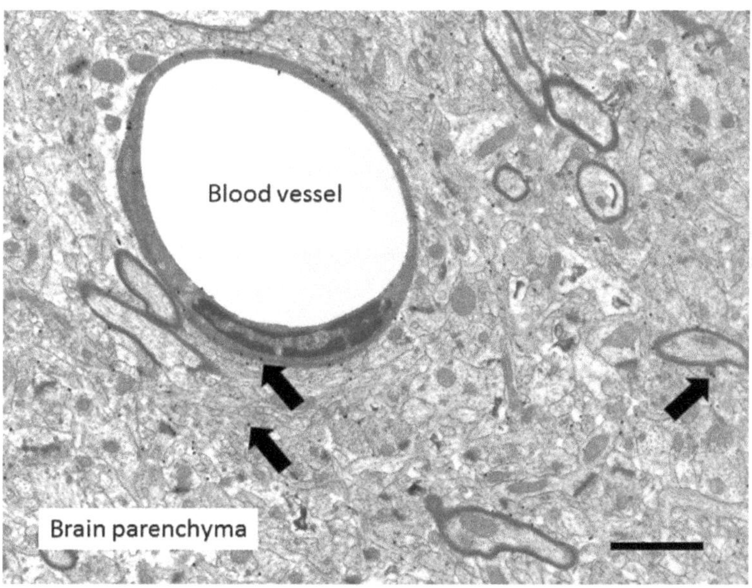

Fig. 4 Gold nanoparticle movement into cortex 2nm gold nanoparticles covalently coated with PEG-amine were infused by intracarotid injection into rats, 10 minutes before animals were perfused/fixed. The 85nm thick, silver-enhanced TEM image shows nanoparticles crossing the endothelium of a microvessel in the cortex. Many nanoparticles (arrowed) are also visible in glia and unmyelinated nerve fibres. Scale bar = 2μM.

smaller nanoparticles (<10 nm) by kidney or (b) reducing uptake of larger nanoparticles (>10 nm) by the mononuclear phagocyte system, particularly in spleen and liver (Liu et al., 2013).

3.2 Toxicity

Gold is normally viewed as nontoxic material. However, nanosized gold has very different physical properties in comparison to its bulk relative (see Section 2). Thus, as the potential reactivity of nanogold increases, the toxicity needs to be assessed before using gold nanoparticles in a biological environment. Toxicological in vitro studies generally confirm low toxicity of gold nanoparticles. Mostly non-CNS cell types have been investigated, although more recent studies have examined relevant cells of the CNS, including brain endothelium (Freese et al., 2012; Gromnicova et al., 2013). Measurement of toxicity is usually made over a short period (eg, using the MTT assay), and these studies have generally confirmed the low toxicity of GNPs. Perhaps the reason for gold nanoparticles to have a toxic potential may be attributed to cellular internalization, a process which differs based on the physicochemical properties of the nanoparticle. These properties are most notably size, charge, surface coating, shape, and even the extracellular environment and composition of tissue fluids as discussed later. A key point is which cellular compartments the nanoparticles pass through and whether they accumulate within the cells.

Size is a major factor determining the ability of gold nanoparticles to traverse the plasma membrane of cells and enter the cytosol; smaller nanoparticles are more effectively internalized than larger NPs by this route (Van Lehn et al., 2013). Depending on their coating, gold nanoparticles can have a similar behavior to macromolecules of similar size. Several studies reveal that many gold nanoparticles that do enter the cell are actively internalized by the endocytic pathways (Caballero-Díaz & Valcárcel, 2014). In a study using gold nanoparticles of varying sizes, Pan et al. (2009) found that nanoparticles of size 1.4 nm were more toxic than 15 nm ones. The authors concluded that the larger particles induced death via necrosis, whereas smaller particles initiated apoptosis. Similarly, 1.4 nm nanoparticles were also toxic in a different study investigating toxicity on normal and cancer cells (Tsoli et al., 2005).

Charge can also affect nanoparticle toxicity and ability to be internalized. Positively charged gold nanoparticles have greater cellular uptake than negatively charged ones, perhaps due to the electrostatic interaction with negatively charged proteoglycans on the cell surface (Alkilany & Murphy, 2010; Fröhlich, 2012; Goodman et al., 2004). Brain endothelium in particular has relatively high levels of sulfated proteoglycans which further contribute to its negative charge (dos Santos et al., 1995). Hence the level of toxicity for different cell type is likely to be affected initially by the level of nanoparticle uptake. Once inside the cell, any toxicity may also be affected by charge. For example, Schaeublin et al. (2011) used thiol stabilized, 1.5 nm gold nanoparticles with positive, negative, or neutral ligand coating. All charged nanoparticles caused disruption to cell morphology in a human keratinocyte cell line (HaCat). Moreover, positively and negatively charged nanoparticles caused mitochondrial stress, whereas neutral nanoparticles appeared to activate pathways of apoptosis (Schaeublin et al., 2011).

Surface coating of gold nanoparticles was initially used to stabilize the nanoparticles and/or enhance their ability to traverse the BBB. Previous work from our group found that glucose- or glutathione-coated gold nanoparticles (<5 nm diameter) enter human brain endothelium and were nontoxic over 24 h at concentrations up to 32 μg/mL (Gromnicova et al., 2013). Subsequent work confirmed lack of cytotoxicity on brain endothelium by this class of nanoparticle, with a variety of covalently bound sugar derivatives at concentrations up to 50 μg/mL. In a study on colloidal gold NPs ranging from 3 to 100 nm in diameter, Chen et al. (2009) found them to be nontoxic; however, when the same range of nanoparticles were injected intraperitoneally into mice, they found that gold nanoparticles of 3, 5, 50, and 100 nm did not show harmful effects, while nanoparticles of 8–37 nm were lethal. This strange distribution of sizes in relation to toxicity is difficult to explain. The larger naked nanoparticles (over 50 nm) appear to be unable to circumvent the cell membrane of healthy cells. The smaller nanoparticles (<50 nm) were able to cross the membrane; however, they aggregated within the cell and produced cytotoxic effects.

Shape of the NP can also have an effect; most gold nanoparticles are spherical but other structures have been observed. In one study the cellular response to 20 nm Au nanospheres (20 nm in diameter and coated with MPS) was compared to Au nanorods (16.7 nm diameter, 43.8 nm long with PEG coating) in a HaCat cell line. Although both formed agglomerates in cell media, the nanospheres were found to be nontoxic, while the nanorods were found to significantly decrease cell viability and this was related to mitochondrial damage (Schaeublin et al., 2012). Nanorods and tubes are generally more toxic than spherical nanoparticles, as they can perforate the plasma membrane and vesicle membranes.

Aggregation can also affect toxicity. For gold nanoparticles the charge/coating is largely responsible for their stability and removal of charged ligands or their neutralization by noncovalently bound protein may cause aggregation, by reducing repulsion between the nanoparticles. Aggregates can often form in cell media as well as biological fluids, such as blood plasma and aggregation can alter the formation of protein–nanoparticles complexes, which affects their ability to be internalized. Using a human fibroblast line it was shown that nonaggregated positively charged gold nanoparticles were fourfold less toxic than positively charged aggregated nanoparticles (Yang, Lohse, & Murphy, 2014). Another study, using transferrin-coated gold nanoparticles on three different cell lines, demonstrated that nonaggregated nanoparticles had increased uptake in two of the cell lines (HeLa and A549 cell lines) however the third cell (MDA-MB-435 cell line) had a higher uptake of the largest aggregates (Albanese & Chan, 2011). Such studies indicate that uptake and potential cytotoxicity are dependent both on the nanoparticle and the cell type.

In some studies the toxic properties of gold are a required characteristic. A number of attempts have been made to treat brain tumors, including glioblastoma multiforme with gold, but in these cases the gold is present for radiosensitization or to promote thermal damage to the target cells, not for the intrinsic cytotoxicity of the gold nanoparticle itself (Hainfeld et al., 2013; Joh et al., 2013).

3.3 Intracellular Effects of Gold Nanoparticles

Gold nanoparticles are generally considered to be nontoxic, but if taken up and stored in cells they can cause damage primarily by interaction with DNA and/or generation of ROS, either by catalytic action of the nanoparticles or by mitochondrial damage (Caballero-Díaz & Valcárcel, 2014; Pan et al., 2009). Positively charged NPs are more likely to be localized in the mitochondria, while negatively charged NPs accumulate in endosomes.

Gold is highly electronegative and has a high binding affinity to DNA. Smaller gold nanoparticles (1.4 nm) can fit into the major groove of DNA and inhibit its replication (Li et al., 2013). In another study 12 nm gold nanoparticles uncoated or coated with hyaluronic acid-induced DNA damage in a fibroblast line (Di Guglielmo et al., 2012).

Gold nanoparticles can generate ROS directly, via reactivity of the metal surface, causing release of free ions and ROS directly or by altering the electron transport chain in the mitochondria (Soenen et al., 2011). The effect is likely to depend on nanoparticle size/composition. For example, 1.4 nm gold nanoparticles coated with triphenylphosphine monosulfate, incubated with reducing agents increased cell survival of HeLa cells, and compared to just nanoparticles alone (Pan et al., 2009). Larger gold nanoparticles ranging in size from 30 to 90 nm were incubated with HepG2 cells, and after incubation they increased in ROS production twofold which also corresponded to depletion of glutathione (Mateo et al., 2014).

Currently, the intracellular routes taken by gold nanoparticles are not well understood, and this factor will affect both their function and potential toxicity (Iversen, Skotland, & Sandvig, 2011). Accumulation of gold nanoparticles in lysosomes, with removal of ligands, is likely to result in reduced solubility, aggregation, and slow clearance. Conversely, nanoparticles that remain in the cytosol or endosomal compartments may exchange or release their cargo molecules relatively quickly, remain soluble and be exocytosed for clearance by kidney and/or liver.

3.4 Effects on the BBB

Several groups have considered whether gold nanoparticles can themselves disrupt the BBB. Earlier work suggested that charged nanoparticles (wax emulsion nanoparticles) might be disruptive (Lockman et al., 2004). In one recent study, gold nanoparticles (40 or 640 nm) caused alterations to the occludins, and ZO-1 in umbilical vein endothelial cells, which lead to an instability in endothelial tight junctions. The relevance of umbilical vein endothelium to the BBB is questionable, but the study also found increased leakage of Evans blue into the brain of treated mice in vivo (Li et al., 2015).

This concept is distinct from the use of gold nanoparticles as imaging reagents to determine whether the barrier has been disrupted (Frigell et al., 2014; Sawyer & Kyriakides, 2013). These studies are based on the observation that gold nanoparticles <10 nm may enter the CNS in areas where the barrier is already disrupted.

4. DETECTING GOLD NANOPARTICLES IN CELLS AND TISSUES

4.1 Electron Microscopy

Electron microscopy has been widely used for observing gold nanoparticles in cells and tissues. The use of this technique is based on long experience of immunogold labeling of molecules of interest in biological samples. During imaging, the electron beam hits the sample resulting in several types of information. The electrons that are transmitted through the sample can be focussed to form a transmission electron microscopy (TEM) image. An electron beam also creates secondary electrons that are used as a readout of the surface topography (information collected by scanning electron microscope, SEM), but other types of scattered electrons are also created that are used for advanced methods of imaging. The electron beam hitting the sample also produces characteristic X-rays that can be used for elemental analysis by energy-dispersive X-ray spectroscopy (EDS or EDX). In general, the gold nanoparticles can be detected on the cell surface by SEM or inside cells and their compartments by TEM, but variations exist if an advanced technique is used (as detailed later). Elemental analysis can then help to distinguish gold from other parts of the image and can be particularly helpful in SEM imaging of biological samples. Occasionally atomic force microscopy has also been used to image gold nanoparticles in cells (Boyoglu et al., 2013). However, this technique provides less detail on the cellular topography and subcellular structure; thus nanoparticle localization is not as easily distinguishable as with TEM or SEM.

Both types of standard electron microscopy involve processing of the sample due to the fact that the sample needs to be dehydrated as the imaging generally happens in a vacuum. Therefore, steps of glutaraldehyde fixation, penetration, osmication (with OsO_4), dehydration, and resin infiltration are involved. In vitro, there are two main ways to handle cell cultures by TEM. First, the cells with nanoparticles can be pelleted (Coulter et al., 2012; Oh & Park, 2014; Shukla et al., 2005; Thomas & Klibanov, 2003) and centrifugation is required for all processing steps. Second, the cells can be grown on a porous plastic membrane in a transwell system (Gromnicova et al., 2013; Ye, Dawson, & Lynch, 2015) and handled similar to a tissue. We find that the latter procedure is technically easier and less time consuming and it is also suitable for quantification of gold nanoparticles. Gold nanoparticles under 10 nm may not be visible by standard TEM or SEM and thus they need to be silver enhanced. Silver enhancement, ie, deposition of silver on gold nanoparticles, results in electron-dense nanoparticles of larger size which can be simply manipulated by the length of the incubation time. Moreover, we found that silver enhancement can be particularly useful in locating higher concentrations of gold nanoparticles in tissues which it stains brown (Fig. 5), thus facilitating light microscope analysis of tissue (Gromnicova et al., 2016).

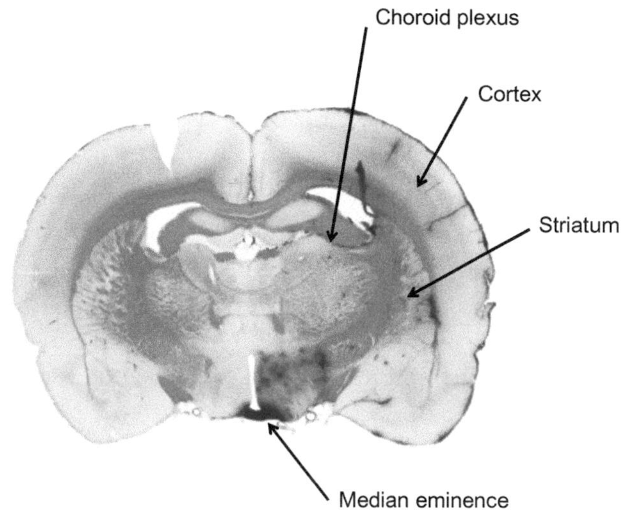

Fig. 5 Silver-enhanced coronal section of rat brain. The rat received an intracarotid injection of PEG-amine-coated gold nanoparticles (*right side*), 10 min before perfusion/fixation. The injected side shows a high amount of brown staining, pointing to regional localization of gold nanoparticles. Areas lacking blood–brain barrier (choroid plexus and median eminence) show high staining. Other areas have diffuse infiltration.

Using TEM for analysis of gold nanoparticle uptake into cells and tissues has many advantages. First, it gives a fixed time image of localization of nanoparticles in subcellular compartments, providing qualitative analysis. Depending on the quality of fixation, compartments such as mitochondria, endosomes, lysosomes, the nucleus, and Golgi apparatus can be distinguished. Moreover, the nanoparticle interactions with the plasma membrane can be also detected. On the extracellular level, the glycocalyx of cells may be visualized with specialized processing techniques (Ebong et al., 2011), but nanoparticles present in this structure are yet to be shown.

Can electron microscopy also be used quantitatively? We have been able to establish a method to use TEM as a quantitative measure for in vitro investigation (Gromnicova et al., 2013; Sreekanthreddy et al., 2015), providing us with a large amount of data on the numbers of nanoparticles localized in different cell compartments. Another advantage of using both SEM and TEM is that they are nondestructive techniques, allowing repeated analysis of the sample and simple long-lasting storage of processed samples. On the other hand, there are some disadvantages to these standard techniques. They are expensive and time-consuming methods. The preparation time for SEM is shorter and may be done within a week. Also, knowledgeable staff is needed for sample preparation and operation of the microscopes. The sample preparation also prohibits real-time imaging. However, some of these hurdles can be overcome by applying advanced or modified techniques of electron microscopy.

In the last 5 years new advanced electron microscopy techniques have allowed greater freedom of choice of sample preparation for imaging of gold nanoparticle in cells or tissues, while reducing time and effort. One of these techniques is high-resolution SEM (HR-SEM) of imaging of nanoparticles in cells and tissue. It has become popular due to shorter time for preparation and less material involved. In standard SEM, the sample needs to be treated with osmium tetroxide to enhance cell contrast; and also coat the surface of the sample to prevent the charging effect—ie, accumulation of negative charge on the sample that may defocus the electron beam. However, this coating may limit the resolution. Recently, a study was shown where the sample preparation technique was modified resulting in detection of small gold nanoparticles (Goldstein et al., 2014). In this study, cells were grown on a glass slide covered with a highly conductive surface of chromium and 10 nm gold nanoparticles that were taken up in the cells were imaged by HR-SEM. Moreover, in a thin 5–10 μm skin section gold nanoparticles down to 5 nm were observed (Goldstein et al., 2014).

Another advanced electron microscopy technique is a synthesis between TEM and SEM at low operation voltage called STEM-in-SEM (STEM for scanning TEM), a technique that has been used to localize nanoparticles in subcellular compartments. Similarly to SEM, secondary electrons are collected and as in TEM, transmitted electrons are also detected. This technique has been used to image relatively thick biological sample (200–300 nm) containing 23 nm gold nanoparticles (García-Negrete et al., 2015). Advantages of STEM-in-SEM over TEM are the availability of elemental analysis to confirm nanoparticle composition and also shorter sample preparation time. STEM can be also employed for nondehydrated biological samples in environmental SEM where water vapor keeps the sample coated with a thin layer of liquid. This technique has been used, for example, to locate and quantify 10–30 nm gold nanoparticles within cells (Peckys & de Jonge, 2014). The same group has also used this technique to visualize 30 nm gold nanoparticles in live cells (Peckys & De Jonge, 2011). The use of nondehydrated/live sample is a great advantage over other electron microscopic techniques even though the subcellular contrast of nanoparticle localization is not as clear as in standard TEM.

4.2 Inductively Coupled Mass Spectrometry

The most sensitive way to measure the amount of nanoparticles in a sample is using ICP-MS (inductively coupled plasma mass spectrometry) as gold concentration is directly related to the amount of nanoparticles in the sample (Krzciuk, 2015; Laborda et al., 2015). ICP-MS is principally used for rapid, precise, and accurate measurements (<1 pg/mL) of metal elements in samples.

Analysis of tissue samples or cells generally requires digestion/separation of the nanoparticles from test media which can be done in a variety of ways including separation by HPLC or chemical digestion. For tissue samples often an aggressive digestion is required, not only to breakdown the tissue but also to dissolve the nanoparticle itself. The methodology usually uses aqua regia, made up of a mixture of nitric acid (HNO_3) and hydrochloric acid (HCl) in a ratio of 1:3. The HNO_3 is a powerful oxidizing agent which can dissolve a small amount of the gold producing Au^{3+} ions these ions are then removed by the Cl^- ions produced by the HCl, and the nitric acid can continue to breakdown the nanoparticles. Tissue breakdown may be accelerated by microwaving the samples. Aqua regia is a highly corrosive solution and must be handled with extreme care in a well ventilated area.

Another option is to use alkaline digestion of the tissue or cellular material by tetramethylammonium hydroxide (TMAH). TMAH can digest the tissue and can cause alterations in the surface coating of the gold nanoparticles but it will not breakdown the gold core. The nanoparticles may then be broken down inside the ICP-MS instrument by the Argon plasma (Gray et al., 2013). Issues can arise from this form of digestion as gold can stick to tubing etc., leading to fluctuations in the signal and variable results. This can be overcome with some of the options mentioned later.

Samples are generally run against a standard curve of gold, with internal iridium or rhodium standards to compensate for any drift in the instrument. To account for the effects of protein and other artifacts on detection, the gold standards should be incorporated into blank tissue/cells matrix.

Gold nanoparticles that do not require digestion can simply be suspended in the matrix and the sample run on the instrument. The technique, media containing the nanoparticles and the digestion reagent, will all influence the time required for the reaction to finish. The sample then can be diluted in a matrix, such as TMAH. The reaction can be stopped by diluting the mixture in a low percentage solution of HCl, often containing iridium tracer.

A number of studies have used inductively coupled plasma atomic emission spectroscopy (ICP-AES) to detect gold (Freese et al., 2012, 2013). This technique nebulizes the sample, introduces it into a plasma, and then analyzes the characteristic emitted spectra. Individual elements are quantitated by comparison of the intensity of emission from the sample with standards. While ICP-AES is faster than ICP-MS, it is much less sensitive (eg, 25 μg/g sample, compared with 0.05 μg/g) and therefore only suitable where very high levels of nanoparticles have been introduced.

4.2.1 Advantages and Disadvantages of ICP-MS

ICP-MS has numerous advantages in measuring the amount of an element in a given sample; these include (a) wide elemental coverage and analytical working range, (b) good control of interferences, and (c) handling relatively small sample volumes (1–3 mL total volume after digest). However, it is the high sensitivity, precision, and accuracy that make it most useful for the quantification of gold or other elements in a given sample.

ICP-MS can have issues with total dissolved solid that leads to deposits on components of the instrument resulting in fluctuations in results. Options to overcome this include using teflon-coated components, replacing piping after each run, cleaning cones, and the nebulizer regularly with strong acid and also alternating the length of time for washes and wash solution. Speed is another issue, which depends on the wash time; the scanning process can take hours for an average run (approx. 30 samples plus standards). This process is also a destructive technique which might not be advantageous compared to other techniques, such as electron microscopy.

A number of authors have analyzed the accuracy of gold determination in tissues, aiming to optimize sensitivity, and reproducibility of the technique for different sizes of nanoparticle (Allabashi et al., 2009; De Jong et al., 2008; Gray et al., 2013).

4.3 Quantitative Analysis by Spectroscopic or Radioactive Techniques

Even though electron microscopy and ICP-MS are the most popular techniques to analyze nanoparticle uptake, other techniques have emerged to overcome the issues of cost and technical expertise required.

Absorption spectroscopy provides a very simple way of quantifying gold concentration as a function of its concentration curve. In this way it is possible to measure the amount of gold present in a cell or medium before and after incubation with gold nanoparticles (Cho, Liu, & Xia, 2010). The advantage is that simple instruments are required. Also the gold does not need to be dissolved so the problem of gold sticking to plastic tubing is avoided (see ICP-MS, Section 4.2). The disadvantage is the low sensitivity of the method even though this may depend on size of nanoparticle used—nanoparticles larger than 5 nm have an absorption peak at around 520 nm (see Section 2.2) and thus can provide a larger signal. Also as the concentration within cells is relatively low in comparison to the medium, the technique may not be sensitive enough to quantify gold content in cells directly.

For tissue analysis, more sensitive types of absorption spectroscopy can be used. Flame atomic absorption spectroscopy is a technique where gold is transformed to the gaseous state during which it can absorb light, allowing detection. In this way, gold nanoparticles within tissues can be dissolved in aqua regia and be quantified (Shilo et al., 2014).

Prior to ICP-MS, radioactive techniques were popular to analyze gold content. For example, in neutron activation analysis, a sample's atomic nuclei are bombarded with neutrons, resulting in a radioactive product that decays, releasing γ-rays. This technique has been used to quantify gold, for example, in rats fed with gold nanoparticles (Hillyer & Albrecht, 2001), to analyze the brain after uptake of transferrin-coated gold nanoparticles (Prades et al., 2012), or in vitro to analyze uptake of gold nanoparticles into different cancer cells (Tsoli et al., 2005). The advantage of this technique is its high sensitivity with no requirement for sample processing; on the other hand, the drawback is the production of radioactive material that needs appropriate handling and disposal.

γ-Rays can be detected also from gold isotope decay. In this case, analysis of gold concentration is performed by γ-spectroscopy. For example, this technique has been used to analyze organs of rats after treatment with gold nanoparticles of different sizes and charges (Hirn et al., 2011; Schleh et al., 2012).

4.4 Other Imaging Techniques in Tissues

Gold nanoparticles can also be observed within animal organs by micro-CT scanning. In this technique, X-rays scan the tissue and are detected in image form. Gold nanoparticles reduce X-rays and thus can be visualized. This technique has been used to detect insulin-coated gold nanoparticles of 20 nm in rat brains (Shilo et al., 2014). Also, gold nanoparticles can be used as labeling for micro-CT, as in the study of Menk et al. (2011) where they demonstrated the use of gold nanoparticles in labeling gliomas and also performed 3D reconstruction of the rat tissues.

In contrast to these specialized techniques, simple fluorescence can also be detected in tissues when a fluorescent ligand is released from gold nanoparticles. As mentioned previously (Section 2.2), gold nanoparticles can quench fluorescence of molecules attached to them. When the molecule is released, the fluorescence can be detected. This property has been used for targeting brain tumors of mice when using a gold nanoparticle with epidermal growth factor peptide and a fluorescent drug (Cheng et al., 2011). The carried drug also functioned as an agent for photothermal therapy to kill cancer cells after irradiation with light, showing the ability of gold nanoparticles to perform both in diagnostics and in therapy, ie, the new field of theranostics.

5. CELLULAR ROUTES FOR NANOPARTICLE TRANSPORT INTO THE CNS

The tight junctions between brain capillary endothelial cells have ~1.5 nm gap and effectively exclude all hydrophilic molecules >1 kDa, including nanoparticles—small gold nanoparticles are >20 kDa and have a molecular diameter about 2 nm. Consequently, if nanoparticles are to be used to transport material into the CNS, they must move through brain endothelium either by a cytosolic route or by a vesicular transcytosis as discussed in this part. The cytosolic route requires the nanoparticle to cross both the apical and basolateral plasma membranes of the endothelium. Vesicular transcytosis may occur via caveolae or via clathrin-coated vesicles, although sorting of vesicles in endosomes may divert them to other intracellular compartments, such as lysosomes. There is an important distinction between promoting uptake by brain endothelial cells and promoting transport across the endothelium.

The thickness of a brain endothelial cell, ie, distance from the apical to the basolateral plasma membranes, is very small. Immortalized brain endothelial cell lines and primary brain endothelium are usually <3 μm thick; in vivo the cells are more flattened and are typically <1 μm thick except at the nucleus. Thus, the distance for cytosolic diffusion or vesicular transport is small and the primary consideration is how the nanoparticles may enter either of these transport pathways.

5.1 Cytosolic Transport

Transport across the cytosol requires the nanoparticle to (1) fuse with the apical plasma membrane, (2) release into the cytosol, (3) diffuse across the cytosol, and then (4) fuse with the basal plasma membrane before (5) being released again into the extracellular environment. This process does not need energy from the cell to take up the nanoparticles in contrast with vesicular transport. Therefore, inhibitors of cellular active transport should not inhibit this process. For nanoparticles with hydrophilic ligands, crossing the plasma membranes is energetically unfavorable and this presents the major barrier to movement across the cell.

A number of surface modifications of gold nanoparticles promote movement across plasma membranes. They include:

- cell-penetrating peptides (Trojan peptides) such as Tat;
- cationic peptides which may perturb the membrane structure;
- amphiphilic molecules;
- bifunctional nanoparticles with mixtures of hydrophobic; and anionic ligands.

Although these surface modifications enhance transport across artificial phospholipid bilayers and/or cell membranes, the mechanism is uncertain. Direct observation and theoretical calculations indicate that the nanoparticles must be at the lower end of the size range—gold core <2 nm (Van Lehn et al., 2013). The focus so far has been on investigating the mechanism of membrane penetration by monolayer-protected amphiphilic nanoparticles. First, lipophilic ligands insert into the core of the phospholipid bilayer, particularly in areas of high membrane curvature (Van Lehn & Alexander-Katz, 2014). The ability of the nanoparticle to enter the membrane may then be dependent on the surface mobility of its ligand groups. The ligands may either distort as the nanoparticle enters the membrane or undergo exchange reactions with other ligands on the nanoparticle surface. These surface rearrangements reduce the energy barrier required to cross the membrane. This theoretical mechanism has been termed as "snorkeling" (Van Lehn & Alexander-Katz, 2014; Van Lehn et al., 2013). Our observation of nanoparticles in vitro by electron microscopy lends support to the idea that small gold nanoparticles with other surface modification (sugars in our experiments) may enter brain endothelium by this mechanism (Fig. 6).

5.1.1 Enhancing Cytosolic Transport

Brain endothelium has a large number of transporters for small molecules, including amino acids, carbohydrates, carboxylic acids, and nucleotides. Some attempts have been made to enhance uptake of nanoparticles exploiting these transporters, by attaching the appropriate ligand to the outside of the nanoparticle, eg, glucose to bind the Glut-1 transporter. However, while these systems have a potentially high transport capacity, they have not been

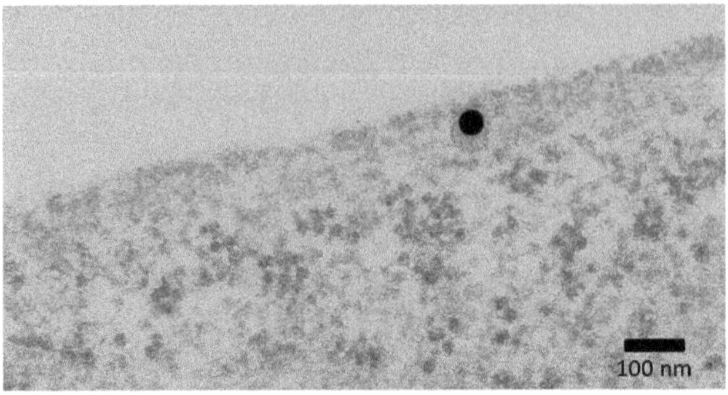

Fig. 6 Transmembrane movement of small gold nanoparticles. Transmission electron micrograph of a 2-nm glucose-coated gold nanoparticle (silver enhanced) crossing the apical plasma membrane of a human brain endothelial cell (hCMEC/D3) in vitro. A proportion of nanoparticles cross the plasma membrane to enter the cytosol.

used successfully to promote passage of gold nanoparticles. There are two intrinsic problems associated with the use of these transporters:

- covalent attachment of the ligand to the nanoparticle may interfere with its ability to bind to its receptor—the ligand must be correctly orientated to engage with its receptor/carrier;
- the solutes are much smaller than nanoparticles, which may not be able to physically fit the transporter mechanism.

Consequently, a correctly orientated ligand might enhance nanoparticle binding to a receptor on the endothelial surface, but that transport system still cannot promote uptake of something as large as a nanoparticle into the cell. There is a distinction between promoting nanoparticle attachment to the endothelial surface and promoting uptake and transcytosis.

Similarly a nanoparticle with, eg, a glucose ligand may move effectively across brain endothelium, but that does not imply that the glucose transporter is involved (Gromnicova et al., 2013). To demonstrate the involvement of a transporter/receptor, it is necessary to show that inhibition or knockdown of that transporter/receptor reduces nanoparticle uptake or transcytosis.

5.2 Vesicular Transport and Transcytosis

In most tissues of the body, a major proportion of nutrient and solute exchange between blood and tissue are by vesicular transcytosis. However, vesicular transport in brain endothelium is limited; brain endothelium has relatively smaller numbers of caveolae and other endocytic vesicles.

However, the number of vesicles is increased in areas of inflammation, around infarcts, and in tumors. This is thought to reflect increased tissue demand for metabolites.

Endocytosis of larger molecules may be mediated by receptors—receptor-mediated transport (RMT), or it may involve nonspecific absorption of the biomolecules to the cell surface—absorptive-mediated transport (AMT). In practice, AMT may show some selectivity, since the initial attachment to the cell surface will depend on molecular size and charge. Moreover, endocytosis is initiated from specific regions of the plasma membrane (lipid rafts), so the ability of the transported molecule to bind to these regions will determine whether they can be transported by AMT.

Vesicles are the best option for bulk transport of nanoparticles across the endothelium. Caveolae are typically 50 nm in diameter and can therefore transport nanoparticles up to 25 nm. Thus, hundreds of smaller gold nanoparticles can be taken up into a single endosome.

Once inside the cell, vesicles are sorted for the apical or basal membranes, via early endosomes. Some molecules are diverted via late endosomes and multivesicular bodies to lysosomes. Brain endothelial cell lines display a relatively high rate of endocytosis and transcytosis; 5 nm gold nanoparticles with different surface coatings are transported across the cell and relatively small amounts accumulate in lysosomes. In vivo there is less evidence for vesicular transcytosis, but this is still thought to be a major route for gold nanoparticle movement into the CNS.

5.2.1 Enhancing Vesicular Transport

Brain endothelium has a number of RMT systems for larger molecules, including the transferrin-receptor, insulin-receptor, and the LDL-receptors LRP-1 and LRP-2. These systems have been used for conventional drug delivery (Gabathuler, 2010), with some success, as they have high transport rates and recycling times, which means that they are difficult to saturate. Moreover, the size of the natural ligands is similar to the size of the smaller nanoparticles.

A number of different targeting approaches have been applied to gold nanoparticles, attempting to enhance brain uptake (reviewed in Velasco-Aguirre et al., 2015). Several studies have targeted the transferrin receptor and transcytosis across brain endothelium was demonstrated in vitro using 6.6 nm gold nanoparticles coupled to transferrin (Velasco-Aguirre et al., 2015). In another study the transferrin receptor was targeted by the addition of a dodecapeptide to the gold nanoparticles (13 nm), which resulted in a large increase in permeability of the nanoparticles across brain endothelium in vitro (Prades et al., 2012). In these experiments the nanoparticles were also linked to a hexapeptide for disruption of β-amyloid. Interestingly the hexapeptide in association with the targeting peptide also appeared to enhance permeability although it produced no increase by itself. The authors argued that the association of serum proteins such as albumin or LDL could associate with the nanoparticle–peptide conjugate, thus enhancing uptake via the brain endothelial albumin or LDL receptor.

Experiments with 5 nm positively charged nanoparticles have also demonstrated enhanced uptake in the presence of serum molecules, although this effect did not occur with serum albumin alone (Fig. 7). This result suggests that the corona of loosely associated serum proteins may have a major effect on relative uptake by different organs in vivo, depending on their receptors.

One study with insulin-coated nanoparticles demonstrated a greater accumulation in brain than for untargeted nanoparticles; however, it was not clear whether the nanoparticles had crossed the BBB, or were accumulating in the vasculature (Shilo et al., 2014). In such studies, it is important to consider how the ligand is attached to the nanoparticle—ie, whether the ligand is attached covalently or noncovalently, and whether this allows the ligand to bind to its receptor (Velasco-Aguirre et al., 2015). For example, in experiments from our laboratory, it was found that nanoparticles with noncovalently bound insulin were taken up very effectively by brain endothelium, however, knockdown of the insulin receptor on the cells by

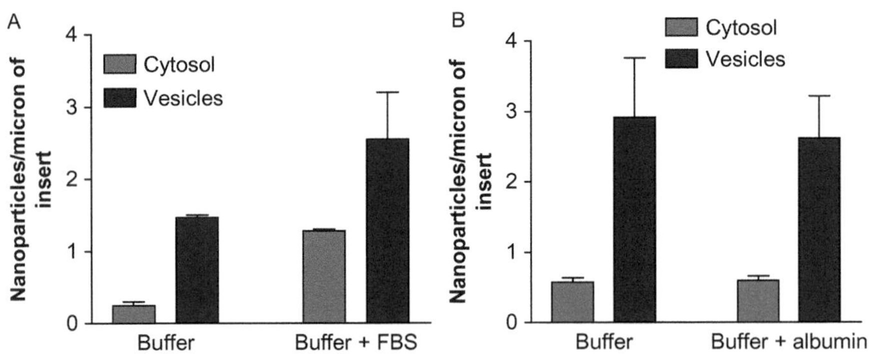

Fig. 7 Effect of protein on uptake of gold nanoparticles (<5 nm) coated with PEG-amine/galactose. Gold nanoparticles were quantified by TEM in brain endothelial cells (hCMEC/D3). (A) Fetal bovine serum (10% FBS) increased the cell uptake of these gold nanoparticles both into cytosol and vesicles. (B) Bovine serum albumin did not affect the uptake of these nanoparticles.

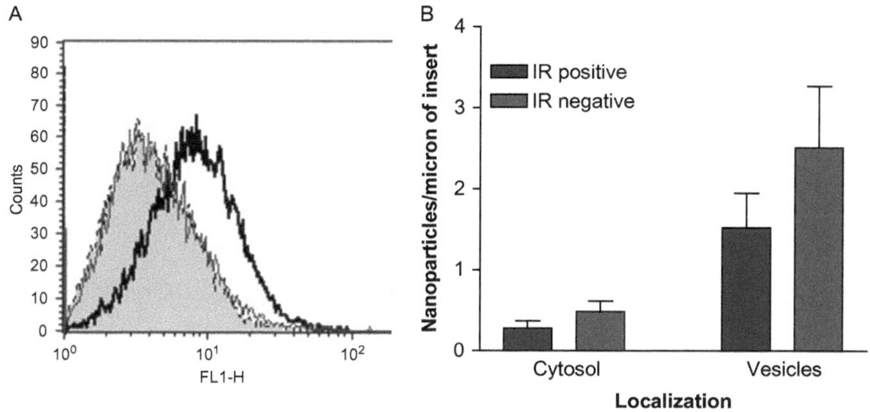

Fig. 8 Effect of insulin-receptor knockdown on the uptake of insulin/PEG-amine nanoparticles. (A) FACS histogram of insulin-receptor expression on hCMEC/D3. *Gray*: control IgG-stained cells; *brown* (*dark gray* in the print version): insulin-receptor siRNA-treated cells demonstrating complete knockdown of the receptor. *Black line*: positive control of untreated cells expressing insulin receptor. (B) The uptake of insulin/PEG-amine nanoparticles into brain endothelial (hCMEC/D3) cells with insulin-receptor knockdown (IR negative) as quantified with TEM. Data represent mean $\pm$ SEM, three independent experiments. Receptor knockdown did not reduce nanoparticle transfer and in practice induced a significant increase in nanoparticle uptake into vesicles.

siRNA, produced no reduction in nanoparticle uptake (in fact a small but significant increase), which implies that the insulin receptor was not involved in endocytosis (Fig. 8).

5.3 Stability of the Nanoparticles During Transcytosis

A successful nanocarrier for the CNS must be relatively stable during transit across the endothelium, unless the endothelium itself is the target. Transit across the cytosol will expose the nanoparticle to reducing conditions—intracytosolic glutathione concentration is typically in the range 1–10 mM. Since many ligand and cargo molecules are bound to the gold core by a thiol linkage, this is potentially unstable during transit across the cells (Fig. 3). The cargo molecules can exchange with cellular glutathione or other molecules with free—SH groups. In practice, the transit time of 5 nm gold nanoparticles across the endothelium can be relatively fast, so that exchange with cellular glutathione is expected to be limited (Gromnicova et al., 2016). Conversely, if the endothelium is the target, it would be advantageous for the nanoparticles to remain in the cytosol for several hours to allow for release of the cargo.

Nanoparticles that transit in vesicles may be subject to acidic conditions down to pH 5. Thiol linkages are relatively stable in these conditions, but cargo molecules bound by electrostatic interactions may dissociate, depending on their isoelectric point (pI). Such conditions also promote uncoupling of targeting molecules from their receptor, which could be advantageous if the nanoparticles have to separate from a recycling receptor or a receptor which is sorted for the lysosomal pathway.

5.4 Cellular Localization of Gold Nanoparticles Within the CNS

Relatively few studies have identified the cellular and subcellular distribution of nanoparticles within the CNS in vivo. Following intravenous administration of 20 nm colloidal gold nanoparticles, they were found in all layers of the retina including 75% in neurons, 17% in endothelial cells, and 8% in glia (Kim et al., 2009). Our own work has identified 5 nm glucose–gold nanoparticles in the cortex in neurons, myelinated and unmyelinated nerve axons, astrocytes, and other glial cells within 10 min of intravascular injection (Gromnicova et al., 2016). Within these cells, the nanoparticles are found primarily in the cytosol, but small numbers are also present in the nuclei. The potential mechanism of transport into the nucleus has not been explored, although nuclear pore complexes are 120 nm in diameter, and sufficiently large to allow passive movement of molecules up to 40 kDa.

In circumventricular organs (choroid plexus, median eminence, etc.), which have fenestrated endothelium and lack a conventional BBB, we observed that 5 nm gold nanoparticles can cross the endothelium relatively easily, but are restricted from moving further by the tight junctions of the choroid plexus epithelium and/or perivascular astrocytes. In this case the nanoparticles appear to cross the endothelium by the paracellular route (Fig. 9).

Once inside the brain, 5 nm gold nanoparticles can move quickly away from the vessels and appear to migrate by crossing cells. They are not found at high levels in the intercellular spaces. Moreover, movement by diffusion through the tortuous intercellular spaces would be too slow to explain the rapid rate of movement observed—up to 1 μm/min (Gromnicova et al., 2016).

There has been very little published on long-term persistence of gold nanoparticles in the CNS, or the mechanisms of clearance. A study with 15 nm albumin-coated nanoparticles suggested that the gold had been cleared from the brain by 7 days, as it was no longer detectable (Sousa

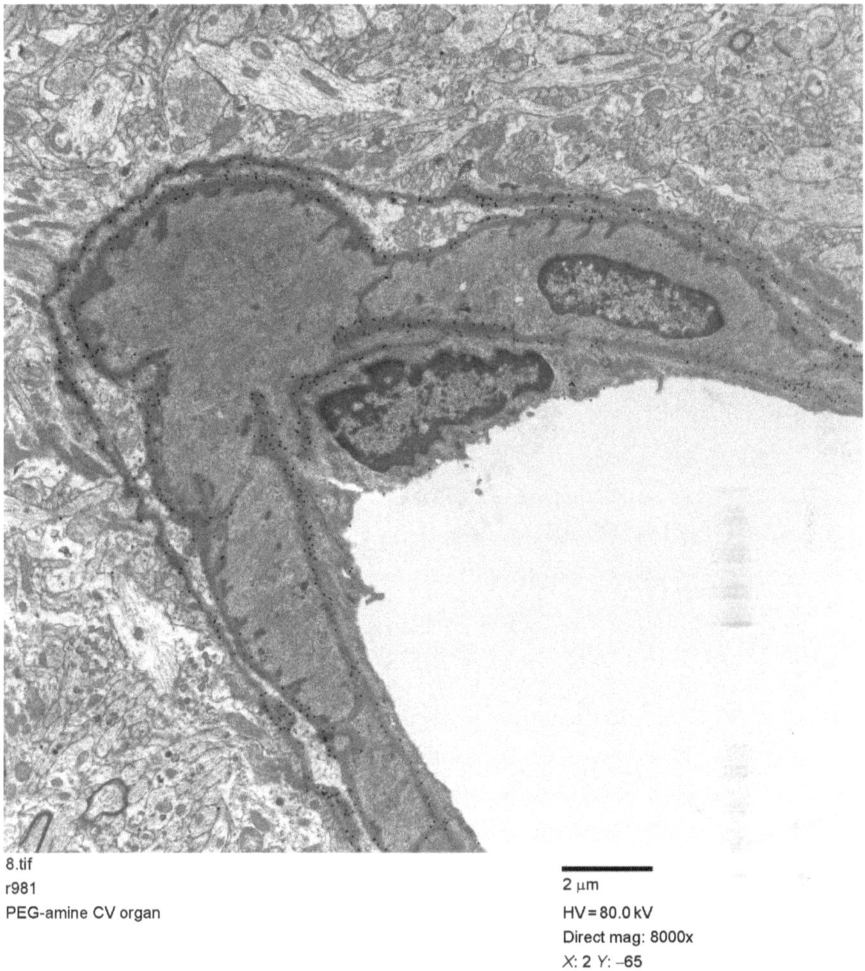

Fig. 9 Gold nanoparticle movement into circumventricular organs. 2 nm gold nanoparticles covalently coated with PEG-amine were infused by intracarotid injection into rats, 10 min before animals were perfused/fixed. The 85-nm thick, silver-enhanced TEM image of a blood vessel in the choroid plexus, which has fenestrated endothelium and lacks a conventional blood–brain barrier. In this case the nanoparticles cross the endothelium rapidly but are mostly retained within the glia limitans.

et al., 2010). It has been assumed that nanoparticles that remain soluble (not aggregated within lysosomes) will be removed according to the properties of the nanoparticle—the gold core itself is biologically inert so breakdown does not occur to any significant extent. For smaller nanoparticles it is thought that long-term clearance will occur by vesicular or cytosolic exocytosis and transcytosis across endothelium, followed by excretion from the kidney. Larger nanoparticles may be taken up by phagocytic cells within the brain, including perivascular macrophages and microglia. Material which cannot be degraded by these cells tends to accumulate in lysosomes. Larger gold nanoparticles could also be removed by perivascular drainage to the CSF, and from there either to cervical lymphatics via the cribriform plate or to venous sinuses via arachnoid granulations in the meninges.

The potential for alternative clearance routes of nanoparticles of different sizes is likely to be an important consideration, if they are to be used as nanocarriers for repeated treatment of chronic conditions. It is important from a therapeutic perspective that the nanoparticles do not accumulate to cytotoxic levels within the CNS.

6. CONCLUSIONS

6.1 Gold Nanocarriers

Gold nanoparticles have a number of distinct advantages for delivery of therapeutic agents to the CNS. Smaller nanoparticles (<10 nm) can cross brain endothelial cells effectively in vitro and in vivo using a combination of cytosolic and vesicular transport routes. The gold core shows minimal interaction with biological molecules and consequently has very low cytotoxicity and immunogenicity.

Gold nanoparticles are relatively easy to synthesize and can be produced in defined size ranges. Consistency is promoted by synthesizing the nanoparticles in continuous flow conditions, rather than batches. Ligands can be attached to the nanoparticles during synthesis via a thiol bond or by an exchange reaction after synthesis with thiolated ligands after synthesis. The first method is simple, but subjects the ligands to highly reducing conditions in nonphysiological buffers as the nanoparticles are produced. However, the exchange reaction allows addition of any thiolated ligand to the nanoparticles, including proteins and nucleic acids. It does have one disadvantage that the products of the exchange reaction are initially formed as an equilibrium mixture over a number of hours/days, and the preparation of a nanoparticle with a stoichiometrically defined ratio of ligand molecules, may therefore require purification.

The carrying capacity of nanoparticles depends on their size, but the smallest ones may have 50% of their molecular mass as cargo molecules or surface-bound ligands. One study with a mouse brain endothelial cell line indicated maximal uptake of gold by 70 nm nanoparticles, but the smallest nanoparticles used in the study (20 nm) had the highest carrying capacity (Shilo et al., 2015). This research highlights the fact that it is the number of cargo molecules that enter the CNS that is important, rather than the amount of gold, unless the aim is to use the gold itself for imaging. At this point in time, most studies have measured gold uptake or entry into the CNS, rather than the amount of drug/ligand delivered.

The stability of the nanoparticles and rate of release depends on the way that the cargo is attached. Noncovalent attachment to charged surface ligands (eg, PEG-amine or oligonucleotides) depends on the charge/pI of the ligand molecule and this would require optimization for each ligand. The advantage of noncovalent attachment is that biological activity is more likely to be retained. Conversely, direct attachment to the gold core is more stable, but may result in reduced biological activity once the cargo is released.

The method of attachment may also be partly determined by whether the cargo is intended to be released in the intercellular space, the cytosol (reducing conditions), or in endosomes (acidic conditions).

6.1.1 Comparison with Silica Nanoparticles and Carbon Nanoparticles

In comparison with gold nanoparticles, mesoporous silica nanoparticles also offer good biocompatibility through surface functionalization and they provide stable structures that can withstand changes in pH and mechanical stress with a large surface area due to their size (Boodhia, 2013). Studies have shown that silica nanoparticles can traverse the BBB in vitro and in vivo and their transport is size dependent, with increased particle size resulting in decreased efficiency (Liu et al., 2014). However, there has been evidence of the negative effect of exposure to silica nanoparticles, one study found them to induce neurotoxicity, and pathological signs of Alzheimer's disease (Yang et al., 2014).

Carbon nanotubes have a similar cargo carrying capacity to gold nanoparticles. However, it has been found that regardless of their modification they produced an inflammatory response in in vivo and produced damage to the lungs comparable to asbestosis (Lam et al., 2006). For this reason the gold nanoparticles are likely to be safer nanocarriers for the CNS and other tissues.

6.1.2 Comparison with Liposomes, Solid Lipid Nanoparticles, and Polymeric Nanoparticles

Liposomes and solid lipid nanoparticles transport their cargo internally, and they can deliver therapeutic agents to target cells through adsorption, endocytosis, or lipid transfer. Liposomes have high solubility and the ability to carry different therapeutics, depending on the size of the liposomes. As with gold nanoparticles, the release of the therapeutic can be controlled by their composition or environmental factors such as pH (Lopes et al., 2013). One advantage of liposomes over gold nanoparticles is that they can be degraded in biological systems; however, the larger size of liposomes limits their ability to cross the BBB.

Solid lipid nanoparticles are stable, lipid-based nanocarriers with a solid hydrophobic lipid core, in which drugs can be attached or encapsulated. They use triglycerides and fatty acids, which mimic the cell membrane of cells, thus displaying good biocompatibility. The matrix of lipids allows for encapsulation of hydrophilic or hydrophobic therapeutic molecules, protecting the cargo from degradation. Solid lipid nanoparticles encasing the anticancer drug, camptothecin, have been used in vitro (Martins et al., 2012). The ability to integrate with the phospholipids of the cell membrane has advantages for transport of smaller hydrophobic drugs, but is less suitable than gold nanoparticles for transport of larger hydrophilic biomolecules.

Polymeric nanoparticles can be produced from natural or synthetic polymers, such as chitosan or poly(lactic-*co*-glycolic acid). They are generally biocompatible and biodegradable and suitable for the delivery of water-soluble drugs, but their entrapment capacities vary for different drugs (Zhu & Liao, 2015).

Dendrimers comprised of a central core and branched polymers capable of forming a layer around the core and are in some ways similar to nanoparticles. The symmetric scaffold structure of the dendrimer has been found to be a suitable carrier for a variety of drugs and siRNA and plasmids; the dendrimer was able to protect the plasmid from attack by DNase I and had increased gene transfer compared with naked DNA (Zhu & Liao, 2015). This indicates that dendrimers can transport larger biomolecules and they can be used for direct transfection of cells, including tumors. However, their ability to cross the BBB has not been evaluated.

6.2 Gold Nanoparticles as Imaging Reagents

Gold nanoparticles are currently used as X-ray contrast reagents (Hainfeld et al., 2013), and they have some advantages over standard contrast agents, such as gadolinium (lower nephrotoxicity) or iodine (higher contrast, better resolution, nonallogeneic, and longer persistence in blood). Typically, however, the gold nanoparticles have been used for imaging the kidney, where they preferentially accumulate or to image tumors. Although gold is excellent as an imaging agent, the low level of accumulation of nanoparticles in the brain has prevented their use for CNS conditions, to date. If a method of targeting nanoparticles to the CNS can be developed, then gold nanoparticles could become a major diagnostic tool for brain diseases.

The weak magnetic properties of gold nanoparticles mean that more attention has been paid to ferromagnetic nanoparticles including iron oxide or nanoparticles doped with iron, for MRI imaging and magnetometry.

Undoubtedly the plasmonic properties of gold nanoparticles are promising for imaging, and they are used for many in vitro studies. However, as this modality depends on light, their potential for use in imaging the brain in vivo is more limited. Potentially nanoparticles could be used to identify the extent of invasive brain tumors or areas of vascular damage in real time, to guide brain surgery. In this case the ability to see exactly where the nanoparticles are located, without additional instrumentation, has great advantages.

To summarize, gold nanoparticles enter the brain by crossing the brain endothelium. They have considerable potential as nanocarriers for therapeutic drugs and biomolecules. They have some potential as imaging reagents, particularly for directing surgeons. Although entry of nanoparticles into the CNS is surprisingly efficient given their size, all of these potential applications may be improved further by selective targeting of the nanoparticles to the brain endothelium.

REFERENCES

Ackerson, C. J., Jadzinsky, P. D., & Kornberg, R. D. (2005). Thiolate ligands for synthesis of water-soluble gold clusters. *Journal of the American Chemical Society*, *127*(18), 6550–6551. Available at: http://www.ncbi.nlm.nih.gov/pubmed/15869273.

Agasti, S. S., et al. (2009). Photoregulated release of caged anticancer drugs from gold nanoparticles. *Journal of the American Chemical Society*, *131*(16), 5728–5729. Available at: http://dx.doi.org/10.1021/ja900591t.

Ahmad, T., et al. (2012). Gold-coated iron oxide nanoparticles as a T2 contrast agent in magnetic resonance imaging. *Journal of Nanoscience and Nanotechnology*, *12*(7), 5132–5137. Available at: http://www.ncbi.nlm.nih.gov/pubmed/22966533. Accessed December 15, 2015.

Albanese, A., & Chan, W. C. W. (2011). Effect of gold nanoparticle aggregation on cell uptake and toxicity. *ACS Nano*, *5*(7), 5478–5489. Available at: http://pubs.acs.org/doi/abs/10.1021/nn2007496.

Alkilany, A. M., & Murphy, C. J. (2010). Toxicity and cellular uptake of gold nanoparticles: What we have learned so far? *Journal of Nanoparticle Research*, *12*(7), 2313–2333. Available at: http://www.pubmedcentral.nih.gov/articlerender.fcgi?artid=2988217&tool=pmcentrez&rendertype=abstract. Accessed May 22, 2013.

Allabashi, R., et al. (2009). ICP-MS: A powerful technique for quantitative determination of gold nanoparticles without previous dissolving. *Journal of Nanoparticle Research*, *11*(8), 2003–2011.

Atukorale, P. U., et al. (2015). Influence of the glycocalyx and plasma membrane composition on amphiphilic gold nanoparticle association with erythrocytes. *Nanoscale*, 7, 11420–11432. Available at: http://xlink.rsc.org/?DOI=C5NR01355K.

Baker, D., & Hankey, D. J. R. (2003). Gene therapy in autoimmune, demyelinating disease of the central nervous system. *Gene Therapy*, *10*(10), 844–853. Available at: http://www.ncbi.nlm.nih.gov/pubmed/12732870. Accessed March 10, 2016.

Baker, D., et al. (2011). Critical appraisal of animal models of multiple sclerosis. *Multiple Sclerosis (Houndmills, Basingstoke, England)*, *17*(6), 647–657. Available at: http://www.ncbi.nlm.nih.gov/pubmed/21372117. Accessed March 10, 2016.

Bhumkar, D. R., et al. (2007). Chitosan reduced gold nanoparticles as novel carriers for transmucosal delivery of insulin. *Pharmaceutical Research*, *24*(8), 1415–1426. Available at: http://link.springer.com/10.1007/s11095-007-9257-9.

Boodhia, K. (2013). *Assessing the potential toxicity of gold nanoparticle carrier systems conjugated with therapeutic peptides*. Dissertation Faculty of Science, South Africa: University of Johannesburg. Available at: http://www.nioh.ac.za/assets/files/Msc%20Kailen%2021213(1).pdf.

Boyoglu, C., et al. (2013). Microscopic studies of various sizes of gold nanoparticles and their cellular localizations. *ISRN Nanotechnology*, Article ID 123838, 13 pages, http://dx.doi.org/10.1155/2013/123838.

Braun, G. B., et al. (2009). Laser-activated gene silencing via gold nanoshell-siRNA conjugates. *ACS Nano*, *3*(7), 2007–2015.

Brust, M., et al. (1994). Synthesis of thiol-derivatised gold nanoparticles in a two-phase liquid-liquid system. *Journal of the Chemical Society, Chemical Communications*, (7), 801–802.

Caballero-Díaz, E., & Valcárcel, M. (2014). Toxicity of gold nanoparticles. *Comprehensive Analytical Chemistry*, *66*, 207–254.

Chen, Y.-S., et al. (2009). Assessment of the in vivo toxicity of gold nanoparticles. *Nanoscale Research Letters*, *4*(8), 858–864. Available at: http://www.pubmedcentral.nih.gov/articlerender.fcgi?artid=2894102&tool=pmcentrez&rendertype=abstract. Accessed June 26, 2013.

Cheng, Y., et al. (2011). Addressing brain tumors with targeted gold nanoparticles: A new gold standard for hydrophobic drug delivery? *Small (Weinheim an der Bergstrasse, Germany)*, 7(16), 2301–2306. Available at: http://www.pubmedcentral.nih.gov/articlerender.fcgi?artid=3837553&tool=pmcentrez&rendertype=abstract. Accessed May 6, 2014.

Cho, E. C., Liu, Y., & Xia, Y. (2010). A simple spectroscopic method for differentiating cellular uptakes of gold nanospheres and nanorods from their mixtures. *Angewandte Chemie (International Edition in English)*, *49*(11), 1976–1980. Available at: http://www.pubmedcentral.nih.gov/articlerender.fcgi?artid=3359148&tool=pmcentrez&rendertype=abstract. Accessed June 12, 2013.

Choi, C. H. J., et al. (2010). Mechanism of active targeting in solid tumors with transferrin-containing gold nanoparticles. *Proceedings of the National Academy of Sciences of the United States of America*, *107*(3), 1235–1240. Available at: http://www.pnas.org/cgi/doi/10.1073/pnas.0914140107.

Coulter, J. A., et al. (2012). Cell type-dependent uptake, localization, and cytotoxicity of 1.9 nm gold nanoparticles. *International Journal of Nanomedicine*, 7, 2673–2685.

Cutler, J. I., et al. (2011). Polyvalent nucleic acid nanostructures. *Journal of the American Chemical Society*, *133*(24), 9254–9257. Available at: http://www.pubmedcentral.nih.gov/articlerender.fcgi?artid=3154250&tool=pmcentrez&rendertype=abstract.

Daniel, M.-C., & Astruc, D. (2004). Gold nanoparticles: Assembly, supramolecular chemistry, quantum-size-related properties, and applications toward biology, catalysis, and nanotechnology. *Chemical Reviews*, *104*(1), 293–346. Available at: http://www.ncbi.nlm.nih.gov/pubmed/14719978.

Day, E. S., et al. (2011). Nanoshell-mediated photothermal therapy improves survival in a murine glioma model. *Journal of Neuro-Oncology*, *104*(1), 55–63. Available at: http://www.pubmedcentral.nih.gov/articlerender.fcgi?artid=3710584&tool=pmcentrez&rendertype=abstract. Accessed December 15, 2015.

De Jong, W. H., & Borm, P. J. A. (2008). Drug delivery and nanoparticles: Applications and hazards. *International Journal of Nanomedicine*, *3*(2), 133–149. Available at: http://www.pubmedcentral.nih.gov/articlerender.fcgi?artid=2527668&tool=pmcentrez&rendertype=abstract. Accessed January 21, 2015.

De Jong, W. H., et al. (2008). Particle size-dependent organ distribution of gold nanoparticles after intravenous administration. *Biomaterials*, *29*(12), 1912–1919. Available at: http://www.sciencedirect.com/science/article/pii/S014296120701071X. Accessed April 30, 2014.

Dhar, S., et al. (2009). Polyvalent oligonucleotide gold nanoparticle conjugates as delivery vehicles for platinum(IV) warheads. *Journal of the American Chemical Society*, *131*(41), 14652–14653. Available at: http://www.pubmedcentral.nih.gov/articlerender.fcgi?artid=2761975&tool=pmcentrez&rendertype=abstract. Accessed May 5, 2014.

Di Guglielmo, C., et al. (2012). In vitro safety toxicology data for evaluation of gold nanoparticles-chronic cytotoxicity, genotoxicity and uptake. *Journal of Nanoscience and Nanotechnology*, *12*(8), 6185–6191. Available at: http://www.ncbi.nlm.nih.gov/pubmed/22962725. Accessed March 10, 2016.

dos Santos, W. L., et al. (1995). Distribution and analysis of surface charge on brain endothelium in vitro and in situ. *Acta Neuropathologica*, *90*(3), 305–311. Available at: http://www.ncbi.nlm.nih.gov/pubmed/8525805. Accessed February 17, 2016.

Dulkeith, E., et al. (2005). Gold nanoparticles quench fluorescence by phase induced radiative rate suppression. *Nano Letters*, *5*(4), 585–589. Available at: http://www.ncbi.nlm.nih.gov/pubmed/15826091.

Ebong, E. E., et al. (2011). Imaging the endothelial glycocalyx in vitro by rapid freezing/freeze substitution transmission electron microscopy. *Arteriosclerosis, Thrombosis, and Vascular Biology*, *31*(8), 1908–1915. Available at: http://www.pubmedcentral.nih.gov/articlerender.fcgi?artid=3141106&tool=pmcentrez&rendertype=abstract. Accessed January 9, 2014.

Elbakry, A., et al. (2009). Layer-by-layer assembled gold nanoparticles for siRNA delivery. *Nano Letters*, *9*(5), 2059–2064.

Etame, A. B., et al. (2011). Design and potential application of PEGylated gold nanoparticles with size-dependent permeation through brain microvasculature. *Nanomedicine: Nanotechnology, Biology and Medicine*, 7(6), 992–1000. Available at: http://www.sciencedirect.com/science/article/pii/S154996341100164X. Accessed May 3, 2014.

Freese, C., et al. (2012). Uptake and cytotoxicity of citrate-coated gold nanospheres: Comparative studies on human endothelial and epithelial cells. *Particle and Fibre Toxicology*, *9*(1), 23. Available at: http://www.pubmedcentral.nih.gov/articlerender.fcgi?artid=3407003&tool=pmcentrez&rendertype=abstract. Accessed March 31, 2014.

Freese, C., et al. (2013). Uptake of poly(2-hydroxypropylmethacrylamide)-coated gold nanoparticles in microvascular endothelial cells and transport across the blood–brain barrier. *Biomaterials Science*, *1*, 824–833. Available at: http://dx.doi.org/10.1039/C3BM60050E.

Frens, G. (1973). Controlled nucleation for the regulation of the particle size in monodisperse gold suspensions. *Nature*, *241*(105), 20–22. Available at: http://www.nature.com/nature-physci/journal/v241/n105/abs/physci241020a0.html. Accessed December 14, 2015.

Frigell, J., et al. (2014). 68Ga-labeled gold glyconanoparticles for exploring blood–brain barrier permeability: Preparation, biodistribution studies, and improved brain uptake via neuropeptide conjugation. *Journal of the American Chemical Society*, *136*(1), 449–457. Available at: http://www.ncbi.nlm.nih.gov/pubmed/24320878. Accessed March 10, 2016.

Fröhlich, E. (2012). The role of surface charge in cellular uptake and cytotoxicity of medical nanoparticles. *International Journal of Nanomedicine*, 7, 5577–5591. Available at: http://www.pubmedcentral.nih.gov/articlerender.fcgi?artid=3493258&tool=pmcentrez&rendertype=abstract. Accessed February 25, 2016.

Gabathuler, R. (2010). Approaches to transport therapeutic drugs across the blood–brain barrier to treat brain diseases. *Neurobiology of Disease*, *37*(1), 48–57. Available at: http://www.ncbi.nlm.nih.gov/pubmed/19664710. Accessed March 10, 2016.

García-Negrete, C. A., et al. (2015). STEM-in-SEM high resolution imaging of gold nanoparticles and bivalve tissues in bioaccumulation experiments. *The Analyst*, *140*(9), 3082–3089. Available at: http://xlink.rsc.org/?DOI=C4AN01643B.

Gibson, J. D., Khanal, B. P., & Zubarev, E. R. (2007). Paclitaxel-functionalized gold nanoparticles. *Journal of the American Chemical Society*, *129*(37), 11653–11661. Available at: http://www.ncbi.nlm.nih.gov/pubmed/17718495.

Giljohann, D. A., et al. (2009). Gene regulation with polyvalent siRNA-nanoparticle conjugates. *Journal of the American Chemical Society*, *131*(6), 2072–2073.

Goldstein, A., et al. (2014). High resolution SEM imaging of gold nanoparticles in cells and tissues. *Journal of Microscopy*, *256*(3), 237–247.

Goodman, C. M., et al. (2004). Toxicity of gold nanoparticles functionalized with cationic and anionic side chains. *Bioconjugate Chemistry*, *15*(4), 897–900. Available at: http://www.ncbi.nlm.nih.gov/pubmed/15264879. Accessed January 13, 2016.

Gray, E. P. E., et al. (2013). Extraction and analysis of silver and gold nanoparticles from biological tissues using single particle inductively coupled plasma mass spectrometry. *Environmental Science & Technology*, *47*(24), 14315–14323.

Gromnicova, R., et al. (2013). Glucose-coated gold nanoparticles transfer across human brain endothelium and enter astrocytes in vitro. *PloS One*, *8*(12), e81043. Available at: http://www.pubmedcentral.nih.gov/articlerender.fcgi?artid=3855187&tool=pmcentrez&render type=abstract. Accessed May 6, 2014.

Gromnicova, R., et al. (2016). Localisation and mobility of glucose-coated gold nanoparticles within the brain. *Nanomedicine (London, England)*, *11*, 617–625. Available at: http://www.ncbi.nlm.nih.gov/pubmed/19437620. Accessed February 17, 2016.

Guo, S., et al. (2010). Enhanced gene delivery and siRNA silencing by gold nanoparticles coated with charge-reversal polyelectrolyte. *ACS Nano*, *4*(9), 5505–5511.

Hainfeld, J. F., et al. (2013). Gold nanoparticle imaging and radiotherapy of brain tumors in mice. *Nanomedicine (London, England)*, *8*(10), 1601–1609. Available at: http://www.pubmedcentral.nih.gov/articlerender.fcgi?artid=3657324&tool=pmcentrez &rendertype=abstract. Accessed May 6, 2014.

Hillyer, J. F., & Albrecht, R. M. (2001). Gastrointestinal persorption and tissue distribution of differently sized colloidal gold nanoparticles. *Journal of Pharmaceutical Sciences*, *90*(12), 1927–1936. Available at: http://www.ncbi.nlm.nih.gov/pubmed/11745751.

Hirn, S., et al. (2011). Particle size-dependent and surface charge-dependent biodistribution of gold nanoparticles after intravenous administration. *European Journal of Pharmaceutics and Biopharmaceutics: Official Journal of Arbeitsgemeinschaft für Pharmazeutische Verfahrenstechnik e.V*, *77*(3), 407–416. Available at: http://www.sciencedirect.com/science/article/pii/S093964111000370X. Accessed April 30, 2014.

Hoft, R. C., et al. (2007). Adsorption of amine compounds on the Au(111) surface: A density functional study. *The Journal of Physical Chemistry C*, *111*(37), 13886–13891. Available at: http://pubs.acs.org/cgi-bin/doilookup/?10.1021/jp072494t.

Hong, R., et al. (2006). Glutathione-mediated delivery and release using monolayer protected nanoparticle carriers. *Journal of the American Chemical Society, 128*(4), 1078–1079.
Hostetler, M. J., et al. (1996). Monolayers in three dimensions: Synthesis and electrochemistry of ω-functionalized alkanethiolate-stabilized gold cluster compounds. *Journal of the American Chemical Society, 118*, 4212–4213.
Hostetler, M. J., et al. (1998). Alkanethiolate gold cluster molecules with core diameters from 1.5 to 5.2 nm: Core and monolayer properties as a function of core size. *Langmuir, 14*(1), 17–30.
Hu, M., et al. (2006). Gold nanostructures: Engineering their plasmonic properties for biomedical applications. *Chemical Society Reviews, 35*(11), 1084–1094. Available at: http://www.ncbi.nlm.nih.gov/pubmed/17057837. Accessed May 23, 2013.
Huang, X., & El-Sayed, M. A. (2010). Gold nanoparticles: Optical properties and implementations in cancer diagnosis and photothermal therapy. *Journal of Advanced Research, 1*(1), 13–28. Available at: http://linkinghub.elsevier.com/retrieve/pii/S2090123210000056. Accessed January 21, 2014.
Huang, X., et al. (2007). Gold nanoparticles: Interesting optical properties and recent applications in cancer diagnostics and therapy. *Nanomedicine (London, England), 2*(5), 681–693. Available at: http://www.ncbi.nlm.nih.gov/pubmed/17976030.
Hulteen, J. C., & Martin, C. R. (1997). A general template-based method for the preparation of nanomaterials. *Journal of Materials Chemistry*, 7(7), 1075–1087.
Hvolbæk, B., et al. (2007). Catalytic activity of Au nanoparticles. *Nanotoday, 2*(4), 14–18.
Iversen, T.-G., Skotland, T., & Sandvig, K. (2011). Endocytosis and intracellular transport of nanoparticles: Present knowledge and need for future studies. *Nano Today, 6*(2), 176–185. Available at: http://linkinghub.elsevier.com/retrieve/pii/S1748013211000181. Accessed October 19, 2013.
Jelveh, S., & Chithrani, D. B. (2011). Gold nanostructures as a platform for combinational therapy in future cancer therapeutics. *Cancers, 3*(1), 1081–1110. Available at: http://www.pubmedcentral.nih.gov/articlerender.fcgi?artid=3756404&tool=pmcentrez&rendertype=abstract. Accessed May 4, 2014.
Joh, D. Y., et al. (2013). Selective targeting of brain tumors with gold nanoparticle-induced radiosensitization. *PloS One, 8*(4), e62425. Available at: http://www.pubmedcentral.nih.gov/articlerender.fcgi?artid=3640092&tool=pmcentrez&rendertype=abstract. Accessed April 29, 2014.
Joshi, H. M., et al. (2006). Gold nanoparticles as carriers for efficient transmucosal insulin delivery. *Langmuir, 22*(17), 300–305.
Kang, K. A., et al. (2011). Fluorescence manipulation by gold nanoparticles: From complete quenching to extensive enhancement. *Journal of Nanobiotechnology, 9*, 16. Available at: http://www.pubmedcentral.nih.gov/articlerender.fcgi?artid=3112388&tool=pmcentrez&rendertype=abstract. Accessed May 26, 2014.
Kanwar, J. R., et al. (2012). Nanoparticles in the treatment and diagnosis of neurological disorders: Untamed dragon with fire power to heal. *Nanomedicine: Nanotechnology, Biology and Medicine, 8*(4), 399–414. Available at: http://www.ncbi.nlm.nih.gov/pubmed/21889479. Accessed March 10, 2016.
Khlebtsov, N., & Dykman, L. (2011). Biodistribution and toxicity of engineered gold nanoparticles: A review of in vitro and in vivo studies. *Chemical Society Reviews, 40*(3), 1647–1671. Available at: http://www.ncbi.nlm.nih.gov/pubmed/21082078. Accessed May 23, 2013.
Kim, J. H., et al. (2009). Intravenously administered gold nanoparticles pass through the blood-retinal barrier depending on the particle size, and induce no retinal toxicity. *Nanotechnology, 20*(50), 505101. Available at: http://www.ncbi.nlm.nih.gov/pubmed/19923650. Accessed March 10, 2016.

Kim, B., et al. (2010). Tuning payload delivery in tumour cylindroids using gold nanoparticles. *Nature Nanotechnology*, *5*(6), 465–472. Available at: http://dx.doi.org/10.1038/nnano.2010.58.

Krzciuk, K. (2015). Intelligent analysis of samples by semiquantitative inductively coupled plasma mass spectrometry (ICP-MS) technique: A review. *Critical Reviews in Analytical Chemistry/CRC*, 1–7. Available at: http://www.ncbi.nlm.nih.gov/pubmed/26517237. Accessed March 10, 2016.

Laborda, F., et al. (2015). Detection, characterization and quantification of inorganic engineered nanomaterials: A review of techniques and methodological approaches for the analysis of complex samples. *Analytica Chimica Acta*, *904*, 10–32. Available at: http://www.ncbi.nlm.nih.gov/pubmed/26724760. Accessed November 30, 2015.

Lam, C.-W., et al. (2006). A review of carbon nanotube toxicity and assessment of potential occupational and environmental health risks. *Critical Reviews in Toxicology*, *36*(3), 189–217.

Lee, S. H., et al. (2008). Amine-functionalized gold nanoparticles as non-cytotoxic and efficient intracellular siRNA delivery carriers. *International Journal of Pharmaceutics*, *364*, 94–101.

Letsinger, R. L., et al. (2000). Use of a steroid cyclic disulfide anchor in constructing gold nanoparticle-oligonucleotide conjugates. *Bioconjugate Chemistry*, *11*(2), 289–291.

Li, K., et al. (2013). Nanoparticles inhibit DNA replication by binding to DNA: Modeling and experimental validation. *ACS Nano*, 7(11), 9664–9674.

Li, C.-H., et al. (2015). Gold nanoparticles increase endothelial paracellular permeability by altering components of endothelial tight junctions, and increase blood–brain barrier permeability in mice. *Toxicological Sciences: An Official Journal of the Society of Toxicology*, *148*(1), 192–203. Available at: http://www.ncbi.nlm.nih.gov/pubmed/26272951. Accessed January 23, 2016.

Libutti, S. K., et al. (2010). Phase I and pharmacokinetic studies of CYT-6091, a novel PEGylated colloidal gold-rhTNF nanomedicine. *Clinical Cancer Research: An Official Journal of the American Association for Cancer Research*, *16*(24), 6139–6149. Available at: http://www.pubmedcentral.nih.gov/articlerender.fcgi?artid=3004980&tool=pmcentrez&rendertype=abstract. Accessed April 29, 2014.

Link, S., et al. (1999). Laser photothermal melting and fragmentation of gold nanorods: Energy and laser pulse-width dependence. *The Journal of Physical Chemistry. A*, *103*(9), 1165–1170. Available at: http://dx.doi.org/10.1021/jp983141k. Accessed March 13, 2014.

Liu, X., et al. (2013). Surface and size effects on cell interaction of gold nanoparticles with both phagocytic and nonphagocytic cells. *Langmuir: The ACS Journal of Surfaces and Colloids*, *29*(29), 9138–9148. Available at: http://www.ncbi.nlm.nih.gov/pubmed/23815604. Accessed March 10, 2016.

Liu, D., et al. (2014). In vitro and in vivo studies on the transport of PEGylated silica nanoparticles across the blood–brain barrier. *ACS Applied Materials and Interfaces*, *6*(3), 2131–2136.

Lockman, P. R., et al. (2004). Nanoparticle surface charges alter blood–brain barrier integrity and permeability. *Journal of Drug Targeting*, *12*(9–10), 635–641. Available at: http://www.ncbi.nlm.nih.gov/pubmed/15621689. Accessed March 10, 2016.

Lopes, S., et al. (2013). Liposomes as carriers of anticancer drugs. In L. Rangel (Ed.), *Cancer treatment: Conventional and innovative approaches* (pp. 85–124). Intech.

Marchesi, V. T. (1965). The role of pinocytic vesicles in the transport of materials across the walls of small blood vessels. *Investigative Ophthalmology*, *4*(6), 1111–1121. Available at: http://www.ncbi.nlm.nih.gov/pubmed/5321528. Accessed March 10, 2016.

Martin, C. R. (1994). Nanomaterials: A membrane-based synthetic approach. *Science (New York, N.Y.)*, *266*(5193), 1961–1966. Available at: http://www.ncbi.nlm.nih.gov/pubmed/17836514. Accessed January 5, 2016.

Martins, S., et al. (2012). Brain delivery of camptothecin by means of solid lipid nanoparticles: Formulation design, in vitro and in vivo studies. *International Journal of Pharmaceutics, 439*(1–2), 49–62.

Mateo, D., et al. (2014). Oxidative stress contributes to gold nanoparticle-induced cytotoxicity in human tumor cells. *Toxicology Mechanisms and Methods, 24*(3), 161–172.

McIntosh, C. M., et al. (2001). Inhibition of DNA transcription using cationic mixed monolayer protected gold clusters. *Journal of the American Chemical Society, 123*(31), 7626–7629.

Menk, R. H., et al. (2011). Gold nanoparticle labeling of cells is a sensitive method to investigate cell distribution and migration in animal models of human disease. *Nanomedicine: Nanotechnology, Biology and Medicine*, 7(5), 647–654. Available at: http://dx.doi.org/10.1016/j.nano.2011.01.010.

Mie, G. (1908). Beiträge zur Optik trüber Medien, speziell kolloidaler Metallösungen. *Annalen der Physik, 330*(3), 377–445. Available at: http://doi.wiley.com/10.1002/andp.19083300302. Accessed July 9, 2014.

Nealon, G. L., et al. (2012). Magnetism in gold nanoparticles. *Nanoscale, 4*(17), 5244. Available at: http://xlink.rsc.org/?DOI=c2nr30640a.

Niikura, K., et al. (2014). Amphiphilic gold nanoparticles displaying flexible bifurcated ligands as a carrier for siRNA delivery into the cell cytosol. *ACS Applied Materials and Interfaces, 6*, 22146–22154.

Nirmala Grace, A., & Pandian, K. (2007). Antibacterial efficacy of aminoglycosidic antibiotics protected gold nanoparticles—A brief study. *Colloids and Surfaces A: Physicochemical and Engineering Aspects, 297*(1–3), 63–70. Available at: http://linkinghub.elsevier.com/retrieve/pii/S092777570600762X.

Oh, N., & Park, J.-H. (2014). Surface chemistry of gold nanoparticles mediates their exocytosis in macrophages. *ACS Nano, 8*(6), 6232–6241. Available at: http://www.ncbi.nlm.nih.gov/pubmed/24836308.

Oishi, M., et al. (2006). Smart PEGylated gold nanoparticles for the cytoplasmic delivery of siRNA to induce enhanced gene silencing. *Chemistry Letters, 35*(9), 1046–1047. Available at: https://www.jstage.jst.go.jp/article/cl/35/9/35_9_1046/_article. Accessed January 5, 2016.

Pan, Y., et al. (2009). Gold nanoparticles of diameter 1.4 nm trigger necrosis by oxidative stress and mitochondrial damage. *Small (Weinheim an der Bergstrasse, Germany), 5*(18), 2067–2076. Available at: http://www.ncbi.nlm.nih.gov/pubmed/19642089. Accessed April 29, 2014.

Peckys, D. B., & De Jonge, N. (2011). Visualizing gold nanoparticle uptake in live cells with liquid scanning transmission electron microscopy. *Nano Letters, 11*(4), 1733–1738.

Peckys, D. B., & de Jonge, N. (2014). Gold nanoparticle uptake in whole cells in liquid examined by environmental scanning electron microscopy. *Microscopy and Microanalysis, 20*(1), 189–197.

Perez-Juste, J., et al. (2005). Gold nanorods: Synthesis, characterization and applications. *Coordination Chemistry Reviews, 249*(17–18), 1870–1901. Available at: http://linkinghub.elsevier.com/retrieve/pii/S0010854505000287.

Prades, R., et al. (2012). Delivery of gold nanoparticles to the brain by conjugation with a peptide that recognizes the transferrin receptor. *Biomaterials, 33*(29), 7194–7205. Available at: http://www.ncbi.nlm.nih.gov/pubmed/22795856. Accessed June 12, 2013.

Rai, A., Prabhune, A., & Perry, C. C. (2010). Antibiotic mediated synthesis of gold nanoparticles with potent antimicrobial activity and their application in antimicrobial coatings. *Journal of Materials Chemistry, 20*(32), 6789.

Rosi, N. L., et al. (2006). Oligonucleotide-modified gold nanoparticles for intracellular gene regulation. *Science, 312*(5776), 1027–1030. Available at: http://www.ncbi.nlm.nih.gov/pubmed/16709779. Accessed April 29, 2014.

Ryou, S.-M., et al. (2010). Delivery of shRNA using gold nanoparticle-DNA oligonucleotide conjugates as a universal carrier. *Biochemical and Biophysical Research Communications*, *398*(3), 542–546. Available at: http://www.sciencedirect.com/science/article/pii/S0006291X10012556. Accessed July 10, 2014.
Ryou, S.-M., et al. (2011). Gold nanoparticle-assisted delivery of small, highly structured RNA into the nuclei of human cells. *Biochemical and Biophysical Research Communications*, *416*(1–2), 178–183. Available at: http://www.sciencedirect.com/science/article/pii/S0006291X11020195. Accessed July 10, 2014.
Sandhu, K. K., et al. (2002). Gold nanoparticle-mediated transfection of mammalian cells. *Bioconjugate Chemistry*, *13*(1), 3–6.
Saris, S. C., et al. (1988). Intravascular streaming and variable delivery to brain following carotid artery infusions in the Sprague-Dawley rat. *Journal of Cerebral Blood Flow and Metabolism*, *8*, 116–120.
Sarkadi, B., et al. (2006). Human multidrug resistance ABCB and ABCG transporters: Participation in a chemoimmunity defense system. *Physiological Reviews*, *86*(4), 1179–1236. Available at: http://www.ncbi.nlm.nih.gov/pubmed/17015488. Accessed December 30, 2015.
Sawyer, A. J., & Kyriakides, T. R. (2013). Nanoparticle-based evaluation of blood–brain barrier leakage during the foreign body response. *Journal of Neural Engineering*, *10*(1), 016013. Available at: http://www.pubmedcentral.nih.gov/articlerender.fcgi?artid=3602133&tool=pmcentrez&rendertype=abstract. Accessed March 10, 2016.
Schaeublin, N. M., et al. (2011). Surface charge of gold nanoparticles mediates mechanism of toxicity. *Nanoscale*, *3*(2), 410–420. Available at: http://www.ncbi.nlm.nih.gov/pubmed/21229159. Accessed January 13, 2016.
Schaeublin, N. M., et al. (2012). Does shape matter? Bioeffects of gold nanomaterials in a human skin cell model. *Langmuir*, *28*(6), 3248–3258.
Schleh, C., et al. (2012). Size and surface charge of gold nanoparticles determine absorption across intestinal barriers and accumulation in secondary target organs after oral administration. *Nanotoxicology*, *6*(1), 36–46. Available at: http://www.pubmedcentral.nih.gov/articlerender.fcgi?artid=3267526&tool=pmcentrez&rendertype=abstract. Accessed July 9, 2014.
Seferos, D. S., et al. (2007). Locked nucleic acid-nanoparticle conjugates. *Chembiochem: A European Journal of Chemical Biology*, *8*(11), 1230–1232. Available at: http://www.ncbi.nlm.nih.gov/pubmed/17562553. Accessed May 5, 2014.
Shilo, M., et al. (2014). Transport of nanoparticles through the blood–brain barrier for imaging and therapeutic applications. *Nanoscale*, *6*(4), 2146–2152. Available at: http://www.ncbi.nlm.nih.gov/pubmed/24362586. Accessed March 27, 2014.
Shilo, M., et al. (2015). The effect of nanoparticle size on the probability to cross the blood–brain barrier: An in-vitro endothelial cell model. *Journal of Nanobiotechnology*, *13*, 19. Available at: http://www.pubmedcentral.nih.gov/articlerender.fcgi?artid=4359781&tool=pmcentrez&rendertype=abstract. Accessed March 8, 2016.
Shukla, R., et al. (2005). Biocompatibility of gold nanoparticles and their endocytotic fate inside the cellular compartment: A microscopic overview. *Langmuir: The ACS Journal of Surfaces and Colloids*, *21*(23), 10644–10654. Available at: http://www.ncbi.nlm.nih.gov/pubmed/16262332.
Sloane, E., et al. (2009). Anti-inflammatory cytokine gene therapy decreases sensory and motor dysfunction in experimental multiple sclerosis: MOG-EAE behavioral and anatomical symptom treatment with cytokine gene therapy. *Brain, Behavior, and Immunity*, *23*(1), 92–100.
Soenen, S. J., et al. (2011). Cellular toxicity of inorganic nanoparticles: Common aspects and guidelines for improved nanotoxicity evaluation. *Nano Today*, *6*, 446–465.

Sonavane, G., Tomoda, K., & Makino, K. (2008). Biodistribution of colloidal gold nanoparticles after intravenous administration: Effect of particle size. *Colloids and Surfaces. B, Biointerfaces*, *66*(2), 274–280. Available at: http://www.ncbi.nlm.nih.gov/pubmed/18722754. Accessed May 22, 2013.

Sousa, F., et al. (2010). Functionalized gold nanoparticles: A detailed in vivo multimodal microscopic brain distribution study. *Nanoscale*, *2*(12), 2826–2834. Available at: http://www.ncbi.nlm.nih.gov/pubmed/20949211. Accessed May 6, 2014.

Sreekanthreddy, P., et al. (2015). A three-dimensional model of the human blood–brain barrier to analyse the transport of nanoparticles and astrocyte/endothelial interactions. *F1000Research*, *4*, 1279. Available at: http://f1000research.com/articles/4-1279/v1. Accessed December 1, 2015.

Swierczewska, M., Lee, S., & Chen, X. (2011). The design and application of fluorophore-gold nanoparticle activatable probes. *Physical Chemistry Chemical Physics (PCCP)*, *13*(21), 9929–9941. Available at: http://www.pubmedcentral.nih.gov/articlerender.fcgi?artid=3632443&tool=pmcentrez&rendertype=abstract. Accessed December 15, 2015.

Templeton, A. C., Wuelfing, W. P., & Murray, R. W. (2000). Monolayer-protected cluster molecules. *Accounts of Chemical Research*, *33*(1), 27–36.

Templeton, A. C., et al. (1998). Reactivity of monolayer-protected gold cluster molecules: Steric effects. *Journal of the American Chemical Society*, *120*(8), 1906–1911. Available at: http://pubs.acs.org/doi/abs/10.1021/ja973863+. Accessed April 28, 2014.

Thomas, M., & Klibanov, A. M. (2003). Conjugation to gold nanoparticles enhances polyethylenimine's transfer of plasmid DNA into mammalian cells. *Proceedings of the National Academy of Sciences of the United States of America*, *100*(16), 9138–9143.

Tsoli, M., et al. (2005). Cellular uptake and toxicity of Au55 clusters. *Small (Weinheim an der Bergstrasse, Germany)*, *1*(8–9), 841–844. Available at: http://www.ncbi.nlm.nih.gov/pubmed/17193536. Accessed August 20, 2013.

Turkevich, J., Stevenson, P. C., & Hillier, J. (1951). A study of the nucleation and growth process in the synthesis of colloidal gold. *Discussions of the Faraday Society*, *11*, 55–75.

Van Lehn, R. C., & Alexander-Katz, A. (2014). Fusion of ligand-coated nanoparticles with lipid bilayers: Effect of ligand flexibility. *The Journal of Physical Chemistry A*, *118*, 5848–5856. Available at: http://www.ncbi.nlm.nih.gov/pubmed/24779418.

Van Lehn, R. C., et al. (2013). Effect of particle diameter and surface composition on the spontaneous fusion of monolayer-protected gold nanoparticles with lipid bilayers. *Nano Letters*, *13*(9), 4060–4067. Available at: http://www.ncbi.nlm.nih.gov/pubmed/23915118.

Velasco-Aguirre, C., et al. (2015). Peptides and proteins used to enhance gold nanoparticle delivery to the brain: Preclinical approaches. *International Journal of Nanomedicine*, *10*, 4919–4936. Available at: https://www.dovepress.com/peptides-and-proteins-used-to-enhance-gold-nanoparticle-delivery-to-th-peer-reviewed-fulltext-article-IJN. Accessed October 29, 2015.

Verma, A., et al. (2004). Tunable reactivation of nanoparticle-inhibited beta-galactosidase by glutathione at intracellular concentrations. *Journal of the American Chemical Society*, *126*(43), 13987–13991. Available at: http://www.ncbi.nlm.nih.gov/pubmed/15506760.

Visaria, R. K., et al. (2006). Enhancement of tumor thermal therapy using gold nanoparticle-assisted tumor necrosis factor-alpha delivery. *Molecular Cancer Therapeutics*, *5*(4), 1014–1020. Available at: http://www.ncbi.nlm.nih.gov/pubmed/16648573. Accessed May 5, 2014.

Wang, Z., et al. (2014). A multi-stimuli responsive gold nanocage-hyaluronic platform for targeted photothermal and chemotherapy. *Biomaterials*, *35*(36), 9678–9688. Available at: http://www.ncbi.nlm.nih.gov/pubmed/25176062.

Wolburg, H., & Lippoldt, A. (2002). Tight junctions of the blood–brain barrier: Development, composition and regulation. *Vascular Pharmacology*, *38*(6), 323–337. Available at: http://www.ncbi.nlm.nih.gov/pubmed/12529927. Accessed March 10, 2016.

Xia, Y., et al. (2011). Gold nanocages: From synthesis to theranostic applications. *Accounts of Chemical Research*, *44*(10), 914–924.

Yang, J. A., Lohse, S. E., & Murphy, C. J. (2014). Tuning cellular response to nanoparticles via surface chemistry and aggregation. *Small*, *10*(8), 1642–1651.

Yang, P.-H., et al. (2005). Transferrin-mediated gold nanoparticle cellular uptake. *Bioconjugate Chemistry*, *16*(3), 494–496.

Yang, X., et al. (2014). Uptake of silica nanoparticles: Neurotoxicity and Alzheimer-like pathology in human SK-N-SH and mouse neuro2a neuroblastoma cells. *Toxicology Letters*, *229*(1), 240–249.

Ye, D., Dawson, K. A., & Lynch, I. (2015). A TEM protocol for quality assurance of in vitro cellular barrier models and its application to the assessment of nanoparticle transport mechanisms across barriers. *The Analyst*, *140*, 83–97. Available at: http://dx.doi.org/10.1039/C4AN01276C.

Zhao, E., et al. (2012). Surface engineering of gold nanoparticles for in vitro siRNA delivery. *Nanoscale*, *4*, 5102.

Zhu, Y., & Liao, L. (2015). Applications of nanoparticles for anticancer drug delivery: A review. *Journal of Nanoscience and Nanotechnology*, *15*(7), 4753–4773.

CHAPTER SIX

Metal Nanoparticles as Targeted Carriers Circumventing the Blood–Brain Barrier

A.C. Sintov*[,1], C. Velasco-Aguirre[†,‡], E. Gallardo-Toledo[†,‡], E. Araya[‡,§], M.J. Kogan[†,‡,1]

*Faculty of Engineering Sciences, Ben Gurion University of the Negev, Be'er Sheva, Israel
[†]Facultad de Ciencias Químicas y Farmacéuticas, Universidad de Chile, Santiago, Chile
[‡]Advanced Center for Chronic Diseases (ACCDiS), Santiago, Chile
[§]Facultad de Ciencias Exactas, Universidad Andrés Bello, Santiago, Chile
[1]Corresponding authors: e-mail address: asintov@bgu.ac.il; mkogan@ciq.uchile.cl

Contents

Abstract

Metal nanoparticles have been proposed as a carrier and a therapeutic agent in biomedical field because of their unique physiochemical properties. Due to these physicochemical properties, they can be used in different fields of biomedicine. In relation to this, plasmonic nanoparticles can be used for detection and photothermal destruction of tumor cells or toxic protein aggregates, and magnetic iron nanoparticles can be used

International Review of Neurobiology, Volume 130
ISSN 0074-7742
http://dx.doi.org/10.1016/bs.irn.2016.06.007

for imaging and for hyperthermia of tumor cells. In addition, both therapy and imaging can be combined in one nanoparticle system, in a process called theranostics. Metal nanoparticles can be synthesized to modulate their size and shape, and conjugated with different ligands, which allow their application in drug delivery, diagnostics, and treatment of central nervous system diseases. This review is focused on the potential applications of metal nanoparticles and their capability to circumvent the blood–brain barrier (BBB). Although many articles have demonstrated delivery of metal nanoparticles to the brain by crossing the BBB after systemic administration, the percentage of the injected dose that reaches this organ is low in comparison to others, especially the liver and spleen. In connection with this drawback, we elaborate the architecture of the BBB and review possible mechanisms to cross this barrier by engineered nanoparticles. The potential uses of metal nanoparticles for treatment of disorders as well as related neurotoxicological considerations are also discussed. Finally, we bring up for discussion a direct and relatively simpler solution to the problem. We discuss this in detail after having proposed the use of the intranasal administration route as a way to circumvent the BBB. This route has not been extensively studied yet for metal nanoparticles, although it could be used as a research tool for mechanistic understanding and toxicity as well as an added value for medical practice.

1. INTRODUCTION

Earlier diagnosis as well as effective treatment of brain diseases is of great relevance. Novel nanoparticles (NPs) have been proposed as new methods for diagnosing and treating central nervous system (CNS) diseases, such as drug delivery systems targeting the CNS, nanoscaffolds for tissue regeneration, and NPs for CNS imaging. Metal nanoparticles can be synthesized, modulating their size and shape, and modified with various chemical functional groups which allow them to be conjugated with different ligands such as drugs, peptides, antibodies, or even polymers, among others. In addition, their physicochemical properties allow them to be used in different fields of biomedicine. In the case of therapy, the surface plasmon resonance (silver and gold) (Afifi, Almaghrabi, & Kadasa, 2015; Day et al., 2012) and the application of magnetic fields on magnetic nanoparticles (MNPs) (iron nanoparticles) (Zhang et al., 2015) can be used for the destruction of cancer cells or protein aggregates; this process is called photothermia. On the other hand, metallic nanoparticles have been studied for their application in diagnostic imaging, mainly due to their high electron density, which allows them to be used as contrast agents. Both therapy and imaging can be combined in one nanoparticle system, a process called theranostics (Dilnawaz & Sahoo, 2015). As pointed out in this review, it is essential first to find new quantitative technological methods regarding how to overcome the blood–brain barrier (BBB) for effectively targeting nanocarriers to the CNS. For this purpose, different polymeric nanoparticles have been explored, some of them in clinical research, as described in an interesting review by Tosi et al. (2013). Metallic NPs, however, have not been widely implemented in clinical area in spite of their extensive scientific research. This review focuses on the potential application of the metal nanoparticles and their capability to circumvent the BBB. In the first section, we describe the architecture of the BBB and the possible mechanisms to cross this barrier by nanoparticles. In the next section, we discuss the potential use of metal nanoparticles for treatment of CNS disorders as well as neurotoxicological considerations. Finally, we have made a revision to elaborate and discuss the use of the intranasal administration route as a direct and easy way to circumvent the BBB.

2. THE BBB AND MECHANISMS OF NANOPARTICLES BY WHICH THEY CROSS THIS BARRIER

The maintenance of CNS homeostasis is critical for its correct function (Abbott, Patabendige, Dolman, Yusof, & Begley, 2010), being achieved by the BBB, which contains the largest interface between blood and CNS. BBB, one of the most important barriers, corresponds to the combination of three membranes, which are the endothelium of the brain microvessels, the epithelium of the choroid plexus, and the epithelium of the arachnoid mater (Becker, Novikoff, & Zimmerman, 1967; Nabeshima, Reese, Landis, & Brightman, 1975). The importance of these membranes resides in the fact that they regulate the transport of substances by generating a mechanism that maintains the homeostasis (Abbott, 1992).

The neurovascular unit of BBB is formed by endothelial cells, microglial cells, pericytes, and astrocytes. The former creates a physical barrier that blocks paracellular transport through BBB due to the generation of complexes of tight and adherent junctions between cells (Nag, 2011). Pericytes sheath approximately one-third of the endothelial tube (37%), and two-thirds of the circumference of the endothelial tube (63%) is covered by astrocytic endfeet (Mathiisen, Lehre, Danbolt, & Ottersen, 2010). Both types of cells are important in the formation and differentiation of CNS vasculature (Mae, Armulik, & Betsholtz, 2011).

The principal role of BBB is to limit the flux of unspecific molecules or xenobiotics into the brain through a strict insulation of CNS. However, there are various mechanisms present in this barrier that allow the passage of essential molecules and nutrients for the correct function of the brain (Fig. 1). Passive mechanisms are used by small gaseous molecules or lipophilic substances in a nonsaturable way. In the case of hydrophilic molecules, these can spread through BBB via saturable mechanisms, which, in most cases, are active mechanisms (energy dependent). Among the different mechanisms of

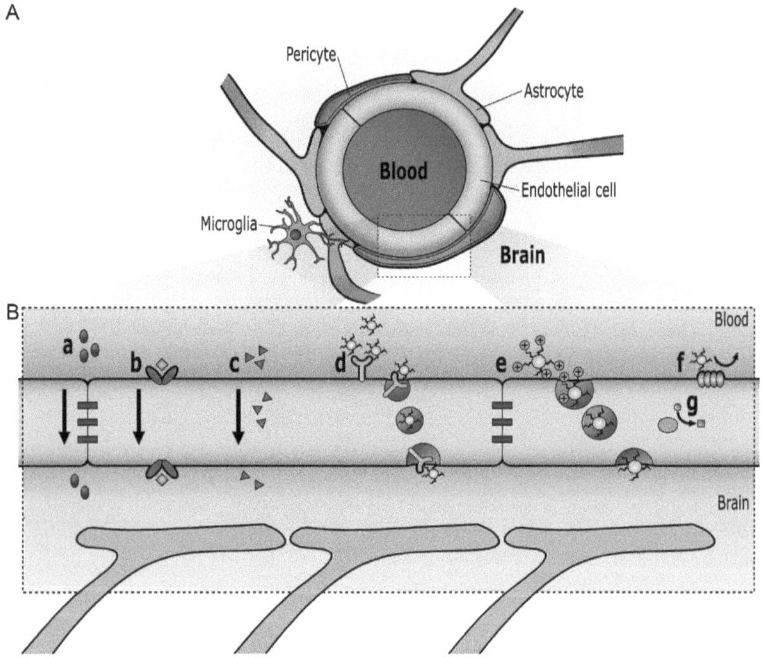

Fig. 1 Blood–brain barrier structure and their principal transport mechanisms. (A) Schematic representation of pericytes and astrocytes, which are part of the neurovascular unit that constitutes the BBB. This unit has a significant role in the development of the CNS vasculature. (B) Main transport mechanisms involved in the crossing of molecules through BBB to the CNS. (a) Tight junctions present in cells that form BBB regulate the entrance of water-soluble compounds and polar drugs. (b) Transport proteins, included in the endothelium, act as carriers for different substances (such as amino acids, glucose, nucleosides, and purine bases, among others). (c) Lipid-soluble molecules cross through the BBB by a diffusive route. (d) Receptor-mediated transcytosis and endocytosis allow the transport of proteins such as transferrin and insulin by specific recognition of these molecules. (e) Molecules that have a positive surface enhance their penetration to CNS by adsorptive-mediated transcytosis and endocytosis. (f) In the endothelium of BBB, efflux pumps are present and expulse the agents from cells to the blood. (g) Representation of cytochrome P450 enzymes. The transport mechanisms (d) and (e) are the most studied routes for nanoparticle delivery to CNS. Abbreviations: *BBB*, blood–brain barrier; *CNS*, central nervous system. *Panel (B): Reprinted with permission from Velasco-Aguirre, C., Morales, F., Gallardo-Toledo, E., Guerrero, S., Giralt, E., Araya, E., & Kogan, M. J. (2015). Peptides and proteins used to enhance gold nanoparticle delivery to the brain: Preclinical approaches.* International Journal of Nanomedicine, 10, *4919–4936.*

saturable transport present in the BBB, these can be divided into carrier-mediated transport (CMT), receptor-mediated transcytosis (RMT), and adsorptive-mediated transcytosis (AMT). The first transport, CMT, carries different kinds of small nutrients or molecules such as ascorbic acid, glucose, glutathione, and tyrosine through highly specialized proteins. In contrast, RMT and AMT are vesicle-based mechanisms which transport large molecules. In addition, a mechanism that relies on monocytes and macrophages has been identified and allows the crossing of viruses such as the human immunodeficiency virus and *Cryptococcus* through the BBB (Velasco-Aguirre et al., 2015).

The main mechanisms studied for metal nanoparticles' delivery to the CNS are AMT and RMT, but not the CMT, which is principally designated for small molecules. AMT depends on the interaction of the negative charges on the plasma membranes (sulfated proteoglycans) with cationic compounds bound to a metal nanoparticle, leading to an endocytic mechanism independent of membrane receptors which effects transcytosis (Pujals, Fernandez-Carneado, Lopez-Iglesias, Kogan, & Giralt, 2006). The lack of specificity of AMT process is the main drawback of this BBB-crossing mechanism. It is also important to consider the potential toxicity and immunogenicity associated with cationic particles (Herve, Ghinea, & Scherrmann, 2008). In contrast, RMT is a highly specific mechanism which consists of receptor–ligand recognition at the luminal side of BBB, followed by exocytosis at the abluminal side. RMT has been widely studied for the transport of large molecules, liposomes, polymeric complexes, and nanoparticles across the BBB (Gabathuler, 2010; Pardridge, 2005, 2007). The main limitations of this strategy are the saturation of the receptors involved in RMT's mechanisms and competition with natural substrates such as insulin, leptin, apolipoprotein, and transferrin. Although many articles have demonstrated the delivery of metal nanoparticles to the brain crossing the BBB by systemic administration, the percentage of injected dose that reaches this organ is low in comparison to others, especially the liver and spleen (Frigell, Garcia, Gomez-Vallejo, Llop, & Penades, 2014; Guerrero et al., 2010, 2012; Prades et al., 2012; Schaffler et al., 2014; Shilo, Motiei, Hana, & Popovtzer, 2014). This phenomenon can be explained by the interactions of nanoparticles with plasma proteins, forming the so-called protein corona (Arvizo et al., 2012; Dell'Orco, Lundqvist, Oslakovic, Cedervall, & Linse, 2010; Lundqvist et al., 2011, 2008; Monopoli et al., 2011; Tenzer et al., 2011). This process allows macrophage cells to recognize and internalize the nanoparticles capped by these proteins, leading to a significant loss of the nanoparticles from circulation and a low delivery to the target organ

(Dobrovolskaia, Aggarwal, Hall, & McNeil, 2008). This loss occurs because of the retention of the nanoparticles by the reticuloendothelial system (RES), resulting in their accumulation in the liver and spleen (Dobrovolskaia et al., 2008). Modification of colloidal particles with poly(ethylene glycol) (PEG) can increase their half-life in the blood, reducing the retention by the RES (Li & Huang, 2008). However, there is little information regarding the composition of protein corona after systemic administration, remaining the main problem of isolating the NPs from the in vivo environment. Specifically, in the case of metal nanoparticles, the composition of the corona was determined using the magnetic properties of superparamagnetic iron oxide NPs (SPIONs), extracting the nanoparticles from sera after systemic administration in rats (Sakulkhu et al., 2014). Moreover, the authors evaluated how the surface charge of the SPIONs affects the composition of the corona and the biodistribution of the nanoparticles, and compared them with the corona composition of particles incubated for the same time in rat serum. The results obtained show that the composition of the protein corona depends on the initial surface charges of the nanoparticles. In general, positively charged particles show the lowest number of different proteins in vivo, neutral particles show the highest number of different proteins both in vitro and in vivo, and particles with a negative charge show the lowest number of different proteins in vitro. These results are very important, because the knowledge of the in vivo protein corona could lead to better understanding and prediction of the fate of NPs in vivo. Moreover, attempts to improve the penetration of metal nanoparticles to the CNS have been studied. One example is the use of focused ultrasound to enhance the permeability of BBB (Etame et al., 2012).

However, it is important to consider that this kind of strategy is not innocuous, since it can allow the entry of other molecules or xenobiotics to the CNS. One more example is the intracranial injection; although this is an invasive strategy, it has been proven to avoid the problem of the nonspecific entrance accompanying systemic administration (Glat, Skaat, Menkes-Caspi, Margel, & Stern, 2013).

3. IMAGING AND TREATMENT OF NEUROLOGICAL DISEASES ASSOCIATED WITH METAL NANOPARTICLES

Nanoparticles are unique in dimension and are able to control drug release. This capacity has primed NPs for the treatment of several diseases, including CNS diseases. However, treatment of CNS disorders is restricted due to the complexity of the BBB, which impedes an adequate exposure to active agents. Nevertheless, NPs that have undergone appropriate surface modifications are capable of reaching the CNS.

Several animal studies have provided direct evidence that NPs could reach and accumulate in the brain parenchyma, including the striatum and hippocampus (Kao et al., 2012; Wiley, Webster, Gale, & Davis, 2013; Wu, Ding, & Sun, 2013). This evidence makes NPs a useful and attractive alternative treatment for neurological diseases, and an important tool to diagnose the disease and to deliver drugs or active compounds to the brain.

3.1 Metal Nanoparticles for CNS Imaging

There are several examples in the literature (summarized in Table 1), which are related to the potential application of metal nanoparticles for brain imaging and drug delivery. The potential uses of MNPs and gold nanoparticles (AuNPs) are mainly documented.

MNPs are used as probes to visualize tumors. Magnetic resonance imaging (MRI) produces high-resolution images of soft tissues down to 1 mm at clinical field strengths. This technique provides detailed anatomic information about structure and physiological conditions and also plays an important role in the oncological imaging of the CNS for surgical localization and intervention, which accurately allow diagnosing, prognosticating treatment effects, and monitoring therapeutic responses (Corot et al., 2006). In applying in vivo brain MRI, it is very important to use an optimally chosen contrast agent since this technique can be insensitive to an exogenous agent. To deal with this issue, manganese nanoparticles have been introduced as a promising T1 contrast neural tracer for MRIs of metastatic brain tumors (Na et al., 2007). However, most of the manganese nanoparticles exhibit a low relaxation rate (Bennewitz et al., 2011), resulting in an insufficient enhancement effect on tiny gliomas. One attempt to avoid this problem is described by Shao et al. (2015), who developed gadolinium-doped $MnCO_3$ nanoparticles as T1 contrast tracer, showing it to be an effective nanoprobe for the diagnostic imaging of tiny brain gliomas due to the high relaxivity, the bimodal imaging capability, and the specificity demonstrated in animal studies. Nevertheless, several studies have reported a cellular toxicity caused by high doses of manganese. It is therefore necessary to reduce the dose required, which is a worse alternative. However, MNPs could be used as an excellent alternative to carry the contrast agent without causing

Table 1 Examples of in vivo Metal NP-Based Carriers for Brain-Targeted Therapy or Drug Delivery

Type of NP	Type of Study	Use	References
MnO nanoparticles	Preclinical	Imaging of brain gliomas	Chen et al. (2015)
Iron oxide nanoparticles	Preclinical	MRI contrast agents to brain tumor and antiretroviral drug delivery	Chertok et al. (2008), Corot, Robert, Idee, and Port (2006), Fiandra et al. (2015), Iv et al. (2015), Kallur, Farr, Bohm-Sturm, Kokaia, and Hoehn (2011), Kong et al. (2012), Muldoon, Sandor, Pinkston, and Neuwelt (2005), Neuwelt et al. (2004), Ramos-Gómez, Seiz, and Martinez-Serrano (2015), Saiyed, Gandhi, and Nair (2010), Simon et al. (2006), Thu et al. (2009), Wu et al. (2013), and Zhang et al. (2015)
Ultrasmall superparamagnetic iron oxide nanoparticles (ferumoxytol and ferumoxtran-10)	Clinical	Brain tumor imaging	Abbott (1992), Gahramanov et al. (2013), Jung and Jacobs (1995), Murillo et al. (2005), Taschner et al. (2005), Varallyay et al. (2013), and Wadghiri et al. (2013)
Gold nanoparticles	Preclinical	Prion disease therapy and brain tumor image	Day et al. (2012), Schultke et al. (2014), Sousa et al. (2010), and Wiley et al. (2013)
Gadolinium metallofullerene nanoparticle	Preclinical	Brain tumor drug delivery and longitudinal imaging	Li et al. (2015) and Shultz et al. (2011)
Ultrasmall gadolinium oxide nanoparticles	Preclinical	Image brain cancer cells in vivo with MRI	Faucher et al. (2011) and Park et al. (2009)

Table 1 Examples of in vivo Metal NP-Based Carriers for Brain-Targeted Therapy or Drug Delivery—cont'd

Type of NP	Type of Study	Use	References
Cerium oxide nanoparticles	Preclinical	Multiple sclerosis treatment	Estevez et al. (2011) and Heckman et al. (2013)
Zinc oxide nanoparticles	Preclinical	Reduce oxidative stress in brain	Afifi et al. (2015), Kao et al. (2012), and Saddick, Afifi, and Abu Zinada (in press)
Silver nanoparticles	Preclinical	Reduce oxidative stress in brain and treatment of glioblastoma	Afifi et al. (2015) and Locatelli et al. (2014)
Platinum nanoparticles	Preclinical	Parkinson's disease	Nellore, Pauline, and Amarnath (2013)

adverse effects (Chen et al., 2015). Chen et al. prepared PEG–Cy5.5 conjugated MnO nanoparticles (MnO–PEG–Cy5.5 NPs) with magnetic resonance and near-infrared fluorescence imaging modalities. They demonstrated the capacity of MnO–PEG–Cy5.5 NPs for the detection of brain gliomas as a dual-modal imaging nanoprobe, seeing an enhancement in the diagnostic obtained with this nanosystem which provides deep and sensitive information on both an anatomical and a cellular level (Chen et al., 2015).

Other studies employing iron oxide nanoparticles have received attention due to their preclinical and clinical potential, especially in theranostics (Iv et al., 2015; Thu et al., 2009). Theranostics is a new term used for the combination of therapy and diagnosis. Due to their magnetic property, the iron oxide nanoparticles are able to be utilized as contrast agents for MRI as well as therapeutic carriers, ie, they are effective theranostic agents, as illustrated in Fig. 2 (Dilnawaz & Sahoo, 2015). This approach enables us to receive real-time information about a CNS disease and at the same time to provide a local treatment (Fig. 2).

In 2012, Kong et al. reported that an external magnetic field applied before systemic administration of Fe_3O_4 nanoparticles significantly enhanced the permeation through the BBB, approaching the perivascular zone of the brain parenchyma in a mouse model (Kong et al., 2012). These iron oxide nanoparticles are classified into two groups based on their mean

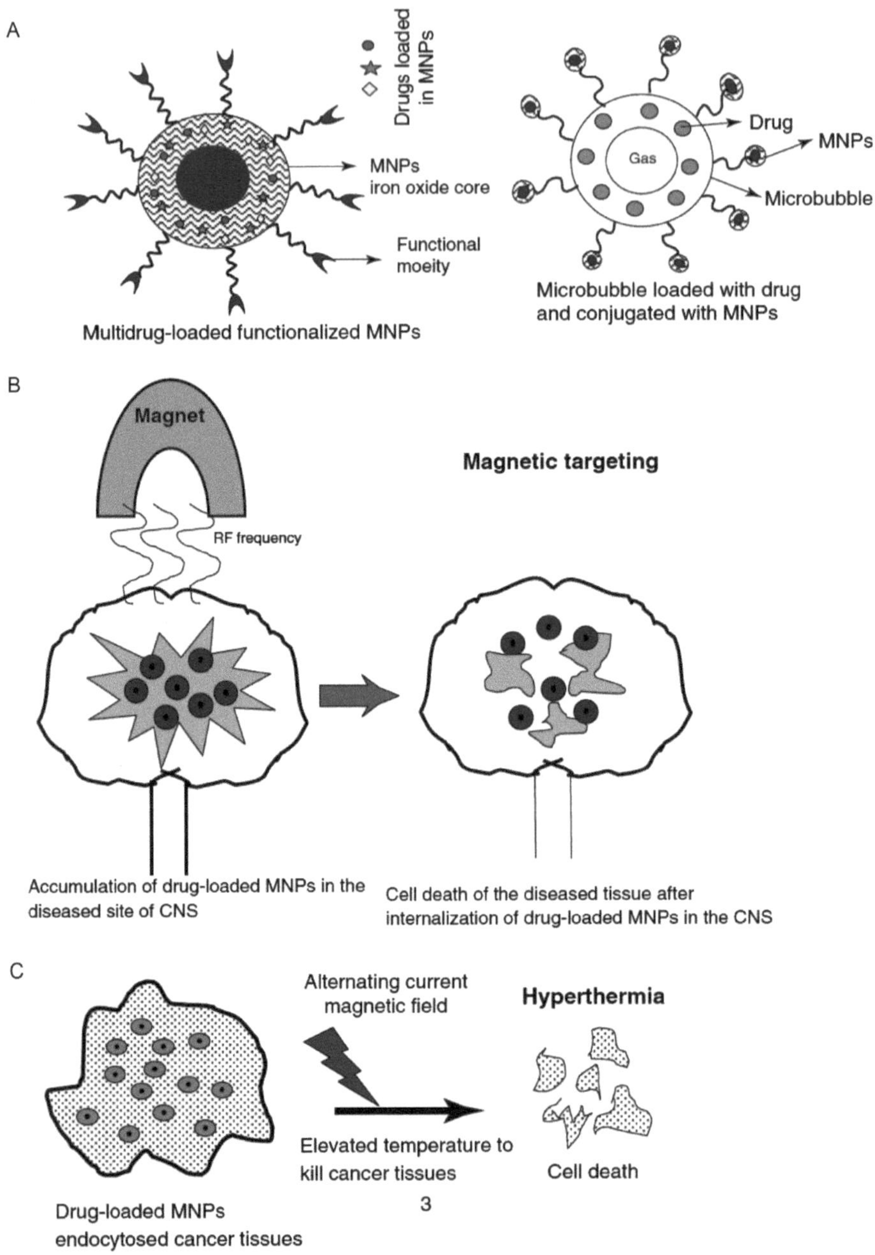

Fig. 2 (A) Drug-loaded functionalized magnetic nanoparticles (MNPs) used for a theranostic approach. (B) By the application of an external magnet, the drug-loaded MNPs will be dragged to the desired site (known as magnetic targeting) for obtaining the significant therapeutic effect. (C) Hyperthermia effect by drug-loaded MNPs, subjected to alternative magnetic field (AMF).

hydrodynamic particle size: superparamagnetic iron oxide and ultrasmall superparamagnetic iron oxide (USPIO) particles. Currently, there are two USPIOs that are in clinical evaluation as an MRI contrast agent: ferumoxtran-10 and ferumoxytol (Corot et al., 2006; Simon et al., 2006). The former is a first-generation USPIO coated with dextran, which has been used in Phase I to Phase III clinical trials mainly for lymph node imaging (Wang et al., 2014). However, the FDA did not approve ferumoxtran-10 as an MRI contrast agent due to its inconclusive impact on patient management. On the other hand, ferumoxytol, a second-generation USPIO with a carboxymethyl dextran coating, was approved by the FDA in 2009 for the treatment of iron deficiency anemia in adults with chronic kidney disease. At the time of writing, both types of iron oxide nanoparticles are in approval by the FDA as MRI contrast agents for CNS clinical application (Neuwelt et al., 2004; Varallyay et al., 2013). Iron oxide nanoparticles can be magnetized in the presence of a magnetic field, so that delivery of drug load can be directed to the specific region, resulting in effective treatment at minimum dosing. In drug delivery, MNPs have also been studied. Saiyed et al. reported the effectiveness of azidothymidine 5′-triphosphate (AZTTP)-loaded magnetic liposomes to cross the BBB through application of an external magnetic field for treatment of neuro-AIDS (Saiyed et al., 2010).

In the field of neural regeneration, MRI provides an excellent noninvasive technique to in vivo cellular tracking of transplanted human neural stem cells (Ramos-Gómez & Martinez-Serrano, 2016). To detect the transplanted cells by MRI with high sensitivity, Kallur et al. labeled human neural stem cells with MNPs and subsequently implanted them. These have been shown to produce sufficient contrast for MRI visualization without disturbance of their functional properties, detecting around 500 labeled cells (Kallur et al., 2011).

Another nanomaterial, AuNP, has been extensively used in biological applications due to its biocompatibility, size, and the extensive knowledge gathered on its surface chemistry. These features make AuNPs very attractive for biomedical applications, including drug delivery and novel diagnostic and therapeutic applications. Guerrero et al. studied ex vivo distribution of the conjugate C(AuNP)-LPFFD for its potential uses in the treatment of Alzheimer's disease. In this study, the researchers covalently labeled the conjugate with [^{18}F]-fluorobenzoate to study the in vivo distribution of the AuNPs by positron emission tomography (PET). This imaging technique allowed the study of the in vivo biodistribution of AuNPs in a sensitive and noninvasive way (Guerrero et al., 2012). A similar work by Frigell et al. intended to synthesize highly sensitive PET probes has yielded novel water-soluble and biocompatible glucose-coated AuNPs conjugated with neuropeptides and a chelator of the positron emitter ^{68}Ga as a PET reporter for in vivo tracking biodistribution (Frigell et al., 2014). Day et al. have developed VEGF and PEG gold nanoshells for vascular-targeted photothermal therapy of glioma. The photothermal ability of nanoshells upon near-infrared laser irradiation was evaluated in vitro. Latter, in vivo studies have therapeutically evaluated the nanoshells in the treatment of mice that had intracerebral glioma tumors. The tumors in the mice were exposed to near-infrared light after systemically delivery of the VEGF-coated nanoshells. The treatment effect resulted in photothermal ablation and in vivo vascular disruption in glioma (Day et al., 2012).

Other materials used for imaging correspond to gadolinium metallofullerene nanoparticles (Li et al., 2015; Shultz et al., 2011) and ultrasmall gadolinium oxide nanoparticles (Faucher et al., 2011; Park et al., 2009), specifically for brain tumor and longitudinal imaging, and imaging of brain cancer cells with MRI, respectively.

3.2 Metal Nanoparticles for the Treatment of Neurodegenerative Disease

Amyloid plaques are the main pathological hallmark of Alzheimer's disease. Wadghiri et al. used bifunctional MNPs that were chemically coupled with $A\beta_{1-42}$ peptide and PEG. They have demonstrated that, following administration of MNPs conjugated with PEG and $A\beta_{1-42}$, the amyloid plaques were detected in vivo in AD transgenic mice using magnetic resonance microimaging (Table 1; Wadghiri et al., 2013). To detect the β-amyloid plaques in a rat model of Alzheimer's disease, Zhang et al. used SPIONs coated with 1,1-dicyano-2-[6-(dimethylamino)-naphthalene-2-yl] propene carboxyl derivative. The researchers have reported a significant decrease in signal intensity detected in the hippocampal area of AD rats after intravenous administration of the nanoparticles, but not in non-AD rats (Zhang et al., 2015).

Another neurodegenerative pathology corresponds to Parkinson's disease (PD). This is one of the most common neurodegenerative disorders that affects the older population, causing tremors, bradykinesia, and akinesia. The hallmark is a selective loss of dopaminergic neurons in the substantia nigra associated with an increase in prooxidant or loss of antioxidants' levels and mitochondrial dysfunction, as well as aggregation of α-synuclein. As a consequence, a difficulty appears in controlling movement. The conventional treatment is increasing the level of dopamine in the brain. However, stem cells have been reported recently as an alternative treatment. Stem cells allow in vivo cell tracking in cell replacement therapies against Parkinson's disease. The Center for Biomedical Technology in Spain has developed an efficient protocol in which the human neural precursor cells are labeled with MNPs that can be detected for up to 5 months by MRI (Ramos-Gómez et al., 2015).

Platinum nanoparticles have also been studied for the treatment of PD (Nellore et al., 2013). Zebrafish was used as a model for this study, because of the high sensitivity of this species to the neurotoxin 1-methyl-4-phenyl-1,2,3,6-tetrahydropyridine (MPTP). MPTP is a disease-inducing agent that produces a similar loss of dopaminergic neurons resembling Parkinsonian symptoms. Pretreatment of zebrafish with platinum nanoparticles significantly reversed the toxic effects produced by MPTP, an effect related to the increment of dopamine levels, its metabolites, glutathione and activities of complex I, superoxide dismutase, glutathione peroxidase and catalase, and a decrease of malondialdehyde levels. These results suggest a dual function of this treatment as mitochondrial complex I and antioxidant activity.

Huntington's disease (HD) is a neurodegenerative disorder created by the expression of a mutated huntingtin protein. This expression leads to the selective death of the medium-size spiny GABAergic neurons in the striatum, resulting in the appearance of involuntary movements. Moraes et al. have developed the SPION-labeled cells for tracking of bone marrow mesenchymal cells, which have been used as a therapy by preventing or retarding the degeneration of striatal neuronal populations in HD. The combination of nanotechnology and stem cells allows the decreases in neurodegeneration and reduces the ventriculomegaly in the lesioned brain in an HD rat model. Moreover, the incorporation of SPION produced MRI signals that were visible for at least 60 days after transplantation, demonstrating the potential of nanotechnology for cell tracking and therapy (Moraes et al., 2012).

In addition, studies of the use of metal nanoparticles for the treatment of neurological diseases have been reported. Shultz et al. (2011) and Locatelli et al. (2014) studied the use of metal nanoparticles for the treatment of brain tumors, whereas Estevez et al. (2011) and Heckman et al. (2013) investigated the use of cerium oxide nanoparticles for multiple sclerosis therapy (Table 1).

4. TRAFFICKING OF METAL NANOPARTICLES TO THE CNS AND NEUROTOXICOLOGICAL CONSIDERATIONS

4.1 Crossing of Nanoparticles Through the BBB

Crossing of metal NPs to the brain and their potential neurotoxicity effects are important topics essential for their potential clinical use. To elaborate these topics further, we recommend two recently published reviews (Feng et al., 2015; Velasco-Aguirre et al., 2015). Depending on the size, shape, composition, surfaces, and aggregation state of the nanoparticles, they are capable of crossing the BBB or producing toxic effects. Because of their very small size, some NPs can easily reach the brain and be taken up by the brain cells, such as neurons and glia. Nanoparticles' uptake by cells includes pinocytosis, endocytosis dependent on caveolae and lipid raft composition, clathrin-dependent endocytosis, and phagocytosis (AshaRani, Hande, & Valiyaveettil, 2009). The intracellular sites of localization of NPs vary depending on the cell type and the uptake mechanism. As mentioned in Section 2, the main mechanisms by which delivery of metal NPs to the CNS is carried out are AMT and RMT. Several animal studies have provided direct evidence that NPs could reach and accumulate in the brain parenchyma, including the striatum and hippocampus. In the case of gold, transferrin-containing AuNPs can enter and accumulate in the brain parenchyma after a systemic administration in mice through an RMT pathway (Wiley et al., 2013). Our group improved the brain delivery of AuNPs by conjugation with the amphipathic peptide CLPFFD (Guerrero et al., 2010). This peptide is a β-breaker peptide designed by Soto et al. (1998), which is based on the original sequence of β-amyloid, and contains a Cys added to the extreme N-terminal portion of the peptide for the chemisorption to the gold surface, resulting in the conjugate AuNP-CLPFFD. In addition, to enhance the permeation of AuNP-CLPFFD across the BBB, we incorporated the peptide THRPPMWSPVWP (which recognizes the transferrin favoring the RMT present in the microvascular endothelial cells of the BBB) to the CLPFFD sequence, causing an increase in the permeability of the conjugate into the brain (Prades et al., 2012). On the other hand, Kao et al. observed translocation of zinc oxide (ZnO) NPs into the brain following in vivo nasal administration in a Sprague Dawley rat model (Kao et al., 2012; Table 1).

4.2 Neurotoxicity of Metal Nanoparticles

As stated by Feng et al. (2015), the main effects of metal NPs on mammalian cells are oxidative stress, immune system dysfunction, and autophagy. The enhanced generation of reactive oxygen species (ROS) affects the mitochondrial respiratory chain and increases CNS toxicity of metallic nanoparticles on the amount of unfolded and misfolded proteins in the endoplasmic reticulum, inducing endoplasmic reticulum stress. Both types of cellular damage lead to further ROS generation, DNA damage, and activation of signaling, resulting in the activation of various cell type-specific pathways for inflammation, apoptosis, or necrotic death (Wu et al., 2011). Moreover, autophagy is a cellular process for elimination of long-lived proteins and damaged organelles components via lysosomal mechanism. The cell autophagy mechanism can be induced by various forms of stress, such as interaction with various types of nanoparticles. Nanoparticles induce autophagy in two ways: one is mediated by an ROS-dependent mechanism and the other by a lysosome-dependent autophagy. Autophagy can play dual roles in cell survival after interaction with NPs. On the one hand, it can be cytoprotective; on the other, it plays a cytotoxic role under different conditions. It is very relevant in the context of this review that the only cell process able to degrade large components, as nanoparticles, is autophagy. Nanoparticles-mediated autophagy may be an adaptive cellular response facilitating the degradation and clearance of nanoparticles, but may also cause harmful cellular dysfunction, as stated by Feng et al. (2015).

There is an increasing amount of evidence indicating that NPs can cause cytotoxicity due to their large surface area, which is capable of inducing ROS, a key factor that generates cellular stress and disease pathogenesis (Nel, Xia, Madler, & Li, 2006). However, the impact of NP-induced ROS on the CNS is unclear. Several types of nanomaterials, such as quantum dots and metal-oxide NPs, have been shown to cause the overexpression of intracellular ROS (Hanley et al., 2009). On the other hand, once nanoparticles enter the CNS, they encounter microglial immune cells and neurons. Once microglia are activated, they can change the form and function. The chronic activation of microglia can cause neurotoxicity. Another effect that has been reported is that nanoparticles can induce and modulate autophagy in different cells (Wu et al., 2011).

One example of toxicity in vivo involves ZnO NPs that were found to attenuate learning and memory ability in rats (Han, Tian, Zhang, Ren, & Yang, 2011). Furthermore, Hu et al. observed an accumulation of TiO_2 nanoparticles in the mouse hippocampus after intragastric administration, and this accumulation led to hippocampal apoptosis and impairment in spatial recognition memory (Hu et al., 2011). Although many nanoparticles studied up to the present time have had dose-dependent neurotoxic effects, some nanoparticles have been found to have protective functions toward the CNS. For example, fullerenol was found to play a neuroprotective role, preventing hydrogen peroxide and cumene hydroperoxide-induced damage in rat hippocampal slices (Tsai, Chen, & Chiang, 1997). In addition, Afifi et al. (2015) observed a protective role of ZnO NPs and silver nanoparticles (AgNPs) by reducing the oxidative stress in the brain. Saddick et al. (in press) reported similar results with ZnO NPs.

In order to consider the future applications of metal NPs, it is necessary to take into account the low bioavailability of NPs to the brain following systemic administration. In the next section, we discuss the capability of intranasal administration to increase brain availability. Nanoparticles delivered through the intranasal route can potentially bypass the BBB, meaning that the retention by the RES can be avoided.

5. THE INTRANASAL ADMINISTRATION: NOSE-TO-BRAIN ROUTE FOR METAL NANOPARTICLE DELIVERY

Nasal delivery of medicaments by sniffing powders or liquids has been practiced since ancient times, and the method has recently experienced a renaissance. The nasal route could provide an attractive easily accessible, needle-free alternative for currently injectable drugs, which may improve patient compliance and allow extended use of self-medication for many chronic diseases or acute conditions. Among the mucosal sites in the human body, the nasal mucosa has the potential to compete with the intravenous route for a fast onset of drug action and a high drug bioavailability. This is due to the vastly vascularized nasal mucosa and the large epithelial surface area (approximately 150 cm^2). These properties present potential benefits for systemic absorption in terms of quick action, an avoidance of any degradation and/or unwanted enterohepatic metabolism of drugs (including peptides and proteins), and an increase in patient compliance. However, the effective delivery of drugs via this route is challenging due to the low delivery payload (which usually necessitates high-volume liquid dosages or appropriate formulations), poor reproducibility, and possible mucosal irritation (Arora, Sharma, & Garg, 2002; Illum, 2003; Merkus, Schipper, & Verhoef, 1996; Sintov, Levy, & Botner, 2010; Ugwoke, Agu, Verbeke, & Kinget, 2005; Wolfe & Bernstone, 2004).

A more challenging and interesting advantage of nasal delivery is the possibility of targeting drugs, contrast agents, and biomarkers to the CNS by bypassing the BBB (Dhuria, Hanson, & Frey, 2010; Hanson & Frey,

2007, 2008; Illum, 2000, 2002, 2003; Mistry, Stolnik, & Illum, 2009). The olfactory region has been marked as a relatively open gate for direct transport from nose to brain (Dhuria et al., 2010; Illum, 2000, 2002, 2003, 2004). Due to the unique anatomical organization of the olfactory nerves, a compound (even hydrophilic and/or high MW) can circumvent the BBB by: (a) neuronal endocytosis followed by intracellular transport and axonal migration to the olfactory bulb with subsequent possible distribution of the drug into distant brain tissues; and (b) paracellular diffusion through the intercellular clefts in the olfactory epithelium and through the channels created between the ensheathing cells surrounding the olfactory axons in the lamina propria toward the cerebrospinal fluid (CSF) and the olfactory bulb (Hanson & Frey, 2007, 2008; Illum, 2000; Thorne, Emory, Ala, & Frey, 1995). Intranasal delivery to targeting active therapeutics into the brain, or the so-called nose-to-brain route, might be efficient enough to achieve therapeutic effects in the CNS of neurological impaired humans by drugs that are not (or less) permeable by the BBB. Concomitant with the enhancement of the therapeutic efficacy, the nasal route may also reduce the systemic toxicity of centrally acting drugs.

5.1 The Nasal Architecture and the Olfactory Region

The human nasal space is divided into two cavities (by the nasal septum), and each consists of three regions: vestibular, olfactory, and respiratory. The respiratory region, containing three turbinates—inferior middle and superior—has a large surface area (Table 2). The respiratory mucosa is well vascularized mainly by the external and the internal carotid arteries from which blood flows through a dense network of capillaries. Due to the large surface area as well as the richly supplied blood, the respiratory region is considered as the major nasal absorption site to the systemic circulation. In contrast, the olfactory region, which is regarded as the main gate to the brain, covers less than one-tenth of the entire surface area of the nasal cavity (10 cm^2 of 150 cm^2; Table 2). The olfactory epithelium consists of three types of cells: neural cells, supporting, and basal cells (Fig. 3). Nonmyelinated axons originated from the olfactory bulb terminate at the columnar olfactory neural cells. The neural cells contain cilia, which are nonmotile and do not play a role in the mucociliary clearance as the common motile cilia in the respiratory region. These nonmotile cilia are designed to detect volatile substances by initiating depolarization of the olfactory axons. Like the whole area of the respiratory mucosa, a mucus layer secreted by Bowman's glands covers the epithelial cells in the olfactory region.

Table 2 Morphological and Physiological Parameters of the Nasal Cavity Associated with Drug Absorption

	Systemic Absorption	**Cerebral Absorption**
Major absorption site	Respiratory region (containing three turbinates)	Olfactory region
Surface area	Approx. 150 cm^2	Approx. 10 cm^2
Volume limits for nasal liquid administration	Variable, usually 0.5–1.0 mL (entire nasal cavity: 15 cm^3)	10–100 μL preferably targeted by using a mucoadhesive technology
Ciliary clearance	Ciliary motion directing the mucus from anterior to posterior	Nonmotile cilia
Mucosal metabolism	Nasal mucosa contains CYP1A, CYP2A, CYP2E, carboxylesterases, and glutathione-*S*-transferases	
Mucosal thickness	100 μm	
Mucosal pH	5.5–6.5	

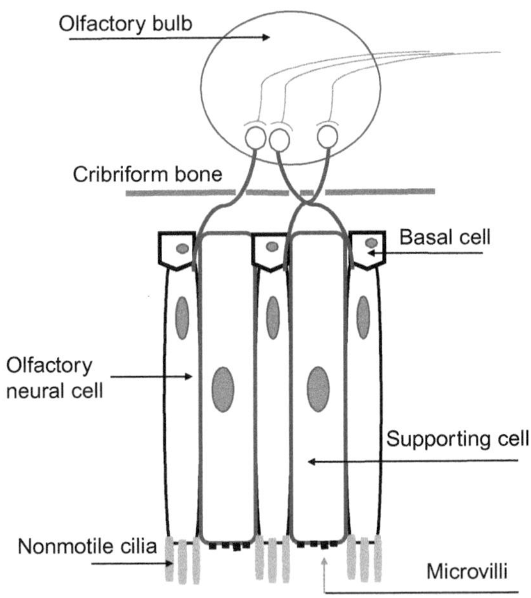

Fig. 3 Schematic representation of the olfactory epithelium and bulb.

5.2 Pathways of Nasal Absorption

When a drug is applied into the nasal cavity in a solution or in a particulate liquid formulation, some portion may be drained into the stomach through the nasopharynx and be absorbed as an oral drug. The magnitude of this oral portion depends on the volume and the form (spray, drops, etc.) by which the liquid is applied. To prevent drainage and irregular absorption, it is recommended to use low application volumes (eg, <1 mL for systemic purposes; <100 μL for brain targeting) (Dhuria et al., 2010), even though the entire volume of the nasal cavity is quite high (15 mL) (Table 2). The possible routes of drug absorption after nasal administration are illustrated in Fig. 4. Drugs that have escaped mucosal metabolism, or drugs that have not been cleared into the mouth and throat through the nasopharynx, can be transported through the large area of respiratory mucosa into the blood. Drugs absorbed into the blood circulation are distributed in the body and can also reach the brain if the drug's chemical properties enable crossing the BBB, although brain levels would be quite low. A drug can alternatively travel directly through the olfactory region into the brain parenchymal tissue and the CSF, from which it can be cleared into the blood circulation.

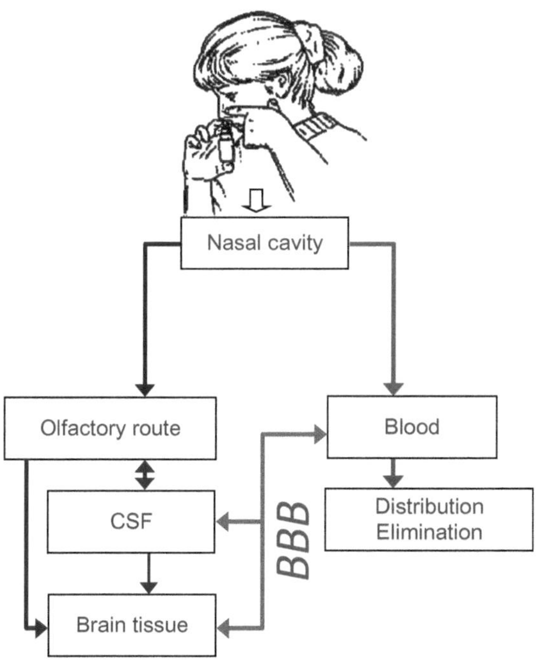

Fig. 4 Pathways of drug absorption from the nasal cavity.

5.3 Mechanism of Nasal Absorption to the Brain

The cellular mechanism of direct nose-to-brain delivery is not exactly known; however, a drug can be principally migrated through three pathways: (1) transcellular transport across the supporting cells in the olfactory region; (2) endocytosis through the olfactory neural cells and transport through the axon to the olfactory bulb; and (3) paracellular diffusion between supporting and neural cells. While the third pathway is a less possible mechanism, especially for nanoparticles (>20 nm), because the tight junctions, the intracellular pathways are more acceptable, although they remain to be elucidated (Mistry et al., 2009).

5.4 Intranasal Administration of Metal Nanoparticles

To date, the strategy to achieve improved delivery of drugs to the brain would be a combination of a nanoparticulate system that specifically targets the olfactory region and inherited bioadhesive properties that retains the nanoparticles at the site by decreasing the effect of the mucociliary clearance, thus enabling time-dependent diffusion (Casettari & Illum, 2014; Mistry et al., 2009; Wen et al., 2011). Among the nanoparticles prepared from biodegradable polymers, those based on chitosan are most desirable due to its cationic nature, biocompatibility, and suitability for high loading of hydrophilic drugs (Casettari & Illum, 2014). For example, AuNPs were produced using chitosan as a reducing agent, and subsequently loaded with insulin. After transmucosal (oral and nasal) administrations of the insulin-loaded chitosan-reduced AuNPs, a significant fall in blood glucose levels occurred (Bhumkar, Joshi, Sastry, & Pokharkar, 2007). Blood glucose lowering at the end of 2 h following the administration to diabetic rats was found to be 30.41% and 20.27% for oral (50 IU/kg) and nasal (10 IU/kg), respectively. It has been noted that the serum levels of insulin after nasal administration of nanoparticles were found to be higher than orally administered nanoparticles (Bhumkar et al., 2007), and also higher than the levels monitored after nasal administration of insulin-loaded aspartic acid-capped AuNPs (Bhumkar et al., 2007).

Unlike studies with polymeric nanoparticles, few studies have been designed so far to examine the direct transfer of *metal* nanoparticles from the nasal cavity to the brain. Forty-five years ago, Czerniawska (1970) applied colloidal ^{198}Au to the mucosa of the olfactory region in 32 rabbits and found the gold in the CSF within 1–2 h. Such activity indicates direct penetration of ^{198}Au from the olfactory membrane into the CSF of the anterior cranial fossa. Concerning environmental toxicity, Elder et al. (2006) have studied manganese accumulation after 12 days of exposure to insoluble manganese oxide nanoparticles (30 nm) with both nostrils patent in rats. Mn concentrations in the olfactory bulbs increased by 3.5 times, whereas lung Mn concentrations doubled. There were also increases in striatum, trigeminal ganglions, midbrain, frontal cortex, and cerebellum. Titanium oxide nanoparticles have also been investigated for brain toxicity since TiO_2 is widely used in paints, food colorant, and sunscreens. Ze et al. (2013) have found oxidative damage in the mouse brain in a study that examined the activation of the P38–Nrf-2 signaling pathway associated with oxidative stress in the mouse brain after intranasal administration of TiO_2 NPs for 90 consecutive days. Oxidative stress in murine brain tissue was also demonstrated after intranasal instillation of copper nanoparticles (Liu et al., 2014). Iron oxide nanoparticles have also neurotoxic potential after intranasal instillation (Wu et al., 2013). It has been shown that Fe_3O_4 NPs were observed in rat brains after the particles were intranasally instilled for 7 days. The particles were found to be deposited at particularly high concentrations in the rat striate and hippocampus. All the results relating to environmental exposure of metal NPs (eg, Mn, TiO_2, Cu, Fe_3O_4) can be used as a good indication for the ability of nanoparticles to translocate directly from the olfactory nerve to the brain.

As AuNPs have turned recently to be potential as imaging agents that could provide a promising future for CNS imaging in various pathologies, there is some disappointment that not many studies have examined AuNPs trafficking to the brain via the nasal route. In addition to Czerniawska's experiments (1970), Hutter et al. (2010) have also reported on intranasal administration of AuNPs. AuNPs of three morphologies—spheres, rods, and urchins, all coated with PEG or cetyltrimethylammonium bromide—were studied after nasal administration to transgenic mice (5 μL in the left nostril) and after treatment of microglial cells in culture. Analysis by dark-field microscopy and by two-photon-induced luminescence indicated that the exposure of neural cells to AuNPs resulted in: (i) AuNPs internalization by both microglial cells and primary hippocampal neurons; (ii) transient toll-like receptor 2 (TLR-2) upregulation in the olfactory bulb, after intranasal administration in transgenic mice in real time; and (iii) differential upregulation in vitro of TLR-2 together with interleukin 1 alpha, granulocyte macrophage colony-stimulating factor, and nitric oxide in microglia. The study demonstrates that the morphology and surface chemistry of AuNPs strongly influence the microglial activation status, and suggests that interactions between GNPs and microglia can be differentially regulated by tuning the nanogeometry of AuNPs. Interestingly, more studies on nasal administration of AgNPs have been published, apparently because of the potential toxicity of nanosilver to industrial workers and the exposed public. AgNPs are generally used as an antibacterial agent applied in wounds and burns, as a coating for hospital beds, as a contraceptive, in textile manufacturing, and as a disinfectant spray, just to name a few. Liu, Guan, Ren, and Yang (2012) investigated the effects of the AgNPs on hippocampal synaptic plasticity and spatial cognition in rats after nasal administrations for 2 weeks. The Morris water maze test was performed for spatial cognition, followed by the long-term potentiation resulting in abnormal data compared with the control. Liu et al. found that the quantity of ROS in hippocampal homogenate was increased significantly, which may be the reason of the neural damage caused by the nanoparticles. Genter, Newman, Shertzer, Ali, and Bolon (2012) have shown by autometallography that AgNPs (25 nm) were accumulated in the olfactory bulb and the lateral brain ventricles after intranasal exposure in mice. Elevated tissue glutathione levels were observed in nasal epithelia and blood at 7 days, indicating additional systemic absorption and reversible oxidative stress. More recent publications by Yin, Yao, Zhou, Faiola, and Jiang (2015) and Yin, Zhang, et al. (2015) have established the neurotoxicity induced by AgNPs after nasal administration. It was found that intranasal instillation of AgNPs in neonatal rats caused activation of neuroglial cells with concomitant destruction of the granular layer of the cerebellum, an effect that was counterbalanced and attenuated by oral administration of vitamin E (Yin, Yao, et al., 2015). In another study, intranasal instillation of AgNPs for 14 weeks to neonatal rats caused cerebellar ataxia-like symptom, evidenced by dysfunction of motor coordination and impairment of locomotor activity, and observation of cerebellum sections revealed that AgNPs can provoke destruction of the cerebellum granular layer with concomitant activation of glial cells (Yin, Zhang, et al., 2015).

6. CONCLUSION AND FUTURE PERSPECTIVE

Nanotechnology is a fast-expanding area of medical science, which is expected to lead to the development of sophisticated drug delivery systems that can challenge body barriers to reach target tissues. One of the important body barriers that needs to be overcome is the BBB. The treatment of neurological and neurodegenerative diseases is hindered by the presence of

BBB, which is impermeable to most active compounds that could have potentially treated the diseases. As evidenced by the new developments, this barrier could be circumvented by advanced nanotechnology and particularly by the forthcoming metal NPs. Metal NPs, such as iron oxide, manganese oxide, silver, platinum, and gold, have been studied as a diagnostic tool by utilizing various imaging techniques (such as MRI or PET). However, they can also be utilized as a therapeutic tool or even preferably as a theranostic tool for neurological disorders. Currently, two general strategies have been proposed for trafficking nanoparticles to the brain: (a) conjugation of NPs to specific peptide moieties that facilitate AMT and RMT in the BBB, thereby being able to travel into the brain after systemic administration; and (b) evading BBB by intranasal administration of NPs. To date, the conjugation strategy has gained more interest than the approach of brain targeting by the direct intranasal route. However, both strategies are still a major challenge for the academic and pharmaceutical communities from both aspects: quantitative delivery and neurotoxicity. It is too early to anticipate neurotoxicity at this stage, but once a breakthrough is achieved, ie, increasing brain availability, neurotoxicity should then be seriously considered. In our view, the efforts that should be made by scientists and medical personnel to eventually achieve effective delivery of drug-loaded metal NPs must focus on simple and practical ideas that ensure the breakthroughs.

ACKNOWLEDGMENTS

M.J.K. is grateful to Fondap15130011, Fondecyt 1130425, and MECESUP UCH-0811. E.A. is grateful to Fondecyt 11130494 and UNAB DI 1309-16/R. C.V.-A. and E.G.-T. give thanks to PhD national CONICYT's scholarship, numbers 21110304 and 21151400, respectively.

REFERENCES

Abbott, N. J. (1992). Comparative physiology of the blood-brain barrier. Physiology and pharmacology of the blood-brain barrier (pp. 371–396). Berlin, Heidelberg: Springer.

Abbott, N. J., Patabendige, A. A., Dolman, D. E., Yusof, S. R., & Begley, D. J. (2010). Structure and function of the blood-brain barrier. *Neurobiology of Disease*, *37*(1), 13–25.

Afifi, M., Almaghrabi, O. A., & Kadasa, N. M. (2015). Ameliorative effect of zinc oxide nanoparticles on antioxidants and sperm characteristics in streptozotocin-induced diabetic rat testes. *BioMed Research International*, *2015*, 153573.

Arora, P., Sharma, S., & Garg, S. (2002). Permeability issues in nasal drug delivery. *Drug Discovery Today*, 7(18), 967–975.

Arvizo, R. R., Giri, K., Moyano, D., Miranda, O. R., Madden, B., McCormick, D. J., … Mukherjee, P. (2012). Identifying new therapeutic targets via modulation of protein corona formation by engineered nanoparticles. *PLoS One*, 7(3), e33650.

AshaRani, P. V., Hande, M. P., & Valiyaveettil, S. (2009). Anti-proliferative activity of silver nanoparticles. *BMC Cell Biology*, *10*, 65.

Becker, N. H., Novikoff, A. B., & Zimmerman, H. M. (1967). Fine structure observations of the uptake of intravenously injected peroxidase by the rat choroid plexus. *The Journal of Histochemistry and Cytochemistry*, *15*(3), 160–165.

Bennewitz, M. F., Lobo, T. L., Nkansah, M. K., Ulas, G., Brudvig, G. W., & Shapiro, E. M. (2011). Biocompatible and pH-sensitive PLGA encapsulated MnO nanocrystals for molecular and cellular MRI. *ACS Nano*, *5*(5), 3438–3446.

Bhumkar, D. R., Joshi, H. M., Sastry, M., & Pokharkar, V. B. (2007). Chitosan reduced gold nanoparticles as novel carriers for transmucosal delivery of insulin. *Pharmaceutical Research*, *24*(8), 1415–1426.

Casettari, L., & Illum, L. (2014). Chitosan in nasal delivery systems for therapeutic drugs. *Journal of Controlled Release*, *190*, 189–200.

Chen, N., Shao, C., Li, S., Wang, Z., Qu, Y., Gu, W., … Ye, L. (2015). Cy5.5 conjugated MnO nanoparticles for magnetic resonance/near-infrared fluorescence dual-modal imaging of brain gliomas. *Journal of Colloid and Interface Science*, *457*, 27–34.

Chertok, B., Moffat, B. A., David, A. E., Yu, F., Bergemann, C., Ross, B. D., & Yang, V. C. (2008). Iron oxide nanoparticles as a drug delivery vehicle for MRI monitored magnetic targeting of brain tumors. *Biomaterials*, *29*(4), 487–496.

Corot, C., Robert, P., Idee, J. M., & Port, M. (2006). Recent advances in iron oxide nanocrystal technology for medical imaging. *Advanced Drug Delivery Reviews*, *58*(14), 1471–1504.

Czerniawska, A. (1970). Experimental investigations on the penetration of ^{198}Au from nasal mucous membrane into cerebrospinal fluid. *Acta Oto-Laryngologica*, *70*(1), 58–61.

Day, E. S., Zhang, L., Thompson, P. A., Zawaski, J. A., Kaffes, C. C., Gaber, M. W., … West, J. L. (2012). Vascular-targeted photothermal therapy of an orthotopic murine glioma model. *Nanomedicine (London, England)*, 7(8), 1133–1148.

Dell'Orco, D., Lundqvist, M., Oslakovic, C., Cedervall, T., & Linse, S. (2010). Modeling the time evolution of the nanoparticle-protein corona in a body fluid. *PLoS One*, *5*(6), e10949.

Dhuria, S. V., Hanson, L. R., & Frey, W. H., 2nd. (2010). Intranasal delivery to the central nervous system: Mechanisms and experimental considerations. *Journal of Pharmaceutical Sciences*, *99*(4), 1654–1673.

Dilnawaz, F., & Sahoo, S. K. (2015). Therapeutic approaches of magnetic nanoparticles for the central nervous system. *Drug Discovery Today*, *20*(10), 1256–1264.

Dobrovolskaia, M. A., Aggarwal, P., Hall, J. B., & McNeil, S. E. (2008). Preclinical studies to understand nanoparticle interaction with the immune system and its potential effects on nanoparticle biodistribution. *Molecular Pharmaceutics*, *5*(4), 487–495.

Elder, A., Gelein, R., Silva, V., Feikert, T., Opanashuk, L., Carter, J., … Oberdorster, G. (2006). Translocation of inhaled ultrafine manganese oxide particles to the central nervous system. *Environmental Health Perspectives*, *114*(8), 1172–1178.

Estevez, A. Y., Pritchard, S., Harper, K., Aston, J. W., Lynch, A., Lucky, J. J., … Erlichman, J. S. (2011). Neuroprotective mechanisms of cerium oxide nanoparticles in a mouse hippocampal brain slice model of ischemia. *Free Radical Biology & Medicine*, *51*(6), 1155–1163.

Etame, A. B., Diaz, R. J., O'Reilly, M. A., Smith, C. A., Mainprize, T. G., Hynynen, K., & Rutka, J. T. (2012). Enhanced delivery of gold nanoparticles with therapeutic potential into the brain using MRI-guided focused ultrasound. *Nanomedicine (London, England)*, *8*(7), 1133–1142.

Faucher, L., Guay-Begin, A. A., Lagueux, J., Cote, M. F., Petitclerc, E., & Fortin, M. A. (2011). Ultra-small gadolinium oxide nanoparticles to image brain cancer cells in vivo with MRI. *Contrast Media & Molecular Imaging*, *6*(4), 209–218.

Feng, X., Chen, A., Zhang, Y., Wang, J., Shao, L., & Wei, L. (2015). Central nervous system toxicity of metallic nanoparticles. *International Journal of Nanomedicine*, *10*, 4321–4340.

Fiandra, L., Colombo, M., Mazzucchelli, S., Truffi, M., Santini, B., Allevi, R., … Corsi, F. (2015). Nanoformulation of antiretroviral drugs enhances their penetration across the blood brain barrier in mice. *Nanomedicine (London, England)*, *11*(6), 1387–1397.

Frigell, J., Garcia, I., Gomez-Vallejo, V., Llop, J., & Penades, S. (2014). 68Ga-labeled gold glyconanoparticles for exploring blood-brain barrier permeability: Preparation, biodistribution studies, and improved brain uptake via neuropeptide conjugation. *Journal of the American Chemical Society*, *136*(1), 449–457.

Gabathuler, R. (2010). Approaches to transport therapeutic drugs across the blood-brain barrier to treat brain diseases. *Neurobiology of Disease*, *37*(1), 48–57.

Gahramanov, S., Muldoon, L. L., Varallyay, C. G., Li, X., Kraemer, D. F., Fu, R., … Neuwelt, E. A. (2013). Pseudoprogression of glioblastoma after chemo- and radiation therapy: Diagnosis by using dynamic susceptibility-weighted contrast-enhanced perfusion MR imaging with ferumoxytol versus gadoteridol and correlation with survival. *Radiology*, *266*(3), 842–852.

Genter, M. B., Newman, N. C., Shertzer, H. G., Ali, S. F., & Bolon, B. (2012). Distribution and systemic effects of intranasally administered 25 nm silver nanoparticles in adult mice. *Toxicologic Pathology*, *40*(7), 1004–1013.

Glat, M., Skaat, H., Menkes-Caspi, N., Margel, S., & Stern, E. A. (2013). Age-dependent effects of microglial inhibition in vivo on Alzheimer's disease neuropathology using bioactive-conjugated iron oxide nanoparticles. *Journal of Nanobiotechnology*, *11*, 32.

Guerrero, S., Araya, E., Fiedler, J. L., Arias, J. I., Adura, C., Albericio, F., … Kogan, M. J. (2010). Improving the brain delivery of gold nanoparticles by conjugation with an amphipathic peptide. *Nanomedicine (London, England)*, *5*(6), 897–913.

Guerrero, S., Herance, J. R., Rojas, S., Mena, J. F., Gispert, J. D., Acosta, G. A., … Kogan, M. J. (2012). Synthesis and in vivo evaluation of the biodistribution of a ^{18}F-labeled conjugate gold-nanoparticle-peptide with potential biomedical application. *Bioconjugate Chemistry*, *23*(3), 399–408.

Han, D., Tian, Y., Zhang, T., Ren, G., & Yang, Z. (2011). Nano-zinc oxide damages spatial cognition capability via over-enhanced long-term potentiation in hippocampus of Wistar rats. *International Journal of Nanomedicine*, *6*, 1453–1461.

Hanley, C., Thurber, A., Hanna, C., Punnoose, A., Zhang, J., & Wingett, D. G. (2009). The influences of cell type and ZnO nanoparticle size on immune cell cytotoxicity and cytokine induction. *Nanoscale Research Letters*, *4*(12), 1409–1420.

Hanson, L. R., & Frey, W. H., 2nd. (2007). Strategies for intranasal delivery of therapeutics for the prevention and treatment of neuroAIDS. *Journal of Neuroimmune Pharmacology*, *2*(1), 81–86.

Hanson, L. R., & Frey, W. H., 2nd. (2008). Intranasal delivery bypasses the blood-brain barrier to target therapeutic agents to the central nervous system and treat neurodegenerative disease. *BMC Neuroscience*, *9*(Suppl. 3), S5.

Heckman, K. L., DeCoteau, W., Estevez, A., Reed, K. J., Costanzo, W., Sanford, D., … Erlichman, J. S. (2013). Custom cerium oxide nanoparticles protect against a free radical mediated autoimmune degenerative disease in the brain. *ACS Nano*, 7(12), 10582–10596.

Herve, F., Ghinea, N., & Scherrmann, J. M. (2008). CNS delivery via adsorptive transcytosis. *The AAPS Journal*, *10*(3), 455–472.

Hu, R., Zheng, L., Zhang, T., Gao, G., Cui, Y., Cheng, Z., … Hong, F. (2011). Molecular mechanism of hippocampal apoptosis of mice following exposure to titanium dioxide nanoparticles. *Journal of Hazardous Materials*, *191*(1–3), 32–40.

Hutter, E., Boridy, S., Labrecque, S., Lalancette-Hebert, M., Kriz, J., Winnik, F. M., & Maysinger, D. (2010). Microglial response to gold nanoparticles. *ACS Nano*, *4*(5), 2595–2606.

Illum, L. (2000). Transport of drugs from the nasal cavity to the central nervous system. *European Journal of Pharmaceutical Sciences*, *11*(1), 1–18.

Illum, L. (2002). Nasal drug delivery: New developments and strategies. *Drug Discovery Today*, 7(23), 1184–1189.
Illum, L. (2003). Nasal drug delivery: Possibilities, problems and solutions. *Journal of Controlled Release*, *87*(1–3), 187–198.
Illum, L. (2004). Is nose-to-brain transport of drugs in man a reality? *The Journal of Pharmacy and Pharmacology*, *56*(1), 3–17.
Iv, M., Telischak, N., Feng, D., Holdsworth, S. J., Yeom, K. W., & Daldrup-Link, H. E. (2015). Clinical applications of iron oxide nanoparticles for magnetic resonance imaging of brain tumors. *Nanomedicine (London, England)*, *10*(6), 993–1018.
Jung, C. W., & Jacobs, P. (1995). Physical and chemical properties of superparamagnetic iron oxide MR contrast agents: Ferumoxides, ferumoxtran, ferumoxsil. *Magnetic Resonance Imaging*, *13*(5), 661–674.
Kallur, T., Farr, T. D., Bohm-Sturm, P., Kokaia, Z., & Hoehn, M. (2011). Spatio-temporal dynamics, differentiation and viability of human neural stem cells after implantation into neonatal rat brain. *The European Journal of Neuroscience*, *34*(3), 382–393.
Kao, Y. Y., Cheng, T. J., Yang, D. M., Wang, C. T., Chiung, Y. M., & Liu, P. S. (2012). Demonstration of an olfactory bulb-brain translocation pathway for ZnO nanoparticles in rodent cells in vitro and in vivo. *Journal of Molecular Neuroscience*, *48*(2), 464–471.
Kong, S. D., Lee, J., Ramachandran, S., Eliceiri, B. P., Shubayev, V. I., Lal, R., & Jin, S. (2012). Magnetic targeting of nanoparticles across the intact blood-brain barrier. *Journal of Controlled Release*, *164*(1), 49–57.
Li, S. D., & Huang, L. (2008). Pharmacokinetics and biodistribution of nanoparticles. *Molecular Pharmaceutics*, *5*(4), 496–504.
Li, T., Murphy, S., Kiselev, B., Bakshi, K. S., Zhang, J., Eltahir, A., ... Dorn, H. C. (2015). A new interleukin-13 amino-coated gadolinium metallofullerene nanoparticle for targeted MRI detection of glioblastoma tumor cells. *Journal of the American Chemical Society*, *137*(24), 7881–7888.
Liu, Y., Gao, Y., Liu, Y., Li, B., Chen, C., & Wu, G. (2014). Oxidative stress and acute changes in murine brain tissues after nasal instillation of copper particles with different sizes. *Journal of Nanoscience and Nanotechnology*, *14*(6), 4534–4540.
Liu, Y., Guan, W., Ren, G., & Yang, Z. (2012). The possible mechanism of silver nanoparticle impact on hippocampal synaptic plasticity and spatial cognition in rats. *Toxicology Letters*, *209*(3), 227–231.
Locatelli, E., Naddaka, M., Uboldi, C., Loudos, G., Fragogeorgi, E., Molinari, V., ... Franchini, M. C. (2014). Targeted delivery of silver nanoparticles and alisertib: In vitro and in vivo synergistic effect against glioblastoma. *Nanomedicine (London, England)*, *9*(6), 839–849.
Lundqvist, M., Stigler, J., Cedervall, T., Berggard, T., Flanagan, M. B., Lynch, I., ... Dawson, K. (2011). The evolution of the protein corona around nanoparticles: A test study. *ACS Nano*, *5*(9), 7503–7509.
Lundqvist, M., Stigler, J., Elia, G., Lynch, I., Cedervall, T., & Dawson, K. A. (2008). Nanoparticle size and surface properties determine the protein corona with possible implications for biological impacts. *Proceedings of the National Academy of Sciences of the United States of America*, *105*(38), 14265–14270.
Mae, M., Armulik, A., & Betsholtz, C. (2011). Getting to know the cast—Cellular interactions and signaling at the neurovascular unit. *Current Pharmaceutical Design*, *17*(26), 2750–2754.
Mathiisen, T. M., Lehre, K. P., Danbolt, N. C., & Ottersen, O. P. (2010). The perivascular astroglial sheath provides a complete covering of the brain microvessels: An electron microscopic 3D reconstruction. *Glia*, *58*(9), 1094–1103.
Merkus, F. W. H. M., Schipper, N. G. M., & Verhoef, J. C. (1996). The influence of absorption enhancers on intranasal insulin absorption in normal and diabetic subjects. *Journal of Controlled Release*, *41*(1–2), 69–75.

Mistry, A., Stolnik, S., & Illum, L. (2009). Nanoparticles for direct nose-to-brain delivery of drugs. *International Journal of Pharmaceutics*, *379*(1), 146–157.

Monopoli, M. P., Walczyk, D., Campbell, A., Elia, G., Lynch, I., Bombelli, F. B., & Dawson, K. A. (2011). Physical-chemical aspects of protein corona: Relevance to in vitro and in vivo biological impacts of nanoparticles. *Journal of the American Chemical Society*, *133*(8), 2525–2534.

Moraes, L., Vasconcelos-dos-Santos, A., Santana, F. C., Godoy, M. A., Rosado-de-Castro, P. H., Jasmin, J., ... Mendez-Otero, R. (2012). Neuroprotective effects and magnetic resonance imaging of mesenchymal stem cells labeled with SPION in a rat model of Huntington's disease. *Stem Cell Research*, *9*(2), 143–155.

Muldoon, L. L., Sandor, M., Pinkston, K. E., & Neuwelt, E. A. (2005). Imaging, distribution, and toxicity of superparamagnetic iron oxide magnetic resonance nanoparticles in the rat brain and intracerebral tumor. *Neurosurgery*, *57*(4), 785–796. discussion 785–796.

Murillo, T. P., Sandquist, C., Jacobs, P. M., Nesbit, G., Manninger, S., & Neuwelt, E. A. (2005). Imaging brain tumors with ferumoxtran-10, a nanoparticle magnetic resonance contrast agent. *Therapy*, *2*, 871–882.

Na, H. B., Lee, J. H., An, K., Park, Y. I., Park, M., Lee, I. S., ... Hyeon, T. (2007). Development of a T1 contrast agent for magnetic resonance imaging using MnO nanoparticles. *Angewandte Chemie (International Ed. in English)*, *46*(28), 5397–5401.

Nabeshima, S., Reese, T. S., Landis, D. M., & Brightman, M. W. (1975). Junctions in the meninges and marginal glia. *The Journal of Comparative Neurology*, *164*(2), 127–169.

Nag, S. (2011). Morphology and properties of brain endothelial cells. *Methods in Molecular Biology*, *686*, 3–47.

Nel, A., Xia, T., Madler, L., & Li, N. (2006). Toxic potential of materials at the nanolevel. *Science*, *311*(5761), 622–627.

Nellore, J., Pauline, C., & Amarnath, K. (2013). Bacopa monnieri phytochemicals mediated synthesis of platinum nanoparticles and its neurorescue effect on 1-methyl 4-phenyl 1,2,3,6 tetrahydropyridine-induced experimental parkinsonism in zebrafish. *Journal of Neurodegenerative Diseases*, *2013*, 972391.

Neuwelt, E. A., Varallyay, P., Bago, A. G., Muldoon, L. L., Nesbit, G., & Nixon, R. (2004). Imaging of iron oxide nanoparticles by MR and light microscopy in patients with malignant brain tumours. *Neuropathology and Applied Neurobiology*, *30*(5), 456–471.

Pardridge, W. M. (2005). Drug and gene targeting to the brain via blood–brain barrier receptor-mediated transport systems. *International Congress Series*, *1277*, 49–62.

Pardridge, W. M. (2007). Drug targeting to the brain. *Pharmaceutical Research*, *24*(9), 1733–1744.

Park, J. Y., Baek, M. J., Choi, E. S., Woo, S., Kim, J. H., Kim, T. J., ... Lee, G. H. (2009). Paramagnetic ultrasmall gadolinium oxide nanoparticles as advanced T1 MRI contrast agent: Account for large longitudinal relaxivity, optimal particle diameter, and in vivo T1 MR images. *ACS Nano*, *3*(11), 3663–3669.

Prades, R., Guerrero, S., Araya, E., Molina, C., Salas, E., Zurita, E., ... Giralt, E. (2012). Delivery of gold nanoparticles to the brain by conjugation with a peptide that recognizes the transferrin receptor. *Biomaterials*, *33*(29), 7194–7205.

Pujals, S., Fernandez-Carneado, J., Lopez-Iglesias, C., Kogan, M. J., & Giralt, E. (2006). Mechanistic aspects of CPP-mediated intracellular drug delivery: Relevance of CPP self-assembly. *Biochimica Et Biophysica Acta-Biomembranes*, *1758*(3), 264–279.

Ramos-Gómez, M., & Martinez-Serrano, A. (2016). Tracking of iron-labeled human neural stem cells by magnetic resonance imaging in cell replacement therapy for Parkinson's disease. *Neural Regeneration Research*, *11*(1), 49–52.

Ramos-Gómez, M., Seiz, E. G., & Martinez-Serrano, A. (2015). Optimization of the magnetic labeling of human neural stem cells and MRI visualization in the hemiparkinsonian rat brain. *Journal of Nanobiotechnology*, *13*, 20.

Saddick, S., Afifi, M., & Abu Zinada, O. A. (in press). Effect of zinc nanoparticles on oxidative stress-related genes and antioxidant enzymes activity in the brain of Oreochromis niloticus and Tilapia zillii. *Saudi Journal of Biological Sciences*.

Saiyed, Z. M., Gandhi, N. H., & Nair, M. P. N. (2010). Magnetic nanoformulation of azidothymidine 5'-triphosphate for targeted delivery across the blood-brain barrier. *International Journal of Nanomedicine*, *5*, 157–166.

Sakulkhu, U., Maurizi, L., Mahmoudi, M., Motazacker, M., Vries, M., Gramoun, A., ... Hofmann, H. (2014). Ex situ evaluation of the composition of protein corona of intravenously injected superparamagnetic nanoparticles in rats. *Nanoscale*, *6*(19), 11439–11450.

Schaffler, M., Sousa, F., Wenk, A., Sitia, L., Hirn, S., Schleh, C., ... Krol, S. (2014). Blood protein coating of gold nanoparticles as potential tool for organ targeting. *Biomaterials*, *35*(10), 3455–3466.

Schultke, E., Menk, R., Pinzer, B., Astolfo, A., Stampanoni, M., Arfelli, F., ... Nikkhah, G. (2014). Single-cell resolution in high-resolution synchrotron X-ray CT imaging with gold nanoparticles. *Journal of Synchrotron Radiation*, *21*(Pt. 1), 242–250.

Shao, C., Li, S., Gu, W., Gong, N., Zhang, J., Chen, N., ... Ye, L. (2015). Multifunctional gadolinium-doped manganese carbonate nanoparticles for targeted MR/fluorescence imaging of tiny brain gliomas. *Analytical Chemistry*, *87*(12), 6251–6257.

Shilo, M., Motiei, M., Hana, P., & Popovtzer, R. (2014). Transport of nanoparticles through the blood-brain barrier for imaging and therapeutic applications. *Nanoscale*, *6*(4), 2146–2152.

Shultz, M. D., Wilson, J. D., Fuller, C. E., Zhang, J., Dorn, H. C., & Fatouros, P. P. (2011). Metallofullerene-based nanoplatform for brain tumor brachytherapy and longitudinal imaging in a murine orthotopic xenograft model. *Radiology*, *261*(1), 136–143.

Simon, G. H., von Vopelius-Feldt, J., Fu, Y., Schlegel, J., Pinotek, G., Wendland, M. F., ... Daldrup-Link, H. E. (2006). Ultrasmall supraparamagnetic iron oxide-enhanced-magnetic resonance imaging of antigen-induced arthritis: A comparative study between SHU 555 C, ferumoxtran-10, and ferumoxytol. *Investigative Radiology*, *41*(1), 45–51.

Sintov, A. C., Levy, H. V., & Botner, S. (2010). Systemic delivery of insulin via the nasal route using a new microemulsion system: In vitro and in vivo studies. *Journal of Controlled Release*, *148*(2), 168–176.

Soto, C., Sigurdsson, E. M., Morelli, L., Kumar, R. A., Castano, E. M., & Frangione, B. (1998). Beta-sheet breaker peptides inhibit fibrillogenesis in a rat brain model of amyloidosis: Implications for Alzheimer's therapy. *Nature Medicine*, *4*(7), 822–826.

Sousa, F., Mandal, S., Garrovo, C., Astolfo, A., Bonifacio, A., Latawiec, D., ... Krol, S. (2010). Functionalized gold nanoparticles: A detailed in vivo multimodal microscopic brain distribution study. *Nanoscale*, *2*(12), 2826–2834.

Taschner, C. A., Wetzel, S. G., Tolnay, M., Froehlich, J., Merlo, A., & Radue, E. W. (2005). Characteristics of ultrasmall superparamagnetic iron oxides in patients with brain tumors. *AJR. American Journal of Roentgenology*, *185*(6), 1477–1486.

Tenzer, S., Docter, D., Rosfa, S., Wlodarski, A., Kuharev, J., Rekik, A., ... Stauber, R. H. (2011). Nanoparticle size is a critical physicochemical determinant of the human blood plasma corona: A comprehensive quantitative proteomic analysis. *ACS Nano*, *5*(9), 7155–7167.

Thorne, R. G., Emory, C. R., Ala, T. A., & Frey, W. H., 2nd. (1995). Quantitative analysis of the olfactory pathway for drug delivery to the brain. *Brain Research*, *692*(1–2), 278–282.

Thu, M. S., Najbauer, J., Kendall, S. E., Harutyunyan, I., Sangalang, N., Gutova, M., ... Aboody, K. S. (2009). Iron labeling and pre-clinical MRI visualization of therapeutic human neural stem cells in a murine glioma model. *PLoS One*, *4*(9), e7218.

Tosi, G., Bortot, B., Ruozi, B., Dolcetta, D., Vandelli, M. A., Forni, F., & Severini, G. M. (2013). Potential use of polymeric nanoparticles for drug delivery across the blood-brain barrier. *Current Medicinal Chemistry*, *20*(17), 2212–2225.

Tsai, M. C., Chen, Y. H., & Chiang, L. Y. (1997). Polyhydroxylated C60, fullerenol, a novel free-radical trapper, prevented hydrogen peroxide- and cumene hydroperoxide-elicited changes in rat hippocampus in-vitro. *The Journal of Pharmacy and Pharmacology*, *49*(4), 438–445.

Ugwoke, M. I., Agu, R. U., Verbeke, N., & Kinget, R. (2005). Nasal mucoadhesive drug delivery: Background, applications, trends and future perspectives. *Advanced Drug Delivery Reviews*, *57*(11), 1640–1665.

Varallyay, C. G., Nesbit, E., Fu, R., Gahramanov, S., Moloney, B., Earl, E., & Neuwelt, E. A. (2013). High-resolution steady-state cerebral blood volume maps in patients with central nervous system neoplasms using ferumoxytol, a superparamagnetic iron oxide nanoparticle. *Journal of Cerebral Blood Flow and Metabolism*, *33*(5), 780–786.

Velasco-Aguirre, C., Morales, F., Gallardo-Toledo, E., Guerrero, S., Giralt, E., Araya, E., & Kogan, M. J. (2015). Peptides and proteins used to enhance gold nanoparticle delivery to the brain: Preclinical approaches. *International Journal of Nanomedicine*, *10*, 4919–4936.

Wadghiri, Y. Z., Li, J., Wang, J., Hoang, D. M., Sun, Y., Xu, H., … Wisniewski, T. (2013). Detection of amyloid plaques targeted by bifunctional USPIO in Alzheimer's disease transgenic mice using magnetic resonance microimaging. *PLoS One*, *8*(2), e57097.

Wang, G., Zhang, X., Skallberg, A., Liu, Y., Hu, Z., Mei, X., & Uvdal, K. (2014). One-step synthesis of water-dispersible ultra-small Fe3O4 nanoparticles as contrast agents for T1 and T2 magnetic resonance imaging. *Nanoscale*, *6*(5), 2953–2963.

Wen, Z., Yan, Z., Hu, K., Pang, Z., Cheng, X., Guo, L., … Lai, R. (2011). Odorranalectin-conjugated nanoparticles: Preparation, brain delivery and pharmacodynamic study on Parkinson's disease following intranasal administration. *Journal of Controlled Release*, *151*(2), 131–138.

Wiley, D. T., Webster, P., Gale, A., & Davis, M. E. (2013). Transcytosis and brain uptake of transferrin-containing nanoparticles by tuning avidity to transferrin receptor. *Proceedings of the National Academy of Sciences of the United States of America*, *110*(21), 8662–8667.

Wolfe, T. R., & Bernstone, T. (2004). Intranasal drug delivery: An alternative to intravenous administration in selected emergency cases. *Journal of Emergency Nursing*, *30*(2), 141–147.

Wu, Y. N., Chen, D. H., Shi, X. Y., Lian, C. C., Wang, T. Y., Yeh, C. S., … Shieh, D. B. (2011). Cancer-cell-specific cytotoxicity of non-oxidized iron elements in iron core-gold shell NPs. *Nanomedicine (London, England)*, 7(4), 420–427.

Wu, J., Ding, T., & Sun, J. (2013). Neurotoxic potential of iron oxide nanoparticles in the rat brain striatum and hippocampus. *Neurotoxicology*, *34*, 243–253.

Yin, N., Yao, X., Zhou, Q., Faiola, F., & Jiang, G. (2015). Vitamin E attenuates silver nanoparticle-induced effects on body weight and neurotoxicity in rats. *Biochemical and Biophysical Research Communications*, *458*(2), 405–410.

Yin, N., Zhang, Y., Yun, Z., Liu, Q., Qu, G., Zhou, Q., … Jiang, G. (2015). Silver nanoparticle exposure induces rat motor dysfunction through decrease in expression of calcium channel protein in cerebellum. *Toxicology Letters*, *237*(2), 112–120.

Ze, Y., Zheng, L., Zhao, X., Gui, S., Sang, X., Su, J., … Hong, F. (2013). Molecular mechanism of titanium dioxide nanoparticles-induced oxidative injury in the brain of mice. *Chemosphere*, *92*(9), 1183–1189.

Zhang, D., Fa, H. B., Zhou, J. T., Li, S., Diao, X. W., & Yin, W. (2015). The detection of beta-amyloid plaques in an Alzheimer's disease rat model with DDNP-SPIO. *Clinical Radiology*, *70*(1), 74–80.

CHAPTER SEVEN

Current Perspective of Carbon Nanotubes Application in Neurology

H. Kafa, J.T.-W. Wang, K.T. Al-Jamal[1]
Institute of Pharmaceutical Science, Faculty of Life Sciences and Medicine, King's College London, London, United Kingdom
[1]Corresponding author: e-mail address: khuloud.al-jamal@kcl.ac.uk

Contents

Abstract

The recent advances in nanotechnology have allowed new fields of research to investigate cutting edge brain-specific therapies and to tackle the complex brain-related disorders. The blood–brain barrier (BBB) is a major obstacle for drug delivery to the brain resulting in only few drugs reaching the market to tackle brain disorders. Nanoparticles (NPs) provide a flexible platform for conjugating drugs and targeting ligands and have been extensively researched to facilitate BBB crossing and effective delivery to the brain. In addition, the inherent properties of NPs are being utilized to facilitate other therapeutic possibilities. One example is carbon nanotubes (CNTs), which exhibit several

International Review of Neurobiology, Volume 130
ISSN 0074-7742
http://dx.doi.org/10.1016/bs.irn.2016.07.001

attractive characteristics allowing their use in the brain environment. The properties include a high aspect ratio, the ability to penetrate biological membranes due to their tubular shape and their infrared absorption properties. In this chapter, we review major advances in using CNTs for treating brain tumor and degenerative diseases with special focus on their abilities to cross the BBB following systemic administration, which is the major obstacle for most other NPs.

1. OVERVIEW OF CARBON NANOTUBES

Carbon nanotube (CNT) is a novel nanomaterial that has attracted great scientific interest in the past decade due to its intrinsic physical, mechanical, and electronic properties (Saito et al., 2009). CNTs can be viewed as rolled up graphene sheets that are formed into single-walled carbon nanotubes (SWNTs) or multiwalled carbon nanotubes (MWNTs) depending on the number of inner layers (Fig. 1). The diameter of CNTs ranges from 0.4 to 2 nm and 1.4 to 100 nm for SWNTs and MWNTs, respectively, while the length varies and can reach several micrometers (Klumpp, Kostarelos, Prato, & Bianco, 2006). The high aspect ratio, results in a large surface, thus allowing multiple and a high degree of functionalization with biocompatible functional groups (Li, Wang, Fan,

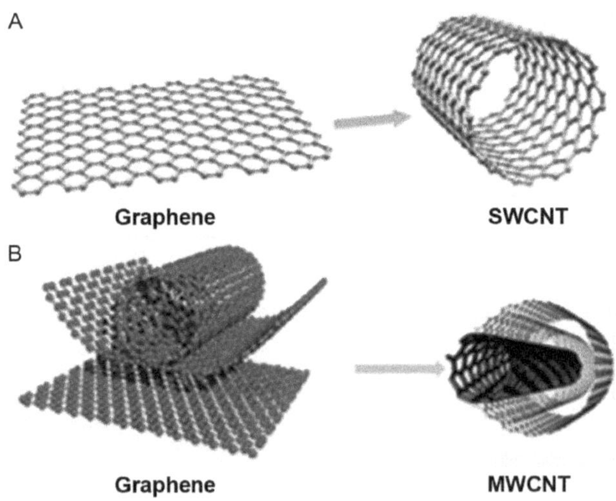

Fig. 1 Wall structure of (A) SWNT and (B) MWNT. SWCNT and MWCNT are referred to as SWNT and MWNT, respectively, described in this chapter. *Vidu, R., Rahman, M., Mahmoudi, M., Enachescu, M., Poteca, T. D., & Opris, I. (2014). Nanostructures: A platform for brain repair and augmentation.* Frontiers in Systems Neuroscience, 8, *91. doi: 10.3389/fnsys.2014.00091. Copyright © 2014.*

Feng, & Cui, 2012). The unique properties of CNTs allow them to be used in a wide array of applications across many scientific fields including composite materials (Qiao, Li, Bao, & Bao, 2007), nanoelectronics (Sgobba & Guldi, 2009), bioimaging (Choi et al., 2007), and drug delivery (Bhirde et al., 2009).

The hydrophobic surface of pristine CNTs makes them prone to bundling in aqueous solution, which limits the biological application of the as-produced CNTs. Functionalization of CNTs is, therefore, required to harness the benefit of CNTs in biological applications. Chemical functionalization is one of the main approaches used to increase CNTs dispersibility, reduce the toxicity, and impart desired characteristics on CNTs in biological environments. The ability to modify the surface of functionalized CNTs (*f*-CNTs) with targeting ligands and drug molecules while being nontoxic (Bardi et al., 2009; Bussy et al., 2015) and nonimmunogenic (Gaillard, Duval, Dumortier, & Bianco, 2011; Medepalli, Alphenaar, Raj, & Sethu, 2011) make them invaluable for designing a drug delivery system capable of specific delivery of high drug payload into living systems (Liu, Chen, et al., 2008).

One of the most interesting characteristics of *f*-CNTs from a drug delivery point of view is their tendency to translocate plasma membranes, and enter the cells either passively by an adsorptive-mediated mechanism or actively via energy-dependent mechanisms (Bianco, Kostarelos, & Prato, 2005). The validity of the former mechanism has been demonstrated in a study conducted by Pantarotto et al. where fluorescently labeled SWNTs were incubated with human and murine fibroblasts, and the uptake was examined by confocal microscopy. The accumulation of SWNTs in the cytoplasm was observed following 60 min of incubation at 37°C (Pantarotto, Briand, Prato, & Bianco, 2004). Accumulation was also observed when the experiment was carried out at 4°C or in the presence of endocytosis inhibitors. The findings demonstrated the ability of SWNTs to passively cross the lipid bilayer of fibroblasts via an energy-independent mechanism (Pantarotto et al., 2004).

By contrast, the cellular uptake of fluorescently labeled SWNTs into human promyelocytic leukemia cells was inhibited by incubation at 4°C, which showed that SWNTs also enter the cells via an endocytosis-dependent pathway (Kam, Jessop, Wender, & Dai, 2004). It is worth mentioning that SWNTs were functionalized with oxidation in this instance compared to the former study where the nanotubes were functionalized with a 1,3-dipolar cycloaddition reaction, a reaction that does not involve shortening or introducing carboxylic groups on the CNTs surface. This illustrates the significance of the functionalization step on the ultimate behavior of *f*-CNTs in living systems. Therefore, while CNTs are able to penetrate plasma membranes and enter the cell via an energy-independent mechanism, this route is highly dependent on the mode of functionalization of CNTs and the cell target involved.

Additionally, the degree of chemical functionalization has a direct impact on the pharmacological profile of CNTs. Singh et al. studied the pharmacological parameters, biodistribution, and excretion profile of amine-functionalized SWNTs with difference surface charges after intravenous administration in mouse. The blood circulation time for *f*-SWNTs was 3 and 3.5 h for the positively charged and neutral *f*-SWNTs, respectively (Singh et al., 2006). Tissue biodistribution showed that the majority of *f*-SWNTs accumulated in the kidneys 30 min after injection. On the other hand, the data showed very low amounts of *f*-SWNTs in the liver, spleen, and lungs indicating the low affinity of *f*-SWNTs for the reticulo-endothelial system (RES). The high accumulation in the kidney was related to the fast urinary excretion. In fact, high amounts of intact *f*-SWNTs were present in the urine, which is important to avoid long-term accumulation in tissues (Singh et al., 2006). By contrast, *f*-MWNTs with different degrees of ammonium functionalization showed marked differences in their biodistribution profiles. *f*-MWNTs with low degree of functionalization exhibited high liver and spleen accumulation after 24 h of intravenous injection. On the other hand, highly functionalized *f*-MWNTs were not detected in liver or spleen. The histological data, therefore, highlights the importance of the degree of chemical functionalization for the biodistribution of *f*-MWNTs following systemic administration (Lacerda et al., 2008).

It is also worth mentioning that uncoated *f*-CNTs results in fast opsonization and elimination by the complement system after encountering serum protein, which reduces blood circulation time, and limits brain exposure to the injected material. Therefore, polyethylene glycol (PEG) coating has been successfully used to mask the surface of circulating *f*-CNT and extend their blood circulation time up to ~22 h allowing longer circulation time for the injection *f*-SWNTs-PEG compared to the uncoated *f*-SWNTs (Prencipe et al., 2009). Notably, blood circulation time of the PEG-coated CNTs depends on the length of the PEG chain. Liu et al. showed that increasing the length of the PEG chain causes a proportional increase in blood circulation time of *f*-SWNTs-PEG (Liu, Davis, et al., 2008).

2. FUNCTIONALIZATION STRATEGIES OF CNTs

Chemical functionalization of CNTs can be achieved either covalently, by conjugating functional groups to the surface of the CNTs using covalent bonds, or noncovalently relying on the close interactions between CNTs and compatible functional groups.

2.1 Covalent Functionalization

Several approaches have been successfully used to covalently conjugate functional groups to the surface of CNTs. The aim is to increase physical stability in aqueous solutions and improve pharmacological profile. Some of the used approaches include halogenation (Khabashesku, Billups, & Margrave, 2002), nucleophilic addition (Holzinger et al., 2001), electrophilic addition (Tagmatarchis, Georgakilas, Prato, & Shinohara, 2002), and radical addition (Strano et al., 2003). However, the two most important reactions that resulted in *f*-CNTs suitable for biomedical applications are oxidation (de Faria et al., 2014) and cycloaddition reactions (Rubio et al., 2015).

Oxidation of CNTs is performed using oxidizing acids such as concentrated nitric acid or a mixture of nitric and sulphuric acids. CNTs oxidation was originally used to shorten the CNTs (Liu et al., 1998), to remove the end caps allowing the filling of the CNT hollow structure for subsequent applications (Tsang, Chen, Harris, & Green, 1994), and to remove amorphous carbon coating to make the surface available for functionalization (Dementev, Osswald, Gogotsi, & Borguet, 2009). The oxidation reaction of CNTs results in shorter fragments with carbon atoms at terminal or defect sites terminated with carboxylic groups, which are then used in subsequent reactions to link other functional groups or molecules, eg, PEG copolymer (Zhao, Hu, Yu, Perea, & Haddon, 2005). By activating the carboxylic groups on the *f*-CNTs, functional groups can be covalently linked to *f*-CNTs using amidation or esterification reactions, without damaging the electronic properties of the *f*-CNTs (Liu et al., 1998) (Fig. 2).

To avoid the disadvantages of the oxidation functionalization that include shortening and introducing surface defects on the side walls of *f*-CNTs, cycloaddition reactions were devised as a method to add functional groups onto the CNTs without disrupting their structural properties (Chen et al., 1998; Jared et al., 2006). This approach preserves the length of the CNTs and does not result in a high amount of surface defects that may affect their electronic properties. Therefore, cycloaddition reactions have been

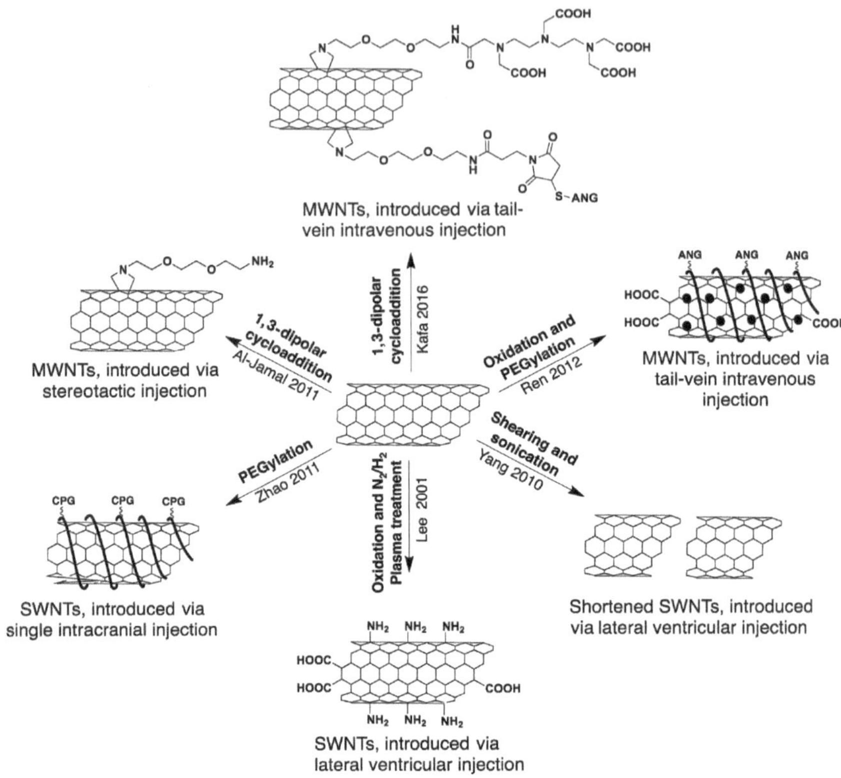

Fig. 2 Common methods used in the functionalization of CNTs for brain-specific applications.

used to produce *f*-CNTs for several applications including nanoelectronics and bioimaging (Wu, Chang, Liu, Zhao, & Zhao, 2010). 1,3-Dipolar cycloaddition reaction is one of the most widely used reactions for the functionalization of CNTs (Fig. 2). It relies on the condensation of amino acids and aldehydes into azomethine ylide that reacts with the side wall of CNTs creating a pyrrolidine ring (Georgakilas, Kordatos, et al., 2002). The 1,3-dipolar cycloaddition produces water-dispersible *f*-CNTs (Georgakilas, Tagmatarchis, et al., 2002). Additionally, the pyrrolidine ring allows the conjugation of other types of molecules for drug delivery and imaging applications (Wu et al., 2010).

2.2 Noncovalent Functionalization

The alternative approach to functionalizing CNTs is to rely on noncovalent interactions between CNTs and dispersing agents such as polymers

(Fujigaya & Nakashima, 2015) and single-stranded DNA (Zheng et al., 2003). The noncovalent interactions can be achieved via van der Waals forces, π–π stacking, or hydrogen bonding. The advantage of noncovalent functionalization is that it protects the structure of the CNTs from surface damage and preserves the electronic properties (Zhao & Stoddart, 2009). Additionally, noncovalent functionalization using PEG coating provides surface protection for CNTs from immunological elimination in the body, and consequently extend blood circulation time (Kotagiri & Kim, 2014). However, due to the weak interactions between the dispersing agents and the CNTs in the noncovalent arrangement, it may not offer the required stability for in vivo applications. The stability of the conjugates in biological environment is an essential prerequisite that needs to be carefully examined prior to undertaking further studies (Liu, Sun, Nakayama-Ratchford, & Dai, 2007).

3. THE SCAFFOLDING EFFECT OF CNTs ON NEURONAL CELLS AS GROWTH SUBSTRATES IN VITRO

Neurodegenerative diseases are generally the result of damage to the neuronal network circuitry due to neuronal loss in the brain. The reason for the neuronal damage, in the case of Alzheimer and Parkinson diseases, is the loss of protein organization in the brain, which eventually leads to loss of memory and motor function in patients. The restoration of healthy neuronal network and stimulating synaptic interfaces could potentially alleviate the symptoms of neuronal loss and stimulate tissue repair (Ellis-Behnke, 2007; Gilmore, Yi, Quan, & Kabanov, 2008).

The structural stability of CNTs and their conserved electronic properties support the rationale for using them for neuronal repair applications. The tubular structure of CNTs combined with their thin diameter and high conductivity allow them to interface with neurons to support neuronal growth and regeneration (Zhang & Webster, 2009). CNTs are, therefore, very promising for neuroprotective applications and could be a key to tackling the complicated neurodegenerative disorders.

The ability of CNTs to support neuronal growth and regeneration was demonstrated by Mattson et al. that showed the ability of MWNTs to support the growth of embryonic rat neurons. MWNTs coated with the bioactive molecule 4-hydroxynonenal led to a significant improvement neurite outgrowth and branching (Fig. 3) (Mattson, Haddon, & Rao, 2000). The use of CNTs to support neuronal growth and differentiation is, therefore, dependent on chemical functionalization and should closely mimic the neurons microenvironment. For example, laminin, found in human extracellular matrix, was used to produce "humanized SWNT" to support the growth of neuronal stem cells. The SWNT–laminin layer allowed neuronal growth and synaptic formation, which was superior to uncoated SWNTs (Kam, Jan, & Kotov, 2009). In another study, nerve growth factor (NGF) was covalently linked to CNTs to stimulate neurites outgrowth in embryonic chick dorsal root ganglion neurons. NGF-coated CNTs facilitated the outgrowth of neurites to a higher extent than the amine-functionlized CNTs (Matsumoto et al., 2007). Furthermore, the use of polymers such as polypyrrole was shown to support the growth of cultured hippocampal neurons

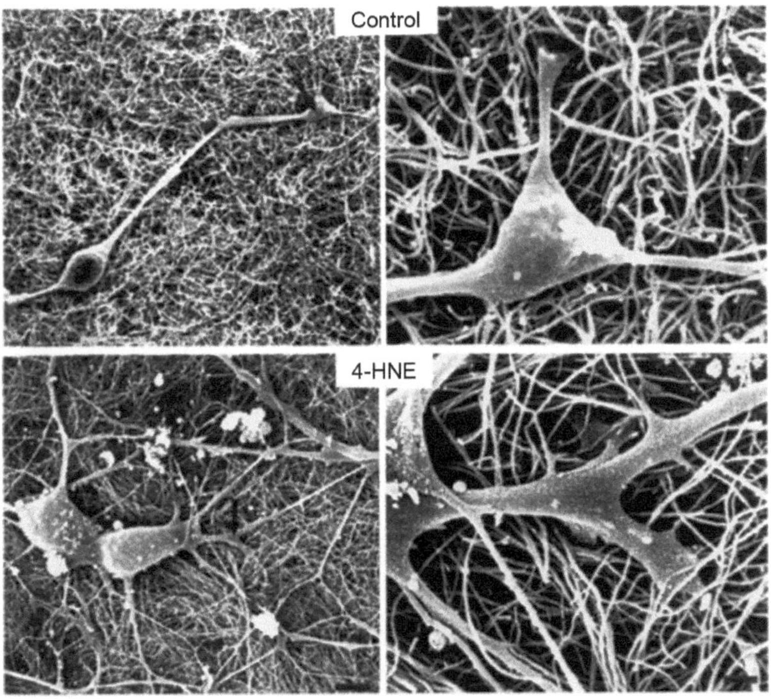

Fig. 3 Neurite outgrowth and branching is enhanced in neurons growing on nanotubes coated with 4-HNE. SEM micrographs showing neurons that had grown for 3 days on unmodified control nanotubes (*upper*) and nanotubes coated with 4-HNE (*lower*). The *right panel* is a high magnification view of the region of the neurite designated by the *arrow* in the *left panel*. Scale bars: *left panels*, 5 μm; *right panels*, 100 nm. *Figure reproduced from Mattson, M. P., Haddon, R. C., & Rao, A. M. (2000). Molecular functionalization of carbon nanotubes and use as substrates for neuronal growth.* Journal of Molecular Neuroscience, 14*(3), 175–182. doi: 10.1385/Jmn:14:3:175 with permission of Springer.*

for 14 days in culture, and allowed neurites growth and branching in addition to the reduction of SWNTs toxic effects on the cells (Hernandez-Ferrer et al., 2014).

In addition to the nature of functional groups, the overall surface charge of CNTs is important for neuronal growth. Hui et al. compared the effect of surface charge of MWNTs on neurite number and branching in rat hippocampal neurons. It was demonstrated that while neurite number was not dependent on the MWNTs surface charge, the length of neurites was most prominent on positively charged MWNTs compared to neutral or negatively charged material (Hu, Ni, Montana, Haddon, & Parpura, 2004) (Fig. 4). Additionally, the type of CNTs also has direct implications on neuronal growth and differentiation. Meng et al. have demonstrated that the concentration of iron catalyst in the CNTs substrate inhibits the differentiation capacity of rat phemocytoma (PC12) cells following exposure to NGF. On the other hand, CNTs with low iron content facilitated the NGF-induced differentiation as observed by the production of neurites in culture (Meng et al., 2013).

Additionally, studies have shown that CNTs substrate enhanced the excitability of cultured neurons by providing electrical shortcuts between neurones. For example, nanomeshworks of SWNTs and MWNTs on glass slides have been shown to support neuronal growth and responsiveness of rat hippocampal neurons. The results showed that CNTs improved the excitability of the neurons by interfacing with the cells creating circuit shortcuts (Cellot et al., 2009). Additionally, it was shown that CNT substrate leads to an early onset of neuronal electrical activity of cultured neurons, which is important for tissue repair and regeneration following nerve damage before astrocyte activation lead to glial scar tissue (Khraiche, Jackson, &

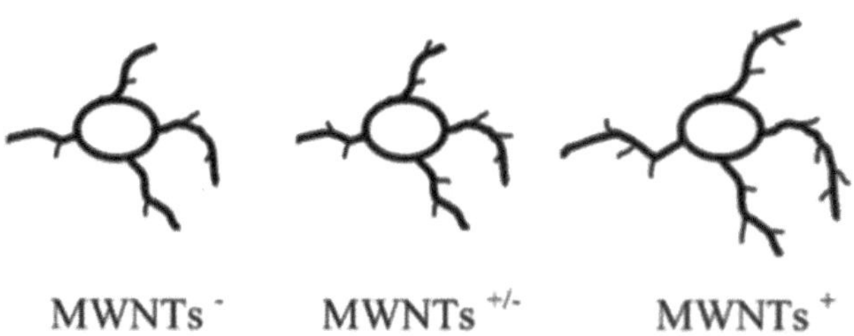

Fig. 4 Drawing summarizing the effects of MWNT charges on the neurite outgrowth and branching. *Reprinted with permission from Hu, H., Ni, Y. C., Montana, V., Haddon, R. C., & Parpura, V. (2004). Chemically functionalized carbon nanotubes as substrates for neuronal growth.* Nano Letters, *4(3), 507–511. doi: 10.1021/nl035193d. Copyright © 2004, American Chemical Society.*

Muthuswamy, 2009). By contrast, Chen et al. showed that MWNTs significantly decreased the hippocampal CA1 glutamatergic postsynaptic transmission in vitro indicating a role in excitotoxicity (Chen et al., 2014). Therefore, while CNTs show great promise for use in prosthesis to repair neuronal tissue damage, more studies are still needed to establish their benefits compared to their potential toxicity to the tissues.

4. CNTs AS HEATING DEVICES FOR HYPERTHERMAL THERAPY APPLICATIONS IN GLIOBLASTOMA IN VITRO

Glioblastoma is the one of the most common forms of brain gliomas, with a median survival time of 15 months in patients (Iacob & Dinca, 2009). Glioblastoma is complicated to manage due to the limited availability of effective therapies, the invasive and resistant nature of the tumor to current therapies, the presence of blood–brain barrier (BBB) that limits access of drugs to the brain and the inability of neuronal tissues to regenerate following tissue damage (Iacob & Dinca, 2009).

Conventional treatment options include chemotherapy, eg, temozolomide and radiotherapy (Moustakas & Kreisl, 2010; Piroth et al., 2007). While modest tumor remission is observed, the side effects of these therapies including neuronal damage, hair loss, skin disorders, nausea, and sterility are difficult to manage. A combination of temozolomide and radiotherapy led to significant increase in survival time in a recent phase III clinical trial compared to radiotherapy alone (Stupp et al., 2009), but new therapies are still needed to avoid the side effects of chemotherapy and radiotherapy.

Hyperthermia relies on creating a local increase in temperature in the tumor site that leads to physiological changes in the malignant cells, which eventually initiates an apoptotic event to eradicate the cells (Hildebrandt et al., 2002). One of the main advantages of hyperthermia over conventional therapies is the lack of systemic side effects. However, there are several limitations to hyperthermia that could hinder its applicability in clinical situation namely, the limited tissue penetration of external heat sources, eg, microwaves or ultrasound, the damaging effect to surrounding tissues and the heat loss in well-vascularized tumors (Verma, Lal, & Van Noorden, 2014).

One of the inherent optical properties of CNTs is their ability to absorb near-infrared radiation (NIR) (Tabakman, Welsher, Hong, & Dai, 2010), which is converted to heat as a function of exposure time and laser power (Ghosh et al., 2009). These optical properties coupled with the recent findings regarding the selective tumor accumulation of MWNTs following systemic administration (Kafa et al., 2016) provide a platform for hyperthermia-based applications. The main advantage of using NIR-based therapy is that biological tissues are naturally transparent to NIR radiation, which eliminates potential damage to surrounding tissues (Iverson et al., 2013).

Markovic et al. have successfully demonstrated the cytotoxic capacity of CNTs incubated with human glioma cell line U251 in vitro. The hyperthermal-related cytotoxic effects were measured as a function of cell viability following exposure to 808 nm NIR light from 30 to 300 s. The data showed a concentration and time-dependent effect following CNTs irradiation in culture compared to untreated cells (Markovic et al., 2011). Similarly, nontargeted SWNTs were also able to induce hyperthermia in U251 human glioma cell line in vitro. In fact, the used SWNTs resulted in a greater cytotoxic effect following 10 min NIR exposure to glioma cells compared to astrocytes or primary endothelial cells. The lack of cytotoxic effects on healthy brain cells, which was due to preferential endocytosis into glioma cells, is very important for future application of hyperthermia in vivo (Santos et al., 2014). Additionally, a combination of intratumoral SWNTs injection and NIR exposure were effective in suppressing tumor growth in athymic nude mice in vivo compared to SWNTs or NIR used separately. The treatment with SWNTs and NIR also inhibited tumor recurrence for up to 80 days (Santos et al., 2014).

One of the advantages of CNTs-based NIR therapy is the ability to target specific subset of tumor cells using targeting ligands. Wang et al. have used SWNTs conjugated with anti-CD133$^+$ monoclonal antibody (anti-CD133-SWNTs) to recognize and eradicate cancer stem-like cells, isolated from human glioblastoma patients, using hyperthermia. The anti-CD133-SWNTs were incubated with mixed culture of CD133$^+$ positive and negative cells, which was then irradiated with 808 nm NIR light for 5 min. Fluorescence imaging confirmed that CD133$^+$ positive cells were eliminated with NIR irradiation, while no effect was observed on the CD133$^+$ negative population (Wang et al., 2011). Furthermore, CD133$^+$ positive cells, pretreated with anti-CD133-SWNTs and irradiated 2 days after xenotransplantation in nude mice, lost their ability to establish a tumor in vivo (Wang et al., 2011). The findings confirmed the ability of SWNTs to target specific subset of cells to achieve local hyperthermia and to inhibit potential tumor remission (Fig. 5).

Furthermore, MWNTs-guided hyperthermia was employed for the elimination of disialoganglioside (GD2)-expressing neuroblastoma cells

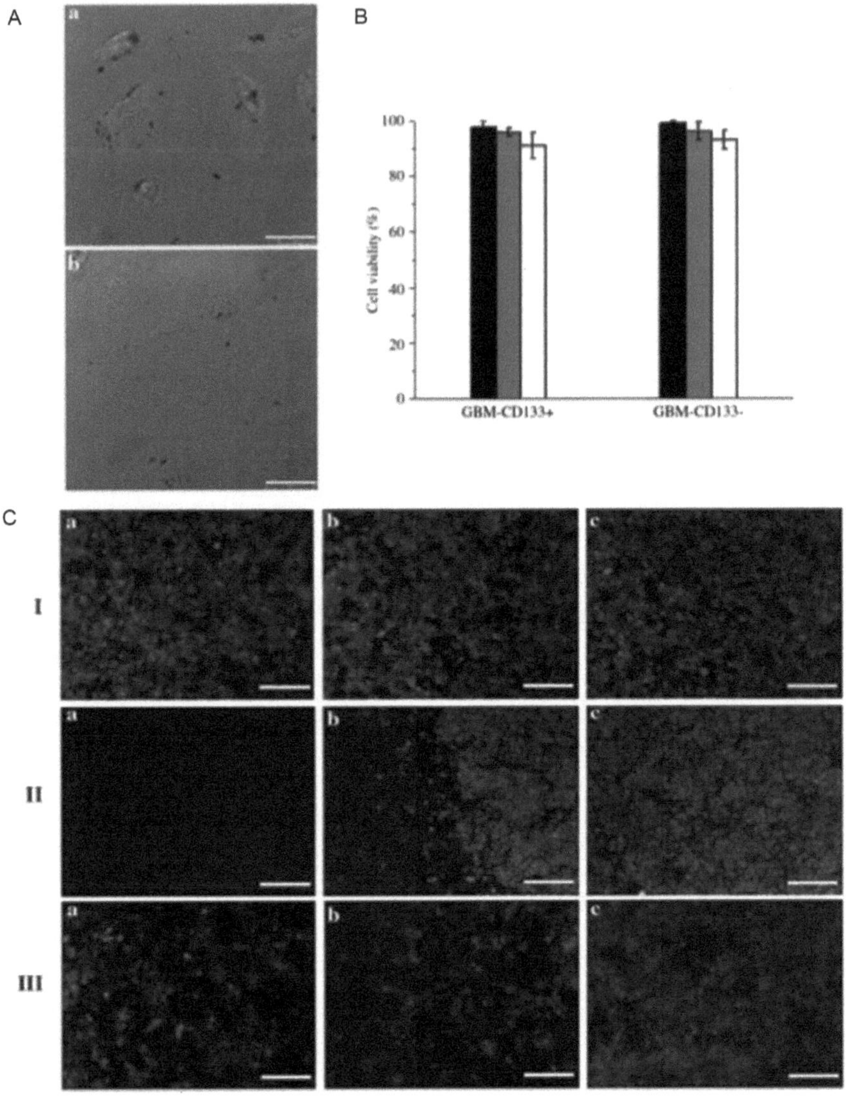

Fig. 5 Endocytosis of anti-CD133-PE-tagged SWNTs and NIR laser-induced photothermal treatment of GBM-CD133$^+$ and GBM-CD133$^-$ cells. (A) Fluorescent images of GBM-CD133$^+$ and GBM-CD133$^-$ cells cocultured with CDSWNTs for 6 h. Scale bar: 10 μm. (B) Cell viability of GBM-CD133$^+$ and GBM-CD133$^-$ cells challenged separately with 5 (*black bar*), 10 (*gray bar*), and 20 (*white bar*) μg/mL of CDSWNTs for 6 h. (C) Photothermal treatment of (I) GBM-CD133$^-$ cells, (II) GBM-CD133$^+$ cells, and (III) mixed population of GBM-CD133$^+$ and GBM-CD133$^-$ with ratio of 1:1. After GBM-CD133$^+$, GBM-CD133$^-$, and mixed cells incubated separately with CDSWNTs for 6 h, 808-nm NIR laser was harnessed to irradiate the cells for 5 min with an intensity of 2 W/cm^2. After staining with calcein-AM dye, fluorescent images of cells were taken (a) within, (b) on the edge of, and (c) outside the NIR laser-shining zone. Scale bar: 100 μm. *Reproduced from Wang, C. H., Chiou, S. H., Chou, C. P., Chen, Y. C., Huang, Y. J., & Peng, C. A. (2011). Photothermolysis of glioblastoma stem-like cells targeted by carbon nanotubes conjugated with CD133 monoclonal antibody.* Nanomedicine: Nanotechnology, Biology and Medicine, *7(1), 69–79. doi: 10.1016/j.nano.2010.06.010.*

in vitro. MWNTs conjugated with anti-GD2 monoclonal antibody (anti-GD2-MWNTs) were incubated with the GD2-positive stNB-V1 neuroblastoma cells, and PC12 cells lacking the GD2 expression. Fluorescence imaging showed the internalization of anti-GD2-MWNTs into stNB-V1 cells, but not the PC12 cells. NIR irradiation using 808 nm light for 10 min induced necrosis in the stNB-V1 culture, which indicated the effectiveness of hyperthermia with anti-GD2-MWNTs (Wang, Huang, Chang, Hsu, & Peng, 2009). The combination of CNTs and hyperthermia provide a promising platform for tackling brain tumors. However, more studies are required to show the effectiveness of CNTs-based hyperthermia in glioblastoma model in vivo as a fundamental step before transitioning this treatment platform to clinical trials.

5. THE INTRINSIC THERAPEUTIC ACTION OF CNTs IN BRAIN TISSUES AS ACTIVE AGENTS

CNTs have shown some therapeutic action in stroke prevention in vivo due to their intrinsic properties, and without the need for deliverable therapeutics. This idea stems from the fact that CNTs proved effective in providing a matrix for neuronal repair in vitro due to their electrical properties (Hernandez-Ferrer et al., 2014; Kam et al., 2009; Mattson et al., 2000), and could therefore be used as scaffolding for stem cell therapy. The preliminary observations by Lee et al. showed that SWNTs scaffolding, even without using stem cells, reduced the size of the infarcted brain areas following ischemic injury suggesting a neuroprotective role in the brain (Lee et al., 2011).

Furthermore, Lee et al. have studied the intrinsic therapeutic effect of amine-functionalized single-walled carbon nanotubes (α-SWNTs) in stroke prevention. A cerebral ischemia injury rat model was used in this study, with SWNTs preinjected into the lateral ventricle of the brain 1 week before the onset of a surgically induced ischemic injury. The extent of neuronal damage following the ischemic injury was significantly lower in the SWNT-treated group compared to the control. The therapeutic effect persisted for 7 days after treatment showing the potential of SWNTs in stroke prevention (Fig. 6) (Lee et al., 2011).

The study also confirmed that SWNTs treatment reduced apoptotic markers as shown with TUNEL assay, and inflammatory mediators following the induced stroke, which is important for neuroprotection from post-ischemic cell death (Higgins, Dawson, & Walters, 2011). The findings also

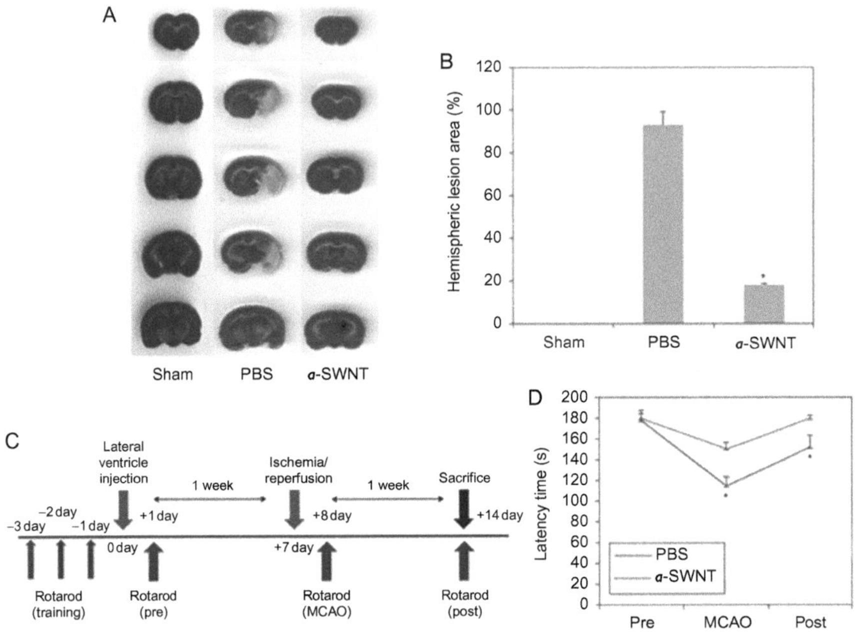

Fig. 6 Pretreatment with α-SWNT-protected animals following ischemia reperfusion. One week before middle cerebral artery occlusion (MCAO), animals received either PBS or a-SWNTs. Animals were killed 1 day after reperfusion. (A) Coronal brain sections (stained with triphenyltetrazolium chloride (TTC) solution) of sham-operated control and PBS- or *a*-SWNT-treated rats. Brain slices were arranged in sequence. *White areas* represent the infarcted region after MCAO. (B) Quantification of the ischemic lesion of brain sections in (A). Data are means ± s.e.m. $*p<0.001$ vs the PBS group. (C) Schematic of injection and training schedule. (D) Graph showing that pretreatment with *a*-SWNTs reduced neurological damage after ischemia. Neurological score was evaluated by the rotarod treadmill test. Data are means ± s.e.m. $*p<0.001$ vs pre-MCAO. *Reprinted by permission from Macmillan Publishers Ltd.: Lee, H. J., Park, J., Yoon, O. J., Kim, H. W., Lee do, Y., Kim do, H., … Kim, S. S. (2011). Amine-modified single-walled carbon nanotubes protect neurons from injury in a rat stroke model.* Nature Nanotechnology, *6(2), 121–125. doi: 10.1038/nnano.2010.281.*

highlight the importance of chemical functionalization for the final therapeutic outcome. In this instance, amine-functionalization, which carries a slight positive charge at physiological pH, showed no inflammatory response following injection. Bardi et al. have previously confirmed that amine-functionlized CNTs did not elicit an inflammatory response compared to oxidized CNTs following stereotactic brain injection in mouse (Bardi et al., 2013).

6. CNTs AS LOCAL DELIVERY TOOLS OF THERAPEUTICS TO THE BRAIN VIA DIRECT INTRACRANIAL ADMINISTRATION IN VIVO

Drug delivery to the brain is a challenging field due to the presence of the BBB. The BBB restricts access for most available therapeutics to the brain, thus hindering the progress to tackle the complex neurodegenerative disorders and brain gliomas. The promise of using CNTs to delivery drugs to the brain, while at the same time providing neuroprotection to the brain encouraged the idea of using them for brain drug delivery. Due to the complexity of crossing the BBB, studies have focused on the direct intracranial delivery route to administer drug-loaded CNTs into the brain cavity to bypass the BBB restriction.

Al-Jamal et al. have shown the successful delivery of caspase-3 siRNA by CNTs to the brain in vivo (Al-Jamal, Gherardini, et al., 2011). MWNTs-NH_3^+ were complexed with caspase-3 siRNA to provide a delivery platform superior to the intracerebral administration of siRNA alone. The perilesional stereotactic administration of MWNTs-NH_3^+:caspase-3 siRNA complexes resulted in reduced neuronal damage and enhanced motor function recovery in an Endothelin-1 rat stroke model (Fig. 7) (Al-Jamal, Gherardini, et al., 2011).

Additionally, the delivery of CpG oligonucleotide to brain glioma was also enhanced using SWNTs. SWNTs-CpG conjugates were injected into the brain using a single intracranial injection and the uptake was evaluated using flow cytometry and confocal microscopy. The data suggested that conjugating CpG to SWNTs significantly enhanced the delivery of SWNTs-CpG into tumor-associated inflammatory cells compared to the free CpG, which resulted in eliminating the brain glioma in the brain and increasing median survival time (Zhao et al., 2011). In fact, the immunostimulatory effect of CPG following the intracranial injection prevented recurrence of glioma following intracranial and systemic challenge (Zhao et al., 2011).

A follow-up study using SWNTs-CpG investigated the difference in antitumor immunity response following intracranial or subcutaneous administration into tumor-bearing mice. The SWNTs-CpG were administered intratumorally in mouse using both intracranial and subcutaneous injections. The aim was to evaluate the stimulatory immunological response

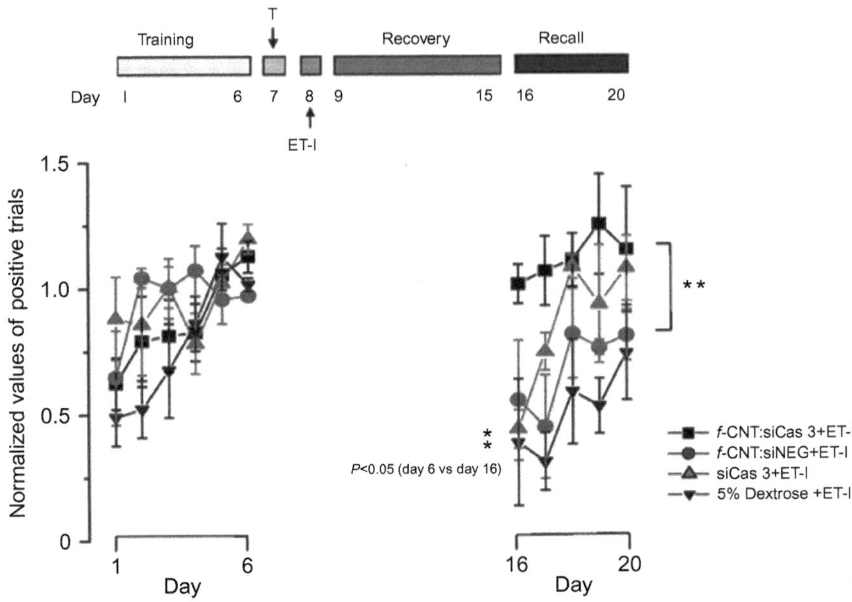

Fig. 7 Behavioral analysis using "skilled reaching" test in rats. Functional improvement in ET-1 ischemic rat forelimb function with or without pretreatment was measured. Rats were pretreated with 5% dextrose, siCas 3 alone (4.7 pmol), *f*-CNT:siCas 3, or *f*-CNT:siNEG complexes (8 : 1 mass ratio) 24 h before exposure to ET-1 (30 pmol) leading to unilateral lesions made contralateral to the preferred (for reaching) forelimb. During recall, rats treated with *f*-CNT:siCas 3 performed a significant higher number of positive trials in food pellet reaching compared to all other treatment groups. (*, $p \leq 0.05$; **, $p \leq 0.01$; $n = 4$) and reaching prelesion levels. *Figure reproduced from Al-Jamal, K. T., Gherardini, L., Bardi, G., Nunes, A., Guo, C., Bussy, C., … Pizzorusso, T. (2011). Functional motor recovery from brain ischemic insult by carbon nanotube-mediated siRNA silencing.* Proceedings of the National Academy of Sciences of the United States of America, 108*(27), 10952–10957. doi: 10.1073/pnas.1100930108.*

from injecting at two different locations, and the effect on brain glioma and melanoma. SWNTs-CpG injected subcutaneously showed only modest tumor elimination in the local injection site. However, SWNTs-CpG injected intracranially elicited antitumor response that was observed in brain tumors and also in subcutaneous melanoma (Fan et al., 2012). Injecting SWNTs-CpG locally into the melanoma site led to abundant myeloid-derived suppressive cells to accumulate in the melanoma site. By contrast, the intracranial injection led to activation of toll-like receptor (TLR)-9-positive microglia, which were abundant in the intracranial tumor site (Fan et al., 2012). The data positively identified the intracranial route to be superior for CpG immunostimulatory response compared to subcutaneous administration. In both cases, SWNTs were the vehicle that facilitates the delivery of CpG to the tumor.

7. CNTs AS DELIVERY TOOLS OF THERAPEUTICS TO THE BRAIN USING SYSTEMIC ADMINISTRATION IN VIVO

While the intracranial delivery of CNTs into the brain cavity showed promising results in vivo, this technique could prove complicated for potential clinical applications due to the invasiveness of the technique and patient compliance. Therefore, the systemic delivery of CNTs into tissues is needed to establish future applications in drug delivery. Systemic administration of CNTs, therefore, remains crucial if a generalized therapeutic effect in the brain is required. Only two studies, detailed later, have investigated the delivery of drugs to the brain following systemic administration.

SWNTs were used for the delivery of acetylcholine (ACh) to the brain of Alzheimer disease mice model (AD mice). SWNTs-ACh were administered using a single gastro-gavage route and the therapeutic effects on learning and memory were tested 8 h after the injection. The data showed a significant improvement in learning and memory capabilities bringing them to normal levels. By contrast, the unloaded SWNTs or the free ACh did not lead to a similar therapeutic outcome. This study was an indirect evidence that SWNTs were able to reach the brain despite the presence of the BBB (Yang et al., 2010). However, the study also showed the importance of SWNTs dose in determining the final outcome. While the lysosomes proved to be the pharmacological target of SWNTs-ACh, the mitochondria were shown to be the target for their toxicity. Therefore, a balance between achieving a therapeutic outcome while avoiding associated toxicity was crucial (Yang et al., 2010).

Angiopep-2 peptide (ANG) is a promising molecule for targeting drugs to the brain via the LRP1 receptor. ANG has shown higher brain uptake and parenchyma accumulation following in situ brain perfusion compared to transferrin (Demeule et al., 2008). Ren et al. have investigated the ability of ANG-targeted *f*-MWNTs to deliver doxorubicin to mouse brain in vivo. PEGylated and carboxylated MWNT (oxMWNT-PEG) were modified with ANG to improve the delivery of doxorubicin to mouse brain. Optical imaging showed a superior brain accumulation for Dox-oxMWNT-PEG-ANG compared to Dox-oxMWNT-PEG or Dox alone after 1 to 6 h. Additionally, Dox-oxMWNT-PEG-ANG significantly increased the median survival time in glioma-bearing mice compared to the saline group

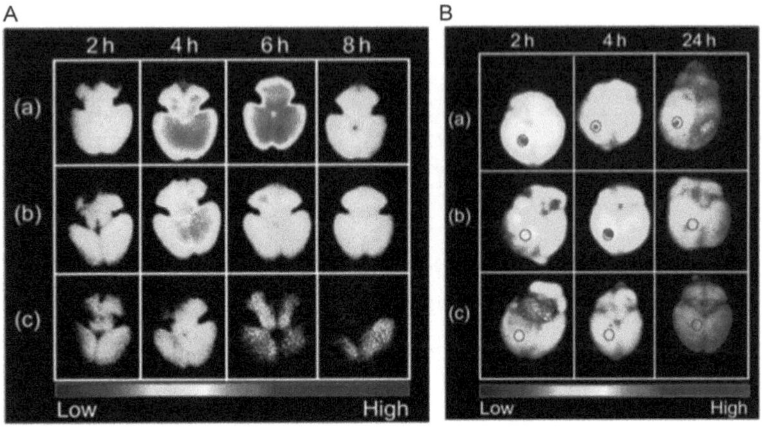

Fig. 8 In vivo fluorescence imaging after intravenous injection at a dose of 5 mg DOX-equiv./kg body weight in mouse. (A) Brains collected from Balb/c mice. (B) Glioma-bearing brains 24 h after injection. DOX-oxMWNTs-PEG-ANG group were shown on the *upper row* (a), DOX-oxMWNTs-PEG group were on the *middle row* (b), and DOX group were on the *lower row* (c). *Reproduced from Ren, J., Shen, S., Wang, D., Xi, Z., Guo, L., Pang, Z., … Jiang, X. (2012). The targeted delivery of anticancer drugs to brain glioma by PEGylated oxidized multiwalled carbon nanotubes modified with angiopep-2.* Biomaterials, 33*(11), 3324–3333. doi: 10.1016/j.biomaterials.2012.01.025.*

(Fig. 8) (Ren et al., 2012). However, the data presented in Ren et al. were based on fluorescence imaging to track the distribution of the doxorubicin, and not the nanocarrier itself. While fluorescence imaging provided qualitative data on the high accumulation of Dox-oxMWNT-PEG-ANG in mouse brain, the lack of quantitative data regarding the distribution of the nanocarrier itself makes it difficult to compare the accumulation rate to other delivery systems.

8. MECHANISTIC STUDIES INVESTIGATING THE BBB CROSSING OF *f*-CNTs

8.1 The Interaction of *f*-MWNTs with BBB In Vitro

While studies so far have focused on delivering therapeutics to the brain using the intracranial or systemic routes, little is known about the interaction of the *f*-CNTs with the BBB. Understanding the behavior of *f*-CNTs in the brain and their interaction with brain endothelial cells is a key for tackling the BBB and is essential for designing drugs and nanocarriers that are capable of crossing the BBB and reaching their brain targets.

CNTs offer several desired characteristics in terms of drug delivery to the brain. The ability of CNTs to penetrate biological membranes due to their unique needle-like structure without perturbing the membrane integrity is promising for BBB crossing (Al-Jamal, Nerl, et al., 2011). Additionally, the high surface area of CNTs allows the loading and delivery of high doses of drugs to local sites, thus maximizing the therapeutic effect while minimizing drug-related cytotoxicity to healthy tissues. This is particularly important when targeting tumor cells where specific cell targeting is a key for therapeutic success. CNTs may also provide protection for the delivered drug from enzymatic degradation or clearance, thus increasing the half-life of the drug within the blood.

We have previously investigated this hypothesis using amine-functionalized multiwalled carbon nanotubes (*f*-MWNTs) (Kafa et al., 2015). The interaction of *f*-MWNTs with the BBB in vitro was investigated using a coculture BBB model comprised of primary porcine brain endothelial cells (PBEC) and primary rat astrocytes. Transmission electron microscopy (TEM) examination revealed the ability of *f*-MWNTs to trigger an endocytosis event, and enter the endothelial cells into vesicles within 4 h. The *f*-MWNTs-filled vesicles were observed near the abluminal side of PBEC within 24–48 h, and fused vesicles allowed the release of *f*-MWNTs in the subluminal space (Fig. 9).

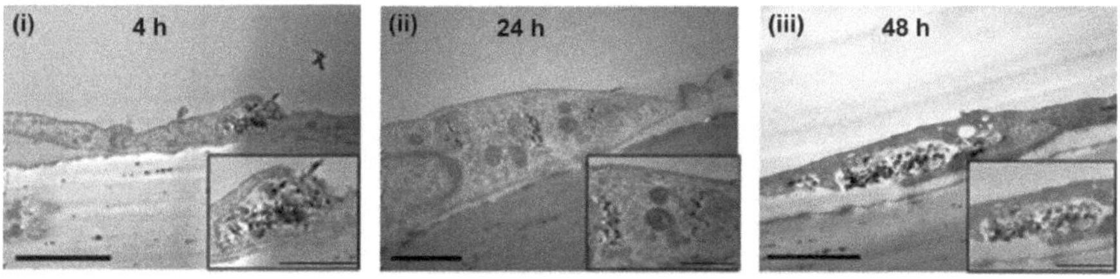

Fig. 9 The transcytosis pattern of *w*-MWNTs across the PBEC monolayer using TEM. *w*-MWNTs (20 μg/mL) were added to the upper chamber and incubated with the PBEC for (i) 4 h, (ii) 24 h, and (iii) 48 h. Bright field TEM images of polyester filters showing the initial interaction of *w*-MWNTs with the PBEC monolayer after 4 h (i). After 24 h (ii), the *w*-MWNTs clusters appeared within EVs or multivesicular bodies. The *w*-MWNTs-containing vesicles showed evidence of fusion with the abluminal plasma membrane and were partly open toward the lower chamber after 48 h (iii). Scale bars: 1 μm and (inset) 500 nm. *Reproduced from Kafa, H., Wang, J. T., Rubio, N., Venner, K., Anderson, G., Pach, E., … Al-Jamal, K. T. (2015). The interaction of carbon nanotubes with an in vitro blood–brain barrier model and mouse brain in vivo.* Biomaterials, *53, 437–452. doi: 10.1016/j.biomaterials.2015.02.083.*

Scanning TEM was used to gain a further understanding into the interaction of *f*-MWNTs with PBEC. Annular dark field (ADF) and high-angular annular dark field (HAADF) images reflected the state and integrity of the plasma membrane and vesicular membranes during the transcytosis stages of *f*-MWNTs. Fig. 10 shows that upon release of *f*-MWNTs from

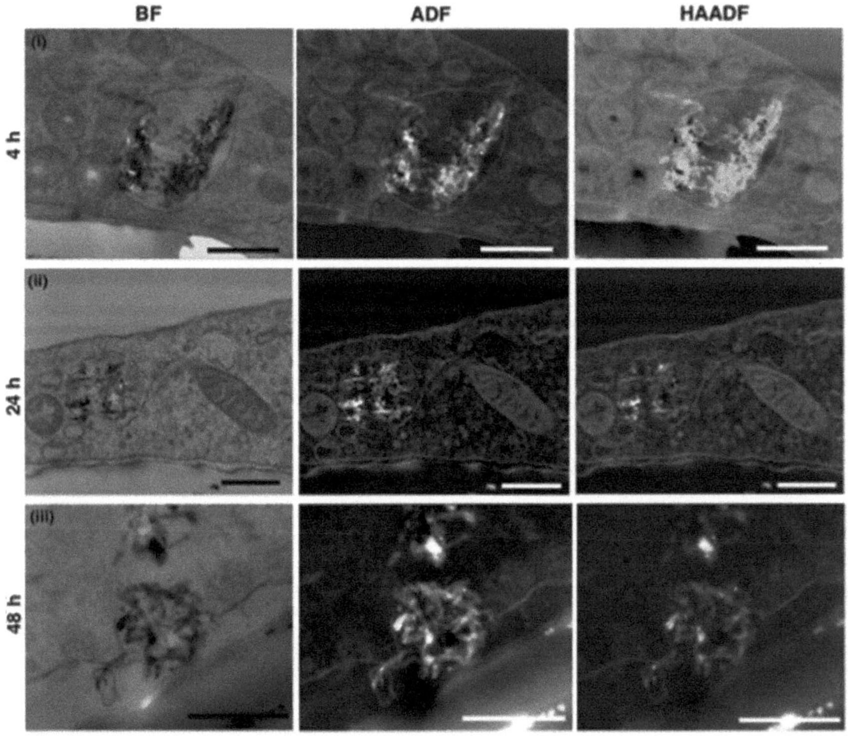

Fig. 10 The transcytosis pattern of *f*-MWNTs across the PBEC monolayer using low voltage STEM. Low voltage STEM images of the polyester filters confirming the uptake and transcytosis of *f*-MWNTs across the PBEC monolayer. *f*-MWNTs appeared within endocytic vesicles after (i) 4 h and (ii) 24 h. The *f*-MWNTs-containing vesicles were then imaged partly open toward the lower chamber allowing the release of *f*-MWNTs after 48 h (iii). The three detection modes: bright field (BF), annular dark field (ADF), and high-angular annular dark field (HAADF) helped identify the electron dense structures of *f*-MWNTs within the cell body and the osmium-rich lipid membranes surrounding the nanostructures. Scale bars: *first* and *second row* 500 nm and *last row* 400 nm. *Reproduced from Kafa, H., Wang, J. T., Rubio, N., Venner, K., Anderson, G., Pach, E., … Al-Jamal, K. T. (2015). The interaction of carbon nanotubes with an in vitro blood–brain barrier model and mouse brain in vivo.* Biomaterials, *53, 437–452. doi: 10.1016/j.biomaterials.2015.02.083.*

PBEC, the vesicular membrane was intact and complete at the edges of the releasing vesicle with no signs of membrane damage. This observation confirmed the involvement of the cell to facilitate the transcytosis event. Furthermore, the tight junctions appeared clear of any interaction with the *f*-MWNTs confirming the dependency of *f*-MWNTs on the transcellular route to cross the BBB. The intact tight junctions add further confirmation that *f*-MWNTs are able to cross the BBB without being toxic to the cells.

The observed transcytosis of *f*-MWNT across PBEC is an energy-dependent mechanism that relies on the triggering vesicle formation at the plasma membrane level to initiate the transcytosis event. Lowering the incubation temperature to 4°C, which inhibits all cellular endocytosis, resulted in a complete uptake inhibition with *f*-MWNTs observed on the luminal side of PBEC unable to enter the cells (Fig. 11).

8.2 The Interaction of *f*-MWNTs with BBB In Vivo

The ability of *f*-MWNTs to accumulate in mouse brain after systemic administration was investigated following intravenous administration of radiolabeled *f*-MWNTs in mouse. Wang et al. have studied the brain accumulation of *f*-MWNTs using single photon emission computed tomography/computed tomography (SPECT/CT) imaging and autoradiography. The SPECT/CT imaging showed that radiolabeled *f*-MWNTs accumulated in the brain cavity especially at the mid-brain region at the early time points between 0.5 and 4 h (Fig. 12). Autoradiography imaging also confirmed that *f*-MWNTs accumulated in the mid regions of the brain at the studied time points (Wang et al., 2016).

Brain accumulation of *f*-MWNTs following i.v. administration was also measured using gamma counting (Kafa et al., 2015). Fig. 13A shows the biodistribution of *f*-MWNTs in mouse from 5 min to 24 h. *f*-MWNTs accumulated mostly in lungs, liver, and spleen. However, there was a detectable brain accumulation in the early time points as well. The highest level was achieved after 5 min of injection reaching 1.1 ± 0.3%ID/g (Fig. 13B). The detected levels of *f*-MWNTs gradually decreased over time until reaching the lowest levels after 24 h. The capillary depletion method was carried out to examine the distribution of *f*-MWNTs in the capillaries and brain parenchyma. Fig. 13C shows that significant amount *f*-MWNTs were detected in the capillary fraction compared to the parenchyma. However, the detected signal in the capillaries decreased over time coupled with

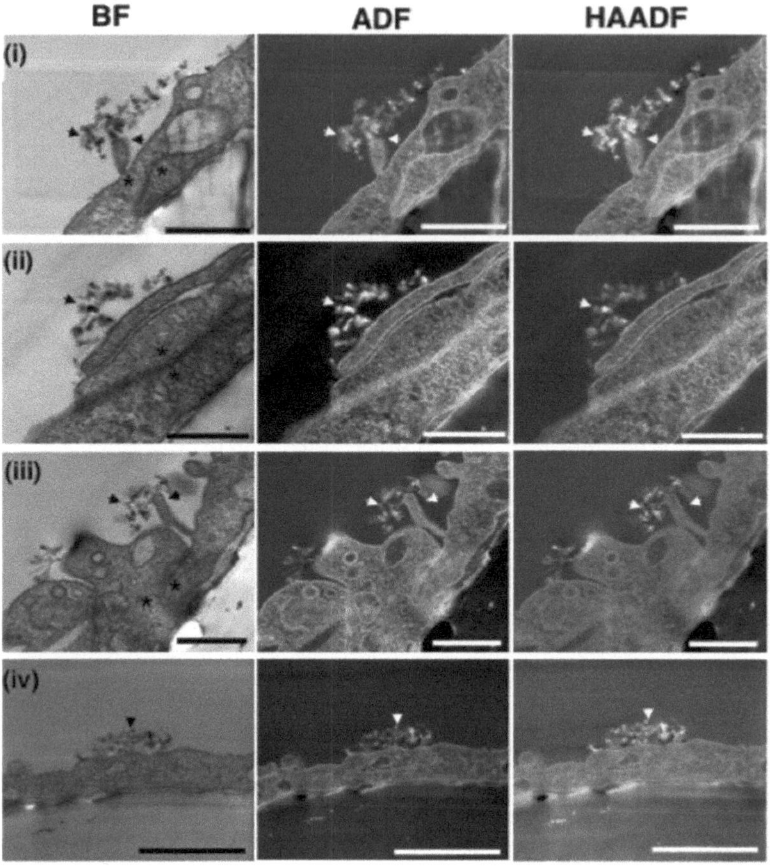

Fig. 11 Inhibition of *f*-MWNTs uptake into endothelial cells following the incubation at 4°C. The transport experiment was carried out at 4°C for 4 h to evaluate the energy dependency of *f*-MWNTs uptake into the endothelial cells. Low voltage STEM images show the accumulation of *f*-MWNTs outside the endothelial cells after 4 h of incubation, but no evidence of vesicular uptake was observed. *f*-MWNTs were mostly observed around endothelial extensions (*arrow heads*) in the vicinity of a tight-junctional zone (*asterisks*) (i–iii). Error bars from top to bottom, 500 nm, 400 nm, and 500 nm, 1 μm. *Reproduced from Kafa, H., Wang, J. T., Rubio, N., Venner, K., Anderson, G., Pach, E., … Al-Jamal, K. T. (2015). The interaction of carbon nanotubes with an in vitro blood–brain barrier model and mouse brain in vivo.* Biomaterials, *53, 437–452. doi: 10.1016/j.biomaterials.2015.02.083.*

a slight increase in the parenchyma fraction indicating brain retention behavior (Kafa et al., 2015), which is important to maximize the *f*-MWNTs bioavailability in the brain to allow for subsequent drug delivery or hyperthermia application.

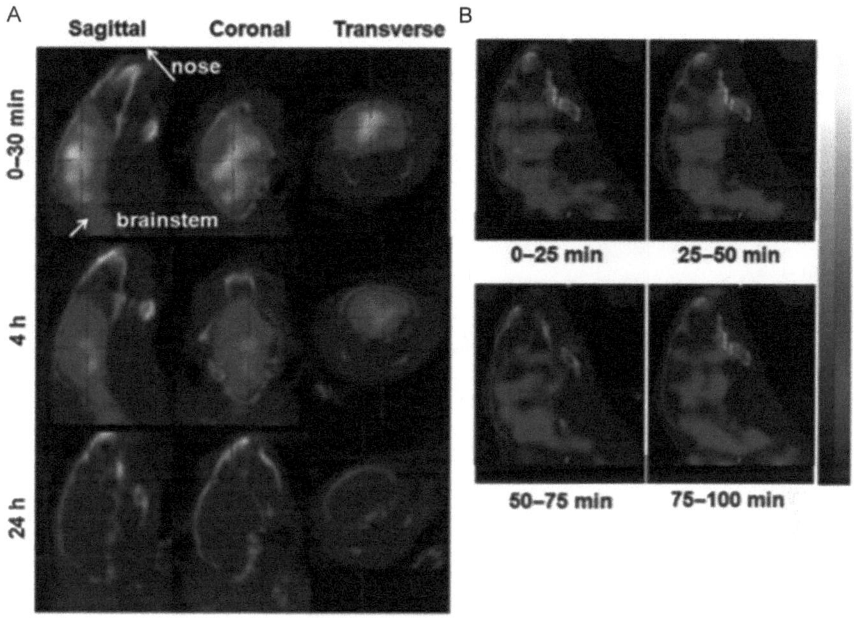

Fig. 12 In vivo SPECT/CT imaging of mouse brains. (A) Sagittal, coronal, and transverse views from 3D reconstructed SPECT/CT images of mouse brains after i.v. injection of [^{111}In]MWNT-Fab′ (50 μg, 5–7 MBq) at 0–30 min, 4 h, and 24 h postinjection. (B) Dynamic SPECT/CT imaging of mouse brain (sagittal views) within 100 min after injection of [^{111}In]MWNT-Fab′. *Reproduced from Wang, J. T., Rubio, N., Kafa, H., Venturelli, E., Fabbro, C., Menard-Moyon, C., … Al-Jamal, K. T. (2016). Kinetics of functionalized carbon nanotube distribution in mouse brain after systemic injection: Spatial to ultra-structural analyses.* Journal of Controlled Release, 224, *22–32. doi: 10.1016/j.jconrel.2015.12.039.*

8.3 The Effect of ANG Targeting on Brain Accumulation of *f*-MWNTs

ANG peptide has previously been used to target therapeutic molecules such as paclitaxel (Demeule et al., 2008) and neurotensin (Demeule et al., 2014) to mouse brain with promising results. Additionally, ANG has been used as a targeting ligand via conjugation to nanocarriers such as PE-PEG micelles (Shao et al., 2010), PEG-modified polyamidoamine dendrimers (Ke et al., 2009), gold nanoparticles (Ruan et al., 2015), and polymersomes (Tian et al., 2015) for brain delivery. Based on previous findings, we have studied the effect of conjugating ANG to *f*-MWNTs to enhance brain uptake and parenchyma accumulation.

Quantitative uptake of the ANG-targeted *f*-MWNTs (*f*-MWNTs-ANG) and nontargeted *f*-MWNTs into the brain was studied in the whole brain

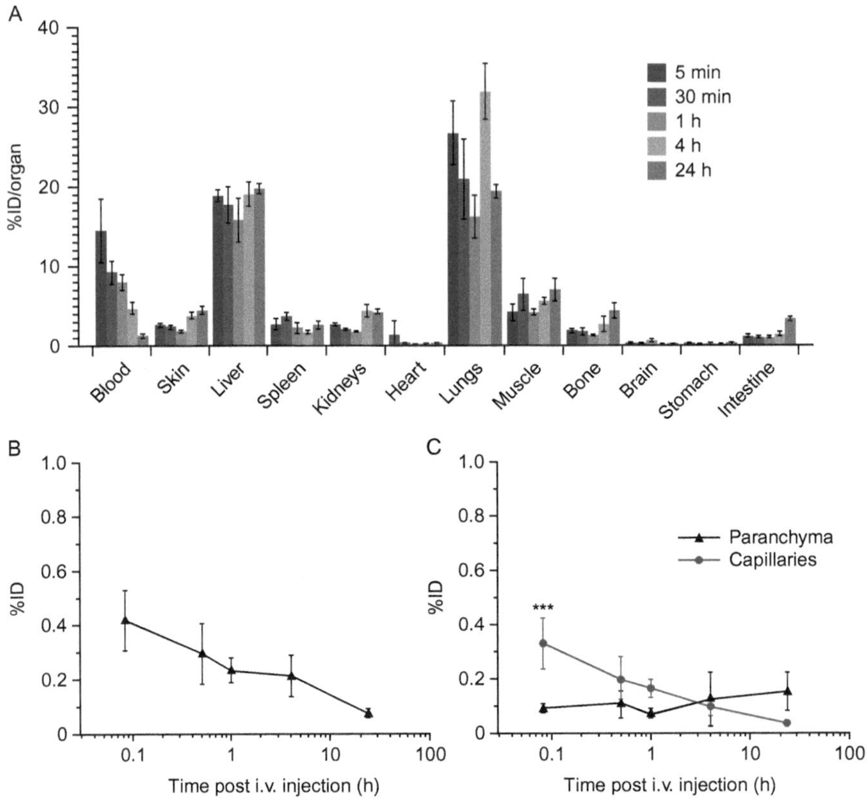

Fig. 13 Organ biodistribution and brain uptake of *f*-MWNTs into mouse following i.v. administration. C57/Bl6 mice were injected with *f*-MWNTs (50 μg/mL, 0.5 MBq) via the tail vein. At each time point (5 min, 30 min, 1 h, 4 h, and 24 h) whole body perfusion with heparinized saline was performed and major organs were harvested for quantitative measurements of radioactivity by γ-scintigraphy. (A) The accumulation of *f*-MWNTs in the major organs after each time point. (B) Brain uptake of *f*-MWNTs over time showing the highest brain accumulation after 5 min. (C) The capillary depletion profile showing the sustained accumulation of *f*-MWNTs in brain parenchyma overtime compared to a gradual reduction in capillary fraction. The data are presented as % injected dose per organ. Results are expressed as mean ± SD, $n = 3$ (*** $p < 0.001$). *Reproduced from Kafa, H., Wang, J. T., Rubio, N., Venner, K., Anderson, G., Pach, E., … Al-Jamal, K. T. (2015). The interaction of carbon nanotubes with an in vitro blood–brain barrier model and mouse brain in vivo.* Biomaterials, *53, 437–452. doi: 10.1016/j.biomaterials.2015.02.083.*

using gamma counting. The highest whole brain uptake of *f*-MWNTs was measured by 5 min postinjection (Fig. 14A). The targeted *f*-MWNTs-ANG achieved significantly higher whole brain uptake (2.0 ± 0.5%ID/g) than non-targeted *f*-MWNTs (1.1 ± 0.3%ID/g) by 5 min postinjection (Kafa et al., 2016).

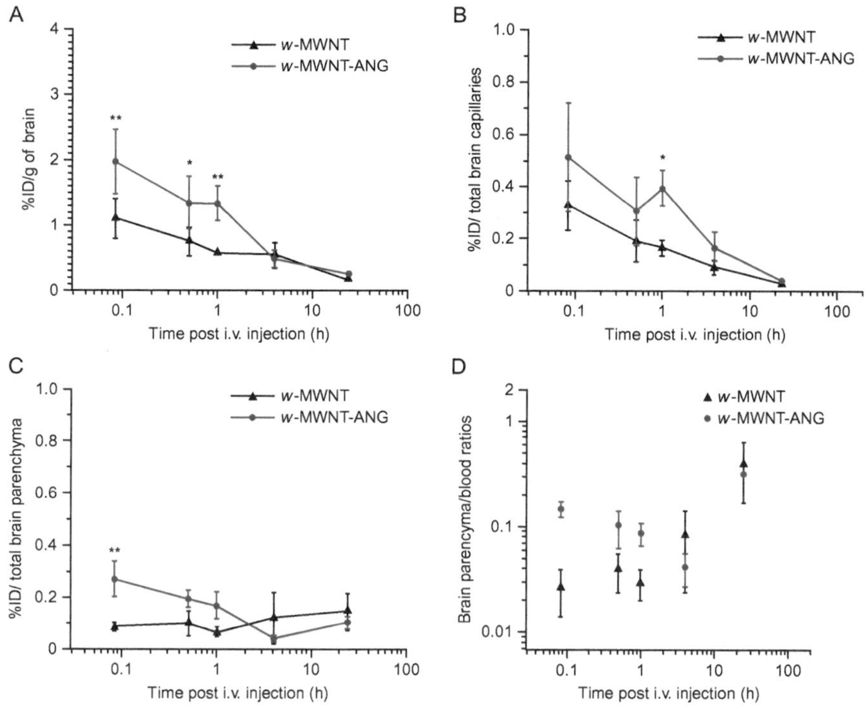

Fig. 14 Uptake of radiolabeled *f*-MWNTs into mouse brain following i.v. administration and capillary depletion. Overall brain uptake of *f*-MWNTs is shown in (A) expressed as % injected dose per gram of brain (%ID/g). The radioactivity of *f*-MWNTs in brain capillaries (B) and parenchyma (C) was measured after capillary depletion with the data presented as %ID per total brain capillaries or parenchyma. Their accumulation in blood (%ID/g) vs parenchyma (%ID/g) is shown in (D). The weight of parenchyma was considered as the weight of the whole brain in this case. *f*-MWNTs were injected (50 μg, 0.5 MBq) into the tail vein of C57/Bl6 mice. Whole body perfusion with heparinized saline was performed to remove residual *f*-MWNTs from circulation. The total brain accumulation was measured by gamma counting followed by capillary depletion at 5 min, 30 min, 1 h, 4 h, and 24 h. Values are expressed as mean ± SD ($n = 3$) and analyzed using One-way ANOVA test. *, $p < 0.05$; **, $p < 0.01$. *Reproduced from Kafa, H., Wang, J. T., Rubio, N., Klippstein, R., Costa, P. M., Hassan, H. A., … Al-Jamal, K. T. (2016). Translocation of LRP1 targeted carbon nanotubes of different diameters across the blood–brain barrier in vitro and in vivo. Journal of Controlled Release, 225, 217–229. doi: 10.1016/j.jconrel.2016.01.031.*

Capillary depletion showed that significant amounts of *f*-MWNTs were detected in the capillaries by 5 min, and the difference between the targeted and nontargeted *f*-MWNTs was significant after 1 h (Fig. 14B). In the parenchyma fraction, *f*-MWNT-ANG showed significantly greater parenchyma accumulation than *f*-MWNTs by 5 min postinjection

(Fig. 14C). The results suggested that the greater overall brain uptake for *f*-MWNTs-ANG than *f*-MWNTs was also associated with enhanced uptake in the brain parenchyma, especially given that levels of *f*-MWNTs associated with capillaries were similar for both *f*-MWNTs types. Additionally, *f*-MWNT-ANG showed a higher parenchyma to blood ratio than *f*-MWNTs by 5 min postinjection (Fig. 14D). This study provided quantitative evidence on the accumulation of *f*-MWNTs in brain parenchyma. Additionally, the data show the positive effect of ANG targeting on increasing *f*-MWNTs-ANG uptake into the brain (Kafa et al., 2016).

Furthermore, all of the experiments discussed earlier were carried out in healthy brain with intact BBB. Under pathological conditions, the brain suffers a leaky BBB with increased paracellular permeability (Abbott, Ronnback, & Hansson, 2006), allowing more molecules to access the brain. Therefore, the brain accumulation of *f*-MWNTs in brain glioma could be even higher under pathological conditions than in healthy brain. Kafa et al. have studied the accumulation of *f*-MWNT-ANG in the brain of a glioma mouse model. Fig. 15 shows the brain accumulation of *f*-MWNTs at 24 h following a tail-veil injection. The data showed that both nontargeted

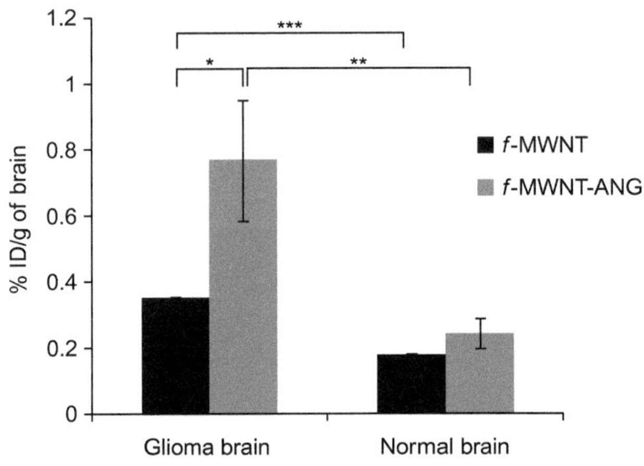

Fig. 15 Brain uptake of *f*-MWNTs in glioma-bearing mice following i.v. administration. *f*-MWNT and *f*-MWNT-ANG were intravenously injected (50 μg, 0.5 MBq) into normal or GL261 glioma-bearing C57/Bl6 mice. Whole body perfusion with heparinized saline was performed to remove residual *f*-MWNTs from circulation. The total brain accumulation was measured by gamma counting at 24 h postinjection. *Reproduced from Kafa, H., Wang, J. T., Rubio, N., Klippstein, R., Costa, P. M., Hassan, H. A., … Al-Jamal, K. T. (2016). Translocation of LRP1 targeted carbon nanotubes of different diameters across the blood–brain barrier in vitro and in vivo.* Journal of Controlled Release, 225, *217–229. doi: 10.1016/j.jconrel.2016.01.031.*

f-MWNT and *f*-MWNT-ANG achieved significantly higher accumulation in glioma brain compared to the healthy brain. This was possibly due to the leakiness of the BBB, and also the unregulated expression of LRP1 receptor in glioma brain, which facilitated the interaction between *f*-MWNT-ANG and the brain endothelium.

The high parenchyma accumulation of *f*-MWNTs coupled with their selective high glioma accumulation is very promising for future drug delivery applications. However, the accumulation of *f*-MWNTs in RES needs to be reduced by using polymer coating *f*-MWNT to allow even higher blood circulation time and increase accumulation in the brain.

9. TOXICITY STUDIES OF CNTs IN BRAIN TISSUES

Toxic side effects for brain delivery systems could be detrimental. The safety of CNTs in brain tissue must, therefore, be established before using them for brain delivery applications. Bardi et al. have demonstrated that pluronic F127-coated MWNTs (MWNT:F127) resulted in no apparent cytotoxicity to neurons in mouse brain cortex following stereotactic microinjection. Histological examination of the brain tissue showed that the observed damage near the injection site was only the result of the mechanical insult of the injection, and no further damage was observed after 18 days (Bardi et al., 2009). Further in vitro testing using primary mouse cortical cultures showed that the incubation of pluronic alone led to high cytotoxicity as measured by the MTT assay. MWNTs:F127, on the other hand, resulted in a significantly reduced cell death compared to pluronic alone. MWNTs in this instance reduced the toxicity of pluronic on primary cultures (Bardi et al., 2009).

The functionalization strategy also contributes to the cytotoxic outcome of CNTs in the brain. We have previously reported that MWNTs functionalized using 1,3-dipolar cycloaddition (MWNT-NH_3^+) resulted in a lower inflammatory response compared to oxidized MWNTs (ox-MWNTs-NH_3^+) following cortical stereotactic injection in mouse brain. Accumulation of ox-MWNTs-NH_3^+ in astrocytes, microglia, and neurons led to a significantly higher release of the proinflammatory cytokines TNF-α and IL-1β compared to MWNTs-NH_3^+, highlighting the impact of functionalization strategy for biomedical applications. A recent follow-up study revealed the role of microglia in the cytotoxic effects of ox-MWNTs brain tissue. Four types of ox-MWNTs were tested in neuronal and mixed cultures isolated either from the rat frontal cortex or striatum. Exposure of ox-MWNTs to neuronal cultures did not lead to detectable cytotoxicity as measured by the modified LDH assay. However, ox-MWNTs resulted in a significant cytotoxicity in mixed glial cultures isolated from striatum, but not from the frontal cortex. The toxicity was, therefore, attributed to the higher density of microglia in the striatum-derived mixed culture compared to the higher amount of astrocytes in the frontal cortex. The study implicated the role of microglia in CNTs-induced cytotoxicity in brain tissue (Bussy et al., 2015).

10. BIODEGRADATION OF CNTs

The high stability and the inert nature of CNTs wall structure raise concerns over their applicability in biological systems. However, biodegradation of CNTs was observed when incubating SWNTs with human myeloperoxidase (MPO) and hydrogen peroxide (H_2O_2). The presence of the MPO enzyme with H_2O_2 resulted in a loss of SWNTs structure as evident by the disappearance of the characteristic D and G bands in Raman spectroscopy (Kagan et al., 2010). The biodegradation mechanism was attributed to the generation of reactive hypochlorite radical intermediate that oxidizes parts of the wall structure of CNTs resulting in eventual degradation (Vlasova, Vakhrusheva, Sokolov, Kostevich, & Ragimov, 2011). Even PEG-coated SWNTs were subject to enzymatic degradation in an in vitro degradation model using recombinant MPO and ex vivo using activated neutrophils. Oxidized and PEG-coated SWNTs (SWNTs-PEG) were incubated with MPO, H_2O_2, and NaCl and biodegradation was observed after 7 days using Raman spectroscopy. Further testing was carried out in an ex vivo conditions using human neutrophils, which secrete MPO along with other proteases. The ex vivo study showed a highly efficient degradation event after only 8 h. The biodegradation was proposed to occur in a two-step process; stripping the PEG coating using the secreted proteases and then SWNTs degradation via surface defects (Bhattacharya et al., 2014). Interestingly, a similar defunctionalization mechanism was previously reported in vivo where the PEG coating on SWNTs-PEG was removed in the liver following intravenous injection. However, no explanation was provided on the mechanism of PEG stripping from the SWNTs surface (Yang et al., 2009). All studies regarding the degradation of CNTs agree that MPO is the main factor in facilitating the biodegradation of the CNTs. In fact, MPO mouse knockouts showed increased inflammation and lower clearance rate of SWNTs following pharyngeal aspiration compared to wild type (Shvedova et al., 2012). Nunes et al. provided further electron microscopic evidence on the degradation of *f*-MWNTs following stereotactic administration into mouse brain cortex. The loss of cylindrical structure of *f*-MWNTs was observed following internalization into microglia (Nunes et al., 2012). It was hypothesized that the oxidative environment of the microglia could be involved in the observed degradation, but further studies are needed to prove the hypothesis.

CNTs hold great potential for future applications in the brain. Their structural and electronic properties provide several advantages for neuronal tissue repair and neuroprotection. Additionally, their optical properties facilitate their use in cancer elimination using hyperthermia. Above all, the high surface area of CNTs and their intrinsic ability to cross the BBB, provide a promising platform for targeted drug delivery to the brain to tackle brain gliomas and neurodegenerative disorders. Additionally, the lack of apparent toxicity coupled with the observed degradability of CNTs in the presence of MPO enzymes are encouraging observations for future applications of CNTs in the brain. However, further studies are required to determine the long-term toxicological outcomes of CNT, and their fate within the living systems before translating their uses into the clinic.

REFERENCES

Abbott, N. J., Ronnback, L., & Hansson, E. (2006). Astrocyte-endothelial interactions at the blood–brain barrier. *Nature Reviews. Neuroscience*, *7*(1), 41–53. http://dx.doi.org/10.1038/Nrn1824.

Al-Jamal, K. T., Gherardini, L., Bardi, G., Nunes, A., Guo, C., Bussy, C., ... Pizzorusso, T. (2011). Functional motor recovery from brain ischemic insult by carbon nanotube-mediated siRNA silencing. *Proceedings of the National Academy of Sciences of the United States of America*, *108*(27), 10952–10957. http://dx.doi.org/10.1073/pnas.1100930108.

Al-Jamal, K. T., Nerl, H., Muller, K. H., Ali-Boucetta, H., Li, S., Haynes, P. D., ... Porter, A. E. (2011). Cellular uptake mechanisms of functionalised multi-walled carbon nanotubes by 3D electron tomography imaging. *Nanoscale*, *3*(6), 2627–2635. http://dx.doi.org/10.1039/c1nr10080g.

Bardi, G., Nunes, A., Gherardini, L., Bates, K., Al-Jamal, K. T., Gaillard, C., ... Kostarelos, K. (2013). Functionalized carbon nanotubes in the brain: Cellular internalization and neuroinflammatory responses. *PLoS One*, *8*(11), e80964. http://dx.doi.org/10.1371/journal.pone.0080964.

Bardi, G., Tognini, P., Ciofani, G., Raffa, V., Costa, M., & Pizzorusso, T. (2009). Pluronic-coated carbon nanotubes do not induce degeneration of cortical neurons in vivo and in vitro. *Nanomedicine*, *5*(1), 96–104. http://dx.doi.org/10.1016/j.nano.2008.06.008.

Bhattacharya, K., Sacchetti, C., El-Sayed, R., Fornara, A., Kotchey, G. P., Gaugler, J. A., ... Fadeel, B. (2014). Enzymatic 'stripping' and degradation of PEGylated carbon nanotubes. *Nanoscale*, *6*(24), 14686–14690. http://dx.doi.org/10.1039/C4nr03604b.

Bhirde, A. A., Patel, V., Gavard, J., Zhang, G., Sousa, A. A., Masedunskas, A., … Rusling, J. F. (2009). Targeted killing of cancer cells in vivo and in vitro with EGF-directed carbon nanotube-based drug delivery. *ACS Nano*, *3*(2), 307–316. http://dx.doi.org/10.1021/nn800551s.

Bianco, A., Kostarelos, K., & Prato, M. (2005). Applications of carbon nanotubes in drug delivery. *Current Opinion in Chemical Biology*, *9*(6), 674–679. http://dx.doi.org/10.1016/j.cbpa.2005.10.006.

Bussy, C., Al-Jamal, K. T., Boczkowski, J., Lanone, S., Prato, M., Bianco, A., & Kostarelos, K. (2015). Microglia determine brain region-specific neurotoxic responses to chemically functionalized carbon nanotubes. *ACS Nano*, *9*(8), 7815–7830. http://dx.doi.org/10.1021/acsnano.5b02358.

Cellot, G., Cilia, E., Cipollone, S., Rancic, V., Sucapane, A., Giordani, S., … Ballerini, L. (2009). Carbon nanotubes might improve neuronal performance by favouring electrical shortcuts. *Nature Nanotechnology*, *4*(2), 126–133. http://dx.doi.org/10.1038/Nnano.2008.374.

Chen, Y., Haddon, R. C., Fang, S., Rao, A. M., Eklund, P. C., Lee, W. H., … Smalley, R. E. (1998). Chemical attachment of organic functional groups to single-walled carbon nanotube material. *Journal of Materials Research*, *13*(09), 2423–2431.

Chen, T., Yang, J., Zhang, H., Ren, G., Yang, Z., & Zhang, T. (2014). Multi-walled carbon nanotube inhibits CA1 glutamatergic synaptic transmission in rat's hippocampal slices. *Toxicology Letters*, *229*(3), 423–429. http://dx.doi.org/10.1016/j.toxlet.2014.06.036.

Choi, J. H., Nguyen, F. T., Barone, P. W., Heller, D. A., Moll, A. E., Patel, D., … Strano, M. S. (2007). Multimodal biomedical imaging with asymmetric single-walled carbon nanotube/iron oxide nanoparticle complexes. *Nano Letters*, 7(4), 861–867. http://dx.doi.org/10.1021/nl062306v.

de Faria, P. C. B., dos Santos, L. I., Coelho, J. P., Ribeiro, H. B., Pimenta, M. A., Ladeira, L. O., … Gazzinelli, R. T. (2014). Oxidized multiwalled carbon nanotubes as antigen delivery system to promote superior CD8(+) T cell response and protection against cancer. *Nano Letters*, *14*(9), 5458–5470. http://dx.doi.org/10.1021/Nl502911a.

Dementev, N., Osswald, S., Gogotsi, Y., & Borguet, E. (2009). Purification of carbon nanotubes by dynamic oxidation in air. *Journal of Materials Chemistry*, *19*(42), 7904–7908. http://dx.doi.org/10.1039/B910217e.

Demeule, M., Beaudet, N., Regina, A., Besserer-Offroy, E., Murza, A., Tetreault, P., … Sarret, P. (2014). Conjugation of a brain-penetrant peptide with neurotensin provides antinociceptive properties. *Journal of Clinical Investigation*, *124*(3), 1199–1213. http://dx.doi.org/10.1172/JCI70647.

Demeule, M., Regina, A., Che, C., Poirier, J., Nguyen, T., Gabathuler, R., & Beliveau, R. (2008). Identification and design of peptides as a new drug delivery system for the brain. *Journal of Pharmacology and Experimental Therapeutics*, *32*(3), 1064–1072. http://dx.doi.org/10.1124/Jpet.107.131318.

Ellis-Behnke, R. (2007). Nano neurology and the four P's of central nervous system regeneration: Preserve, permit, promote, plasticity. *Medical Clinics of North America*, *91*(5), 937–962. http://dx.doi.org/10.1016/j.mcna.2007.04.005.

Fan, H., Zhang, I., Chen, X., Zhang, L., Wang, H., Da Fonseca, A., … Badie, B. (2012). Intracerebral CpG immunotherapy with carbon nanotubes abrogates growth of subcutaneous melanomas in mice. *Clinical Cancer Research*, *18*(20), 5628–5638. http://dx.doi.org/10.1158/1078-0432.CCR-12-1911.

Fujigaya, T., & Nakashima, N. (2015). Non-covalent polymer wrapping of carbon nanotubes and the role of wrapped polymers as functional dispersants. *Science and Technology of Advanced Materials*, *16*(2), 024802, http://dx.doi.org/10.1088/1468-6996/16/2/024802.

Gaillard, C., Duval, M., Dumortier, H., & Bianco, A. (2011). Carbon nanotube-coupled cell adhesion peptides are non-immunogenic: A promising step toward new biomedical devices. *Journal of Peptide Science*, *17*(2), 139–142. http://dx.doi.org/10.1002/psc.1290.

Georgakilas, V., Kordatos, K., Prato, M., Guldi, D. M., Holzinger, M., & Hirsch, A. (2002). Organic functionalization of carbon nanotubes. *Journal of the American Chemical Society*, *124*(5), 760–761.

Georgakilas, V., Tagmatarchis, N., Pantarotto, D., Bianco, A., Briand, J. P., & Prato, M. (2002). Amino acid functionalisation of water soluble carbon nanotubes. *Chemical Communications (Cambridge, England)*, (24), 3050–3051. http://dx.doi.org/10.1039/B209843A.

Ghosh, S., Dutta, S., Gomes, E., Carroll, D., D'Agostino, R., Jr., Olson, J., … Gmeiner, W. H. (2009). Increased heating efficiency and selective thermal ablation of malignant tissue with DNA-encased multiwalled carbon nanotubes. *ACS Nano*, *3*(9), 2667–2673. http://dx.doi.org/10.1021/nn900368b.

Gilmore, J. L., Yi, X., Quan, L., & Kabanov, A. V. (2008). Novel nanomaterials for clinical neuroscience. *Journal of Neuroimmune Pharmacology*, *3*(2), 83–94. http://dx.doi.org/10.1007/s11481-007-9099-6.

Hernandez-Ferrer, J., Perez-Bruzon, R. N., Azanza, M. J., Gonzalez, M., Del Moral, R., Anson-Casaos, A., … Martinez, M. T. (2014). Study of neuron survival on polypyrrole-embedded single-walled carbon nanotube substrates for long-term growth conditions. *Journal of Biomedical Materials Research. Part A*, *102*(12), 4443–4454. http://dx.doi.org/10.1002/jbm.a.35110.

Higgins, P., Dawson, J., & Walters, M. (2011). Nanomedicine: Nanotubes reduce stroke damage. *Nature Nanotechnology*, *6*(2), 83–84. http://dx.doi.org/10.1038/nnano.2011.5.

Hildebrandt, B., Wust, P., Ahlers, O., Dieing, A., Sreenivasa, G., Kerner, T., … Riess, H. (2002). The cellular and molecular basis of hyperthermia. *Critical Reviews in Oncology/Hematology*, *43*(1), 33–56.

Holzinger, M., Vostrowsky, O., Hirsch, A., Hennrich, F., Kappes, M., Weiss, R., & Jellen, F. (2001). Sidewall functionalization of carbon nanotubes. This work was supported by the European Union under the 5th Framework Research Training Network 1999, HPRNT 1999-00011 FUNCARS. *Angewandte Chemie International Edition in English*, *40*(21), 4002–4005.

Hu, H., Ni, Y. C., Montana, V., Haddon, R. C., & Parpura, V. (2004). Chemically functionalized carbon nanotubes as substrates for neuronal growth. *Nano Letters*, *4*(3), 507–511. http://dx.doi.org/10.1021/nl035193d.

Iacob, G., & Dinca, E. B. (2009). Current data and strategy in glioblastoma multiforme. *Journal of Medicine and Life*, *2*(4), 386–393.

Iverson, N. M., Barone, P. W., Shandell, M., Trudel, L. J., Sen, S., Sen, F., … Strano, M. S. (2013). In vivo biosensing via tissue-localizable near-infrared-fluorescent single-walled carbon nanotubes. *Nature Nanotechnology*, *8*(11), 873–880. http://dx.doi.org/10.1038/nnano.2013.222.

Jared, M. A., Keith, B. H., Yuri, M., Cristina, H., Sean, P., Lawrence, B. A., & Lon, J. W. (2006). Functionalization of individual ultra-short single-walled carbon nanotubes. *Nanotechnology*, *17*(20), 5033.

Kafa, H., Wang, J. T., Rubio, N., Klippstein, R., Costa, P. M., Hassan, H. A., … Al-Jamal, K. T. (2016). Translocation of LRP1 targeted carbon nanotubes of different diameters across the blood–brain barrier in vitro and in vivo. *Journal of Controlled Release*, *225*, 217–229. http://dx.doi.org/10.1016/j.jconrel.2016.01.031.

Kafa, H., Wang, J. T., Rubio, N., Venner, K., Anderson, G., Pach, E., … Al-Jamal, K. T. (2015). The interaction of carbon nanotubes with an in vitro blood–brain barrier model and mouse brain in vivo. *Biomaterials*, *53*, 437–452. http://dx.doi.org/10.1016/j.biomaterials.2015.02.083.

Kagan, V. E., Konduru, N. V., Feng, W., Allen, B. L., Conroy, J., Volkov, Y., ... Shvedova, A. A. (2010). Carbon nanotubes degraded by neutrophil myeloperoxidase induce less pulmonary inflammation. *Nature Nanotechnology*, *5*(5), 354–359. http://dx.doi.org/10.1038/nnano.2010.44.

Kam, N. W., Jan, E., & Kotov, N. A. (2009). Electrical stimulation of neural stem cells mediated by humanized carbon nanotube composite made with extracellular matrix protein. *Nano Letters*, *9*(1), 273–278. http://dx.doi.org/10.1021/nl802859a.

Kam, N. W. S., Jessop, T. C., Wender, P. A., & Dai, H. J. (2004). Nanotube molecular transporters: Internalization of carbon nanotube-protein conjugates into mammalian cells. *Journal of the American Chemical Society*, *126*(22), 6850–6851. http://dx.doi.org/10.1021/Ja0486059.

Ke, W. L., Shao, K., Huang, R. Q., Han, L., Liu, Y., Li, J. F., ... Jiang, C. (2009). Gene delivery targeted to the brain using an Angiopep-conjugated polyethyleneglycol-modified polyamidoamine dendrimer. *Biomaterials*, *30*(36), 6976–6985. http://dx.doi.org/10.1016/J.Biomaterials.2009.08.049.

Khabashesku, V. N., Billups, W. E., & Margrave, J. L. (2002). Fluorination of single-wall carbon nanotubes and subsequent derivatization reactions. *Accounts of Chemical Research*, *35*(12), 1087–1095.

Khraiche, M. L., Jackson, N., & Muthuswamy, J. (2009). Early onset of electrical activity in developing neurons cultured on carbon nanotube immobilized microelectrodes. *Conference Proceedings Annual International Conference of the IEEE Engineering in Medicine and Biology Society*, *2009*, 777–780. http://dx.doi.org/10.1109/IEMBS.2009.5333590.

Klumpp, C., Kostarelos, K., Prato, M., & Bianco, A. (2006). Functionalized carbon nanotubes as emerging nanovectors for the delivery of therapeutics. *Biochimica Et Biophysica Acta-Biomembranes*, *1758*(3), 404–412. http://dx.doi.org/10.1016/J.Bbamem.2005.10.008.

Kotagiri, N., & Kim, J. W. (2014). Stealth nanotubes: Strategies of shielding carbon nanotubes to evade opsonization and improve biodistribution. *International Journal of Nanomedicine*, *9*(Suppl. 1), 85–105. http://dx.doi.org/10.2147/IJN.S51854.

Lacerda, L., Ali-Boucetta, H., Herrero, M. A., Pastorin, G., Bianco, A., Prato, M., & Kostarelos, K. (2008). Tissue histology and physiology following intravenous administration of different types of functionalized multiwalled carbon nanotubes. *Nanomedicine (London, England)*, *3*(2), 149–161. http://dx.doi.org/10.2217/17435889.3.2.149.

Lee, H. J., Park, J., Yoon, O. J., Kim, H. W., Lee do, Y., Kim do, H., ... Kim, S. S. (2011). Amine-modified single-walled carbon nanotubes protect neurons from injury in a rat stroke model. *Nature Nanotechnology*, *6*(2), 121–125. http://dx.doi.org/10.1038/nnano.2010.281.

Li, X. M., Wang, L., Fan, Y. B., Feng, Q. L., & Cui, F. Z. (2012). Biocompatibility and toxicity of nanoparticles and nanotubes. *Journal of Nanomaterials*. *2012*. http://dx.doi.org/10.1155/2012/548389. Article ID 548389, 19 pp.

Liu, Z., Chen, K., Davis, C., Sherlock, S., Cao, Q., Chen, X., & Dai, H. (2008). Drug delivery with carbon nanotubes for in vivo cancer treatment. *Cancer Research*, *68*(16), 6652–6660. http://dx.doi.org/10.1158/0008-5472.CAN-08-1468.

Liu, Z., Davis, C., Cai, W., He, L., Chen, X., & Dai, H. (2008). Circulation and long-term fate of functionalized, biocompatible single-walled carbon nanotubes in mice probed by Raman spectroscopy. *Proceedings of the National Academy of Sciences of the United States of America*, *105*(5), 1410–1415. http://dx.doi.org/10.1073/pnas.0707654105.

Liu, J., Rinzler, A. G., Dai, H., Hafner, J. H., Bradley, R. K., Boul, P. J., ... Smalley, R. E. (1998). Fullerene pipes. *Science*, *280*(5367), 1253–1256.

Liu, Z., Sun, X., Nakayama-Ratchford, N., & Dai, H. (2007). Supramolecular chemistry on water-soluble carbon nanotubes for drug loading and delivery. *ACS Nano*, *1*(1), 50–56. http://dx.doi.org/10.1021/nn700040t.

Markovic, Z. M., Harhaji-Trajkovic, L. M., Todorovic-Markovic, B. M., Kepic, D. P., Arsikin, K. M., Jovanovic, S. P., … Trajkovic, V. S. (2011). In vitro comparison of the photothermal anticancer activity of graphene nanoparticles and carbon nanotubes. *Biomaterials*, *32*(4), 1121–1129. http://dx.doi.org/10.1016/j.biomaterials.2010.10.030.

Matsumoto, K., Sato, C., Naka, Y., Kitazawa, A., Whitby, R. L., & Shimizu, N. (2007). Neurite outgrowths of neurons with neurotrophin-coated carbon nanotubes. *Journal of Bioscience and Bioengineering*, *103*(3), 216–220. http://dx.doi.org/10.1263/jbb.103.216.

Mattson, M. P., Haddon, R. C., & Rao, A. M. (2000). Molecular functionalization of carbon nanotubes and use as substrates for neuronal growth. *Journal of Molecular Neuroscience*, *14*(3), 175–182. http://dx.doi.org/10.1385/Jmn:14:3:175.

Medepalli, K., Alphenaar, B., Raj, A., & Sethu, P. (2011). Evaluation of the direct and indirect response of blood leukocytes to carbon nanotubes (CNTs). *Nanomedicine*, 7(6), 983–991. http://dx.doi.org/10.1016/j.nano.2011.04.002.

Meng, L., Jiang, A., Chen, R., Li, C. Z., Wang, L., Qu, Y., … Chen, C. (2013). Inhibitory effects of multiwall carbon nanotubes with high iron impurity on viability and neuronal differentiation in cultured PC12 cells. *Toxicology*, *313*(1), 49–58. http://dx.doi.org/10.1016/j.tox.2012.11.011.

Moustakas, A., & Kreisl, T. N. (2010). New treatment options in the management of glioblastoma multiforme: A focus on bevacizumab. *Onco Targets Therapy*, *3*, 27–38.

Nunes, A., Bussy, C., Gherardini, L., Meneghetti, M., Herrero, M. A., Bianco, A., … Kostarelos, K. (2012). In vivo degradation of functionalized carbon nanotubes after stereotactic administration in the brain cortex. *Nanomedicine (London, England)*, 7(10), 1485–1494. http://dx.doi.org/10.2217/nnm.12.33.

Pantarotto, D., Briand, J. P., Prato, M., & Bianco, A. (2004). Translocation of bioactive peptides across cell membranes by carbon nanotubes. *Chemical Communications (Cambridge, England)*, (1), 16–17. http://dx.doi.org/10.1039/b311254c.

Piroth, M. D., Gagel, B., Pinkawa, M., Stanzel, S., Asadpour, B., & Eble, M. J. (2007). Postoperative radiotherapy of glioblastoma multiforme: Analysis and critical assessment of different treatment strategies and predictive factors. *Strahlentherapie und Onkologie*, *183*(12), 695–702. http://dx.doi.org/10.1007/s00066-007-1739-5.

Prencipe, G., Tabakman, S. M., Welsher, K., Liu, Z., Goodwin, A. P., Zhang, L., … Dai, H. (2009). PEG branched polymer for functionalization of nanomaterials with ultralong blood circulation. *Journal of the American Chemical Society*, *131*(13), 4783–4787. http://dx.doi.org/10.1021/ja809086q.

Qiao, Y., Li, C. M., Bao, S. J., & Bao, Q. L. (2007). Carbon nanotube/polyaniline composite as anode material for microbial fuel cells. *Journal of Power Sources*, *170*(1), 79–84. http://dx.doi.org/10.1016/j.jpowsour.2007.03.048.

Ren, J., Shen, S., Wang, D., Xi, Z., Guo, L., Pang, Z., … Jiang, X. (2012). The targeted delivery of anticancer drugs to brain glioma by PEGylated oxidized multi-walled carbon nanotubes modified with angiopep-2. *Biomaterials*, *33*(11), 3324–3333. http://dx.doi.org/10.1016/j.biomaterials.2012.01.025.

Ruan, S., Yuan, M., Zhang, L., Hu, G., Chen, J., Cun, X., … Gao, H. (2015). Tumor microenvironment sensitive doxorubicin delivery and release to glioma using angiopep-2 decorated gold nanoparticles. *Biomaterials*, *37*, 425–435. http://dx.doi.org/10.1016/j.biomaterials.2014.10.007.

Rubio, N., Hirvonen, L. M., Chong, E. Z., Wang, J. T., Bourgognon, M., Kafa, H., … Al-Jamal, K. T. (2015). Multiphoton luminescence imaging of chemically functionalized multi-walled carbon nanotubes in cells and solid tumors. *Chemical Communications (Cambridge, England)*, *51*(45), 9366–9369. http://dx.doi.org/10.1039/c5cc02675j.

Saito, N., Usui, Y., Aoki, K., Narita, N., Shimizu, M., Hara, K., … Endo, M. (2009). Carbon nanotubes: Biomaterial applications. *Chemical Society Reviews*, *38*(7), 1897–1903. http://dx.doi.org/10.1039/b804822n.

Santos, T., Fang, X., Chen, M. T., Wang, W., Ferreira, R., Jhaveri, N., ... Chen, T. C. (2014). Sequential administration of carbon nanotubes and near-infrared radiation for the treatment of gliomas. *Frontiers in Oncology*, *4*, 180. http://dx.doi.org/10.3389/fonc.2014.00180.

Sgobba, V., & Guldi, D. M. (2009). Carbon nanotubes—Electronic/electrochemical properties and application for nanoelectronics and photonics. *Chemical Society Reviews*, *38*(1), 165–184. http://dx.doi.org/10.1039/b802652c.

Shao, K., Huang, R., Li, J., Han, L., Ye, L., Lou, J., & Jiang, C. (2010). Angiopep-2 modified PE-PEG based polymeric micelles for amphotericin B delivery targeted to the brain. *Journal of Controlled Release*, *147*(1), 118–126. http://dx.doi.org/10.1016/j.jconrel.2010.06.018.

Shvedova, A. A., Kapralov, A. A., Feng, W. H., Kisin, E. R., Murray, A. R., Mercer, R. R., ... Kagan, V. E. (2012). Impaired clearance and enhanced pulmonary inflammatory/fibrotic response to carbon nanotubes in myeloperoxidase-deficient mice. *PLoS One*, 7(3), e30923. http://dx.doi.org/10.1371/journal.pone.0030923.

Singh, R., Pantarotto, D., Lacerda, L., Pastorin, G., Klumpp, C., Prato, M., ... Kostarelos, K. (2006). Tissue biodistribution and blood clearance rates of intravenously administered carbon nanotube radiotracers. *Proceedings of the National Academy of Sciences of the United States of America*, *103*(9), 3357–3362. http://dx.doi.org/10.1073/pnas.0509009103.

Strano, M. S., Dyke, C. A., Usrey, M. L., Barone, P. W., Allen, M. J., Shan, H., ... Smalley, R. E. (2003). Electronic structure control of single-walled carbon nanotube functionalization. *Science*, *301*(5639), 1519–1522. http://dx.doi.org/10.1126/science.1087691.

Stupp, R., Hegi, M. E., Mason, W. P., van den Bent, M. J., Taphoorn, M. J. B., Janzer, R. C., ... Mirimanoff, R.-O. (2009). Effects of radiotherapy with concomitant and adjuvant temozolomide versus radiotherapy alone on survival in glioblastoma in a randomised phase III study: 5-year analysis of the EORTC-NCIC trial. *The Lancet Oncology*, *10*(5), 459–466. http://dx.doi.org/10.1016/S1470-2045(09)70025-7.

Tabakman, S. M., Welsher, K., Hong, G., & Dai, H. (2010). Optical properties of single-walled carbon nanotubes separated in a density gradient; length, bundling, and aromatic stacking effects. *The Journal of Physical Chemistry C Nanomater Interfaces*, *114*(46), 19569–19575. http://dx.doi.org/10.1021/jp106453v.

Tagmatarchis, N., Georgakilas, V., Prato, M., & Shinohara, H. (2002). Sidewall functionalization of single-walled carbon nanotubes through electrophilic addition. *Chemical Communications (Cambridge, England)*, (18), 2010–2011.

Tian, X., Nyberg, S., Sharp, P. S., Madsen, J., Daneshpour, N., Armes, S. P., ... Battaglia, G. (2015). LRP-1-mediated intracellular antibody delivery to the central nervous system. *Scientific Reports*, *5*, 11990. http://dx.doi.org/10.1038/srep11990.

Tsang, S. C., Chen, Y. K., Harris, P. J. F., & Green, M. L. H. (1994). A simple chemical method of opening and filling carbon nanotubes. *Nature*, *372*(6502), 159–162.

Verma, J., Lal, S., & Van Noorden, C. J. (2014). Nanoparticles for hyperthermic therapy: Synthesis strategies and applications in glioblastoma. *International Journal of Nanomedicine*, *9*, 2863–2877. http://dx.doi.org/10.2147/IJN.S57501.

Vlasova, I. I., Vakhrusheva, T. V., Sokolov, A. V., Kostevich, V. A., & Ragimov, A. A. (2011). Peroxidase-induced degradation of single-walled carbon nanotubes: Hypochlorite is a major oxidant capable of in vivo degradation of carbon nanotubes. *Journal of Physics Conference Series*, *291*(1), 012056.

Wang, C. H., Chiou, S. H., Chou, C. P., Chen, Y. C., Huang, Y. J., & Peng, C. A. (2011). Photothermolysis of glioblastoma stem-like cells targeted by carbon nanotubes conjugated with CD133 monoclonal antibody. *Nanomedicine: Nanotechnology, Biology and Medicine*, 7(1), 69–79. http://dx.doi.org/10.1016/j.nano.2010.06.010.

Wang, C. H., Huang, Y. J., Chang, C. W., Hsu, W. M., & Peng, C. A. (2009). In vitro photothermal destruction of neuroblastoma cells using carbon nanotubes conjugated with GD2 monoclonal antibody. *Nanotechnology*, *20*(31), 315101, http://dx.doi.org/10.1088/0957-4484/20/31/315101.

Wang, J. T., Rubio, N., Kafa, H., Venturelli, E., Fabbro, C., Menard-Moyon, C., … Al-Jamal, K. T. (2016). Kinetics of functionalised carbon nanotube distribution in mouse brain after systemic injection: Spatial to ultra-structural analyses. *Journal of Controlled Release*, *224*, 22–32. http://dx.doi.org/10.1016/j.jconrel.2015.12.039.

Wu, H.-C., Chang, X., Liu, L., Zhao, F., & Zhao, Y. (2010). Chemistry of carbon nanotubes in biomedical applications. *Journal of Materials Chemistry*, *20*(6), 1036–1052. http://dx.doi.org/10.1039/B911099M.

Yang, S. T., Wang, H., Meziani, M. J., Liu, Y., Wang, X., & Sun, Y. P. (2009). Biodefunctionalization of functionalized single-walled carbon nanotubes in mice. *Biomacromolecules*, *10*(7), 2009–2012. http://dx.doi.org/10.1021/bm900263z.

Yang, Z., Zhang, Y., Yang, Y., Sun, L., Han, D., Li, H., & Wang, C. (2010). Pharmacological and toxicological target organelles and safe use of single-walled carbon nanotubes as drug carriers in treating Alzheimer disease. *Nanomedicine*, *6*(3), 427–441. http://dx.doi.org/10.1016/j.nano.2009.11.007.

Zhang, L. J., & Webster, T. J. (2009). Nanotechnology and nanomaterials: Promises for improved tissue regeneration. *Nano Today*, *4*(1), 66–80. http://dx.doi.org/10.1016/j.nantod.2008.10.014.

Zhao, D., Alizadeh, D., Zhang, L., Liu, W., Farrukh, O., Manuel, E., … Badie, B. (2011). Carbon nanotubes enhance CpG uptake and potentiate antiglioma immunity. *Clinical Cancer Research*, *17*(4), 771–782. http://dx.doi.org/10.1158/1078-0432.CCR-10-2444.

Zhao, B., Hu, H., Yu, A., Perea, D., & Haddon, R. C. (2005). Synthesis and characterization of water soluble single-walled carbon nanotube graft copolymers. *Journal of the American Chemical Society*, *127*(22), 8197–8203. http://dx.doi.org/10.1021/ja042924i.

Zhao, Y. L., & Stoddart, J. F. (2009). Noncovalent functionalization of single-walled carbon nanotubes. *Accounts of Chemical Research*, *42*(8), 1161–1171. http://dx.doi.org/10.1021/ar900056z.

Zheng, M., Jagota, A., Semke, E. D., Diner, B. A., Mclean, R. S., Lustig, S. R., … Tassi, N. G. (2003). DNA-assisted dispersion and separation of carbon nanotubes. *Nature Materials*, *2*(5), 338–342. http://dx.doi.org/10.1038/Nmat877.

INDEX

Note: Page numbers followed by "*f*" indicate figures, and "*t*" indicate tables.

C

D

E

F

G

N

O

P

Q

R

S

CONTENTS OF RECENT VOLUMES

Volume 39

Volume 40

Volume 41

Volume 42

Volume 43

Volume 44

Volume 45

Volume 46

Volume 47

Volume 48

Volume 49

Volume 50

Volume 51

Volume 52

Volume 53

Volume 54

Volume 55

Volume 56

Volume 57

Volume 58

Volume 59

Volume 60

Volume 61

Volume 62

Volume 63

Volume 64

Volume 65

Volume 66

Volume 67

Volume 68

Volume 69

Volume 70

Volume 71

Volume 72

Volume 73

Volume 74

Volume 75

Volume 76

Volume 77

Volume 78

Volume 79

Volume 80

Volume 81

Volume 82

Volume 83

Volume 84

Volume 85

Volume 86

Volume 87

Volume 88

Volume 89

Volume 90

Volume 91

Volume 92

Volume 93

Volume 94

Volume 95

Volume 96

Volume 97

Volume 98

Volume 99

Volume 100

Volume 101

Volume 102

Volume 103

Volume 104

Volume 105

Volume 106

Volume 107

Volume 108

Volume 109

Volume 110

Volume 111

Volume 112

Volume 113

Volume 114

Volume 115

Volume 116

Volume 117

Volume 118

Volume 119

Volume 120

Volume 121

Volume 122

Volume 123

Volume 124

Volume 125

Volume 126

Volume 127

Volume 128

Volume 129

FSC
www.fsc.org
MIX
Papier | Fördert
gute Waldnutzung
FSC® C083411

Kolibri 360 GmbH
Ferdinand-Jühlke-Straße 7
99095 Erfurt, Deutschland
produktsicherheit@kolibri360.de